THE SUPERIOR COLLICULUS

NEW APPROACHES FOR STUDYING SENSORIMOTOR INTEGRATION

METHODS & NEW FRONTIERS IN NEUROSCIENCE

Series Editors
Sidney A. Simon, Ph.D.
Miguel A.L. Nicolelis, M.D., Ph.D.

Published Titles

Apoptosis in Neurobiology
Yusuf A. Hannun, M.D., Professor of Biomedical Research and Chairman/Department of Biochemistry and Molecular Biology, Medical University of South Carolina
Rose-Mary Boustany, M.D., tenured Associate Professor of Pediatrics and Neurobiology, Duke University Medical Center

Methods for Neural Ensemble Recordings
Miguel A.L. Nicolelis, M.D., Ph.D., Professor of Neurobiology and Biomedical Engineering, Duke University Medical Center

Methods of Behavioral Analysis in Neuroscience
Jerry J. Buccafusco, Ph.D., Alzheimer's Research Center, Professor of Pharmacology and Toxicology, Professor of Psychiatry and Health Behavior, Medical College of Georgia

Neural Prostheses for Restoration of Sensory and Motor Function
John K. Chapin, Ph.D., Professor of Physiology and Pharmacology, State University of New York Health Science Center
Karen A. Moxon, Ph.D., Assistant Professor/School of Biomedical Engineering, Science, and Health Systems, Drexel University

Computational Neuroscience: Realistic Modeling for Experimentalists
Eric DeSchutter, M.D., Ph.D., Professor/Department of Medicine, University of Antwerp

Methods in Pain Research
Lawrence Kruger, Ph.D., Professor of Neurobiology (Emeritus), UCLA School of Medicine and Brain Research Institute

Motor Neurobiology of the Spinal Cord
Timothy C. Cope, Ph.D., Professor of Physiology, Emory University School of Medicine

Nicotinic Receptors in the Nervous System
Edward D. Levin, Ph.D., Associate Professor/Department of Psychiatry and Pharmacology and Molecular Cancer Biology and Department of Psychiatry and Behavioral Sciences, Duke University School of Medicine

Methods in Genomic Neuroscience
Helmin R. Chin, Ph.D., Genetics Research Branch, NIMH, NIH
Steven O. Moldin, Ph.D, Genetics Research Branch, NIMH, NIH

Methods in Chemosensory Research
Sidney A. Simon, Ph.D., Professor of Neurobiology, Biomedical Engineering, and Anesthesiology, Duke University
Miguel A.L. Nicolelis, M.D., Ph.D., Professor of Neurobiology and Biomedical Engineering, Duke University

The Somatosensory System: Deciphering the Brain's Own Body Image
Randall J. Nelson, Ph.D., Professor of Anatomy and Neurobiology,
University of Tennessee Health Sciences Center

The Superior Colliculus: New Approaches for Studying Sensorimotor Integration
William C. Hall, Ph.D., Department of Neuroscience, Duke University
Adonis Moschovakis, Ph.D., Institute of Applied and Computational Mathematics, Crete

New Concepts in Cerebral Ischemia
Rick C.S. Lin, Ph.D., Professor of Anatomy, University of Mississippi Medical Center

DNA Arrays: Technologies and Experimental Strategies
Elena Grigorenko, Ph.D., Technology Development Group, Millennium Pharmaceuticals

Methods for Alcohol-Related Neuroscience Research
Yuan Liu, Ph.D., National Institute of Neurological Disorders and Stroke, National Institutes of Health
David M. Lovinger, Ph.D., Laboratory of Integrative Neuroscience, NIAAA

In Vivo Optical Imaging of Brain Function
Ron Frostig, Ph.D., Associate Professor/Department of Psychobiology,
University of California, Irvine

Primate Audition: Behavior and Neurobiology
Asif A. Ghazanfar, Ph.D., Primate Cognitive Neuroscience Lab, Harvard University

Methods in Drug Abuse Research: Cellular and Circuit Level Analyses
Dr. Barry D. Waterhouse, Ph.D., MCP-Hahnemann University

Functional and Neural Mechanisms of Interval Timing
Warren H. Meck, Ph.D., Professor of Psychology, Duke University

Biomedical Imaging in Experimental Neuroscience
Nick Van Bruggen, Ph.D., Department of Neuroscience Genentech, Inc.,
South San Francisco
Timothy P.L. Roberts, Ph.D., Associate Professor, University of Toronto

The Primate Visual System
John H. Kaas, Department of Psychology, Vanderbilt University
Christine Collins, Department of Psychology, Vanderbilt University

THE SUPERIOR COLLICULUS

NEW APPROACHES FOR STUDYING SENSORIMOTOR INTEGRATION

Edited by
William C. Hall
Adonis Moschovakis

CRC PRESS

Boca Raton London New York Washington, D.C.

Library of Congress Cataloging-in-Publication Data

The superior colliculus : new approaches for studying sensorimotor integration / edited by William C. Hall and Adonis Moschovakis.
p. ; cm. — (Methods & new frontiers in neuroscience)
Includes bibliographical references and index.
ISBN 0-8493-0097-5 (alk. paper)
1. Sensorimotor integration. 2. Superior colliculus. I. Hall, William C., Ph. D. II. Moschovakis, Adonis. III. Methods & new frontiers in neuroscience series.
[DNLM: 1. Superior Colliculus—physiology. 2. Saccades. WL 310 S959 2003]
QP454.S87 2003
612.7′6—dc22 2003058471

Direct all inquiries to CRC Press LLC, 2000 N.W. Corporate Blvd., Boca Raton, Florida 33431.

Visit the CRC Press Web site at www.crcpress.com

International Standard Book Number 0-8493-0097-5
Library of Congress Card Number 2003058471
Printed in the United States of America 1 2 3 4 5 6 7 8 9 0
Printed on acid-free paper

Series Preface

Our goal in creating the **Methods & New Frontiers in Neuroscience** series is to present the insights of experts on emerging experimental techniques and theoretical concepts that are, or will be, at the vanguard of neuroscience. Books in the series cover topics ranging from methods to investigate apoptosis, to modern techniques for neural ensemble recordings in behaving animals. The series also covers new and exciting multidisciplinary areas of brain research, such as computational neuroscience and neuroengineering, and describes breakthroughs in classical fields like behavioral neuroscience. We want these books to be the books every neuroscientist will use in order to get acquainted with new methodologies in brain research. These books can be given to graduate students and postdoctoral fellows when they are looking for guidance to start a new line of research.

Each book is edited by an expert and consists of chapters written by the leaders in a particular field. The books are richly illustrated and contain comprehensive bibliographies. Chapters provide substantial background material relevant to the particular subject. Hence, they are not only "methods books," but they also contain detailed "tricks of the trade" and information as to where these methods can be safely applied. In addition, they include information about where to buy equipment and about web sites helpful in solving both practical and theoretical problems.

We hope that as the volumes become available, the effort put in by us, by the publisher, by the book editors, and by individual authors will contribute to the further development of brain research. The extent to which we achieve this goal will be determined by the utility of these books.

Sidney A. Simon, Ph.D.
Miguel A.L. Nicolelis, M.D., Ph.D.
Series Editors

Preface

To explore the world, the brains of humans and other animals orchestrate movements of several effectors (eyes, head, trunk, upper and lower limbs) to orient the direction of gaze toward interesting features of the visual field. Such orientation is one example of general neural mechanisms that allow organisms to acquire information about the external world and to process the information so that it can be utilized by motor systems to generate adaptive behavior. What are the biological mechanisms, that is, the sequences of neural events, that allow sensory systems to interface and meaningfully engage their motor counterparts? To answer this fundamental question, experts have focused for more than 30 years, on a small, phylogenetically ancient midbrain structure, the superior colliculus. The close proximity within its layers of sensory cells that encode the spatial loci of external stimuli and premotor cells that command corresponding movements of the eyes and head toward these stimuli provides investigators with an almost unique opportunity to study how sensory signals are translated into adaptive motor commands within a single structure. Because of these features and also because of the simplifying benefits of the fixed relationship between stimulus location and the metrics of an appropriate orienting movement, progress in understanding the structure and function of the superior colliculus has reached the point that sophisticated quantitative models can be proposed and tested experimentally. For all of these reasons, investigators in this area appear to be on the threshold of unveiling, for the first time in the mammalian brain, the neural mechanisms responsible for a specific example of sensorimotor integration. The time, therefore, seems opportune for offering, in a single volume, a summary of the progress accomplished to date and a view of the vistas opened up for future exploration.

For this volume, authors were chosen to represent the wide variety of approaches that have contributed to this level of understanding of neural mechanisms in the superior colliculus. To encourage perspectives on what has been accomplished during the past 3 decades as well as the expression of views concerning the future course of research in this field, the volume includes contributions from outstanding scientists at different stages of their careers. In all cases, the authors are recognized experts on the frontier of research in this productive field. The first three chapters discuss the role of the superior colliculus in the control of gaze shifts and, more specifically, the rapid eye movements called saccades. Authors Platt, Lau, and Glimcher consider the place of the superior colliculus in the complex hierarchy of nuclei that control the direction of gaze. Stanford summarizes classical experiments that implicate the superior colliculus in the control of saccades, while Munoz and Schall compare its role to that of the frontal eye field in saccade preparation and initiation.

In the first of several chapters devoted to a detailed description of structural features of the superior colliculus as they relate to its functional properties, Harting describes certain medium-scale morphological features that have intrigued anatomists over the years. Then, Alexej Grantyn and Adonis Moschovakis describe the morphology of collicular neurons and the nature of the signals they carry to their targets both inside and outside of the colliculus. Next, Özen, Helms, and Hall describe the properties of circuits intrinsic to the deeper, premotor layers of the superior colliculus, and the means by which they generate behaviorally relevant signals. Isa, Kobayashi, and Saito consider the interaction of the deeper layer cells with information originating in the superficial, visuosensory layers.

The next two chapters are devoted to developmental issues; Rosemarie Grantyn, Jüttner, and Meier emphasize the development of the visual layers while Mize and Salt focus on mechanisms responsible for the development of the retinal input to these layers.

The next two chapters emphasize computational issues. Guitton, Bergeron, and Choi argue in favor of a transcollicular feedback loop implementing a gaze controller, while Keller summarizes progress in understanding the operation of the highly distributed neural processing that takes place inside the superior colliculus.

In a fitting epilogue to this volume, Sparks summarizes certain methodological and conceptual problems of the field; he considers how collicular efferent signals could be processed by the structures that receive input from the superior colliculus, whether feedback signals must be rerouted to it, and the types of external physical variables that each structure could encode.

The editors of this volume are grateful to the series editors, Sid Simon and Miguel Nicolelis, for providing us with the opportunity and the encouragement to undertake this venture. We also are indebted to the publisher of the life sciences group at CRC, Barbara Norwitz, for her constant attention and expert advice and especially for her persistent, and necessary, prodding in motivating us to complete this project in a timely fashion.

About the Editors

William C. Hall is a Professor in the Department of Neurobiology at Duke University Medical Center in Durham, North Carolina. He did his undergraduate and graduate studies at Duke University, where he trained and conducted his thesis research under the tutelage of Dr. Irving T. Diamond in the Department of Psychology. Since 1970, he has been a faculty member in the Duke University Medical Center. His research interests are focused on the application of *in vitro* methods for analyzing the structure and function of intrinsic circuitry in the superior colliculus.

Adonis Moschovakis is a professor of physiology at the University of Crete, group leader at the Institute of Applied and Computational Mathematics of the Foundation for Research and Technology — Hellas, and section editor for *Neurophysiology and Computational Neuroscience of Brain Research Bulletin*. He has studied medicine at the University of Athens where he obtained his M.D. in 1979, psychiatry at the St. Antoine Hospital in Paris (1979 to 1981), and neuroscience at Washington University in St. Louis, Missouri, where he obtained his Ph.D. in 1987.

He has served as a staff fellow at the Laboratory of Neural Control of the National Institute of Neurological Disorders and Stroke at the National Institutes of Health, and chairman of the Department of Basic Sciences of the Faculty of Medicine at the University of Crete, and director of its graduate program in neurosciences.

Contributors

André Bergeron, Ph.D.
Montreal Neurological Institute
McGill University
Montreal, Quebec, Canada

Woo Young Choi
Montreal Neurological Institute
McGill University
Montreal, Quebec, Canada

Paul W. Glimcher, Ph.D.
Center for Neural Science and Department of Psychology
New York University
New York, New York

Alexej A. Grantyn, M.D., Ph.D.
Laboratoire de Physiologie de la Perception et de l' Action
CNRS–Collège de France
Paris, France

Rosemarie Grantyn, M.D., Ph.D.
Johannes Müller Institute for Physiology
Humboldt University of Berlin
Berlin, Germany

Daniel Guitton, Ph.D.
Montreal Neurological Institute
McGill University
Montreal, Quebec, Canada

William C. Hall, Ph.D.
Department of Neurobiology
Duke University Medical Center
Durham, North Carolina

John K. Harting, Ph.D.
Department of Anatomy
University of Wisconsin Medical School
Madison, Wisconsin

Matthew C. Helms, Ph.D.
Department of Neurobiology
Duke University Medical Center
Durham, North Carolina

Tadashi Isa, M.D., Ph.D.
Department of Integrative Physiology
National Institute of Physiological Sciences
Myodaiji, Okazaki, Japan

René Jüttner, Ph.D.
Max-Delbruck-Centrum für Molekulare Medizin
Berlin, Germany

Edward L Keller, Ph.D.
Smith-Kettlewell Eye Research Institute
San Francisco, California

Yasushi Kobayashi, Ph.D.
Neuroscience Laboratories
Visual Neuroscience Group
Osaka University Graduate School of Frontier Biosciences
Osaka, Japan

Brian Lau
Center for Neural Science
New York University
New York, New York

Jochen Meier, Ph.D.
Johannes Müller Institute for Physiology
Humboldt University of Berlin
Berlin, Germany

R. Ranney Mize, Ph.D.
Department of Cell Biology and Anatomy
Louisiana State University
New Orleans, Louisiana

Adonis K. Moschovakis, M.D., Ph.D.
Institute of Applied and Computational Mathematics, F.O.R.T.H.
Heraklion, Crete, Greece

Douglas P. Munoz, Ph.D.
Centre for Neuroscience Studies
Department of Physiology
Queens University
Kingston, Ontario, Canada

Gülden Özen, M.D., Ph.D.
Department of Neurobiology
Duke University Medical Center
Durham, North Carolina

Michael L. Platt
Department of Neurobiology
Center for Cognitive Neuroscience and
Department of Biological Anthropology and Anatomy
Duke University Medical Center
Durham, North Carolina

Yasuhiko Saito, Ph.D.
Department of Physiology
Gumma University School of Medicine
Maebashi, Japan

Thomas E. Salt, Ph.D.
Institute of Ophthalmology
University College London
London, England

Jeffrey D. Schall, Ph.D.
Center for Integrative and Cognitive Neuroscience
Vanderbilt Vision Research Center
Department of Psychology
Vanderbilt University
Nashville, Tennessee

David L. Sparks, Ph.D.
Division of Neuroscience
Baylor College of Medicine
Houston, Texas

Terrence R. Stanford, Ph.D.
Department of Neurobiology and Anatomy
Wake Forest University School of Medicine
Winston-Salem, North Carolina

Contents

1 Situating the Superior Colliculus within the Gaze Control Network

Michael L. Platt, Brian Lau, and Paul W. Glimcher

CONTENTS

1.1 INTRODUCTION

A wealth of evidence, much of it reviewed in this book, suggests that the superior colliculus (SC) plays a crucial role in the control of orienting eye and head movements. Activity in the superior colliculus, for example, predicts both the metrics and the timing of orienting movements. Electrical microstimulation of the SC, to take another example, can be used to elicit saccadic eye movements that are virtually indistinguishable from naturally occurring saccades. Based upon data like these, many neurobiologists have come to regard the SC as a command center from which nearly all saccades are generated. Thus, the SC has come to be viewed as a *final common path* for the generation of rapid eye movements in the same sense that Sherrington used the term final common path to describe alpha-motor neurons in the spinal cord. For most oculomotor scientists, activity in the SC is assumed to reflect the cumulative output of the neural systems that select and execute orienting movements.

Although some data raise questions about the validity of this framework (e.g., References 1 and 2), the SC is in fact almost ideally situated to perform this proposed role (Figure 1.1). We now know that the SC receives dense inputs from the frontal

0-8493-0097-5/04/$0.00+$1.50

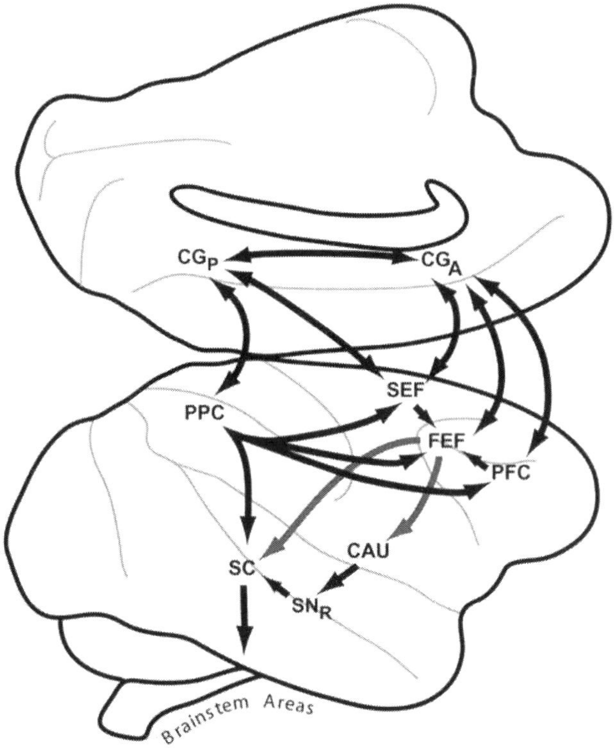

FIGURE 1.1 Some, but not all, of the major connections between elements in the orienting movement control network in the macaque brain, in which the superior colliculus (SC) serves as the final common path. Top: medial surface; bottom: lateral surface. Sensory input pathways are not diagrammed.

eye fields (FEF), which serve as cortical centers for the regulation of orienting eye movements. It also receives a strong and physiologically important input from the substantia nigra pars reticulata (SNr), which serves as the principal oculomotor output structure for the entire basal ganglia. Finally, the SC receives projections from a number of other cortical areas that link it with neural groups which process sensory data, participate in sensorimotor transformations, and compute decisions selecting movements for execution. Together, these observations provide a compelling rationale for suspecting that the SC, at least under some conditions, may indeed serve as a final common pathway for the generation of orienting movements.

In this chapter, we provide an overview of several of the important areas that appear to generate or modulate neural signals, such as those carried by the SC, associated with orienting movements. It begins by examining the frontal cortical areas, which appear to regulate the SC more directly than any other areas. The FEF is the structure probably most closely allied to the SC and it has been known to play a critical role in the generation of orienting eye movements since the time of Ferrier.[3] More than any other structure, the circuitry of the FEF seems to closely parallel the functional properties of the SC circuits to which it projects. Indeed, Schiller and colleagues[1] demonstrated that after collicular ablation a monkey's ability to make saccadic eye movements recovers if the FEF is left intact. The FEF thus appears to represent a pathway by which frontal cortical areas can program saccades for execution by influencing either the SC or its brainstem targets. The FEF, in turn, seems to be strongly influenced by the supplementary eye fields (SEF), a higher order cortical structure that may participate in decisions about which movement to produce. A review of these two areas forms the first part of this chapter.

These frontal areas, however, are not the only cortical regions that influence the colliculus. Parietal cortex also includes what appear to be saccadic control areas, as do regions of the cingulate and prefrontal cortices. While it is not yet entirely clear what role these areas play in eye movement production, there is compelling evidence that they participate in sensorimotor transformations, control the allocation of attention, aid in the identification of erroneously executed movements, and participate in the assessment of whether a recently completed movement was of value. Together, these areas probably provide some of the highest-level cognitive input that the SC receives and they form the subject of the second part of this review.

Finally, it has been known for some time that the SC receives strong and direct inhibitory projections from the basal ganglia via the SNr. Hikosaka and Wurtz were the first to study these areas and to demonstrate how crucial these inhibitory projections are to the normal functioning of the systems regulating orienting movements. Subsequent work by Hikosaka and others has both reinforced this early conclusion and expanded the field within which the basal ganglia appear to make computations. We now believe that the basal ganglia play a critical evaluative role in assessing the value of each orienting movement to the animal, predicting the value of upcoming orienting movements based on those assessments, and updating representations of value after movement completion, using both inhibitory and excitatory signals.

These three sets of areas represent most, but not all, of the primary and secondary orienting inputs that reach the SC. Other areas, some of which are mentioned briefly within this review, also make important connections with the SC but are not yet well enough understood to be reviewed here. These three groups of areas should thus be viewed as a current, and certainly incomplete, description of the important inputs to the SC. Many of the areas described here are also strongly interconnected with each other, another important observation that is only briefly discussed here. However incomplete, current knowledge does suggest that these three groups of areas carry critical signals that eventually coalesce in neuronal activation within the map of orienting movements in the SC. These inputs thus provide a flexible and adaptive network for controlling the phylogenetically ancient tectal orienting system, and, as demonstrated in this review, this network is increasingly, although not yet completely, understood.

1.2 FRONTAL EYE FIELDS

The FEF have been recognized as a critical cortical structure for controlling eye movements since Ferrier's[3] observation that electrical stimulation of the surface of the brain overlying this structure elicited saccades. Subsequent work using intracortical microstimulation demonstrated that fixed-vector saccades can be evoked using currents in the microampere range, and that the direction and amplitude of these saccades is topographically mapped in the FEF.[4,5] These saccades are evoked with short latencies, and are likely mediated by projections from the FEF to the SC,[6–9] although direct connections with brainstem oculomotor areas may also be important.[7,10,11] Hanes and Wurtz[12] recently showed that when the SC is temporarily inactivated, using muscimol or lidocaine, saccades could no longer be elicited through microstimulation in the FEF. This suggests that in the intact animal the FEF most likely exerts its influence via projections to the SC, although studies by Schiller and colleagues[1] demonstrated that monkeys eventually produce saccadic eye movements following recovery from permanent lesions of either the SC or the FEF.

A wealth of anatomical and physiological evidence indicates that the FEF is not simply a motor structure. Rather, it is well situated to link visual information to eye movements (reviewed by Schall and Thompson[13]). Single-unit recordings in the FEF have revealed distinct cell classes that fire in relation to the presentation of visual stimuli, saccades, fixation, and pursuit eye movements.[14–19] Anatomically, the FEF receives dense inputs from the dorsal and ventral visual streams,[20,21] and is interconnected with the oculomotor lateral intraparietal area (LIP),[20,22] the SEF [23,24] and the dorsolateral prefrontal cortex.[25,26] In this section, we review recent work on how different cell classes in the FEF participate in the transformation of visual target information into a saccadic motor command.

Visually guided eye movements require the conversion of visual information into a motor response. This conversion can be viewed as a sequence of two decisions: first deciding where to look and then deciding when to look. Deciding where to look is the result of sensory processing and specifies the goal of an eye movement. Deciding when to look is the result of motor preparation and is revealed by commitment to a particular eye movement. Schall and colleagues developed a visual search task to explore the role of the FEF in decisions about where to look. They trained monkeys to shift gaze to an oddball target that differed in color or shape from an array of distractors. A subset of FEF neurons responded with an increase in activity following the appearance of the visual array. Although the initial neural activity did not distinguish between whether the oddball or a distractor fell within the cell's receptive field, later activity clearly discriminated between the two. If the oddball was in the cell's receptive field, its firing rate remained elevated; otherwise it became sharply attenuated. Schall and colleagues hypothesized that the time at which FEF neurons reliably discriminate the oddball from the distractors marks the completion of sensory processing, or the decision computing where to look.[27] The idea that visually responsive neurons are involved with sensory processing predicts that manipulating the difficulty of sensory discrimination should affect reaction times by changing the time it takes to discriminate the oddball from the distractors. Schall and colleagues tested this prediction by making search more difficult by either increasing the number of distractors[28] or increasing the similarity between the target and the distractors.[29] These studies showed that the amount of time it took for visually responsive FEF neurons to reliably discriminate between the oddball and the distractors was prolonged as the difficulty of the task increased. This delay was accompanied by a proportional increase in saccadic reaction time, indicating that prolonged sensory processing leads to delayed motor responses.

Although the decisions that compute where and when to look are conceptually separable, it is possible that the two processes blend seamlessly together, and the activity of visually responsive neurons may also reflect motor preparation or the commitment to a response. Schall and colleagues argued against this by showing that (1) the time to reliable discrimination by visually responsive neurons is uncorrelated with saccadic reaction time for a fixed level of task difficulty,[27] and (2) these neurons continue to signal the oddball even if the monkey is required to maintain fixation,[30] or makes a movement to another location.[31] These experiments demonstrate that the activity of visually responsive neurons can be dissociated from the overt motor response. However, the commitment to a particular eye movement can itself be distinguished from motor preparation; many potential actions can be prepared despite the fact that only one can be produced. Therefore it remains possible that the activity of these neurons reflects covert motor preparation.

The distinction between commitment and motor preparation is supported by two recent studies of how the output of sensory processing is transferred to the neural elements controlling motor preparation. In considering these studies, it is worth keeping two possibilities in mind. One possibility is that transfer occurs after target selection; the system generating the motor response simply receives the final outcome of sensory processing. Alternatively, transfer could be continuous; the motor response is prepared as sensory processing proceeds. Bichot and colleagues[32] tested these alternatives by recording from movement-related cells in the FEF. In contrast to visually responsive neurons, these cells exhibit a burst of activity during saccades of a particular amplitude and direction, and respond little if at all to visual stimuli.[17,19] Bichot and colleagues recorded from these movement-related neurons while monkeys performed a visual search task where the oddball was defined by a conjunction of shape and color. They reasoned that if transfer followed target selection, then the activity of movement-related neurons should reflect the metrics of the upcoming saccade. If, on the other hand, there is partial transfer of sensory information, they predicted that a neuron's activity should depend on the properties of the distractor falling within its movement field. They found that movement-related cells were differentially active before saccades that shifted gaze to locations outside of a cell's movement field; these neurons tended to be more active for distractors that shared a feature with the oddball compared to distractors that did not share any features. This result clearly demonstrates that movement-related neurons receive sensory information. Bichot and colleagues [32] also noted that the metrics of the eye movements to a particular oddball target were unaffected by the activity of neurons responding to the surrounding distractors; saccade trajectories were not curved towards similar distractors, as might be expected if saccades to distractors were concurrently programmed (e.g., McPeek and Keller[33]). Thus, the differential activity of the movement-related neurons to distractors can be dissociated from the overt motor response.

Both the presence of sensory activity in movement-related cells and the dissociation of this activity from the final motor outcome can be explained if the commitment to a particular eye movement is achieved by a winner-take-all rule, a process where selection is based on which neuron (or population of related neurons) achieves the highest level of activity. One possible mechanism for achieving a winner-take-all rule is a simple threshold; the final motor outcome is specified by whichever neuron attains a particular level of activity first (see References 34 and 35). According to this model, the differential activity observed by Bichot and colleagues represents covert motor planning (also see work by McPeek and Keller[36] for related results in the SC). However, in their task, the monkeys made eye movements after committing to that action, and it is impossible to know whether the increased activation for similar distractors represents covert motor preparation.

Gold and Shadlen[37] determined the ongoing degree of motor preparation by using electrical microstimulation to probe the state of the FEF *prior* to commitment. They trained monkeys to shift gaze to one of two targets depending on the direction of a random-dot motion stimulus. By varying the fraction of coherently moving dots as well as the duration of motion viewing, Gold and Shadlen manipulated the monkey's ability to determine the correct direction of motion. On some trials, they microstimulated a site in the FEF immediately after the offset of the motion stimulus, before the monkey voluntarily initiated a saccade to a choice target. Microstimulation evoked a saccade orthogonal to the axis of motion, which was followed by a voluntary corrective saccade to the desired target. Gold and Shadlen reasoned that if the motor elements of the FEF continuously converted sensory information into motor preparation, then the direction of the evoked saccades should depend on motion coherence and viewing duration. They confirmed this prediction by showing that the endpoints of the evoked saccades deviated in the direction of the monkey's eventual choice, and that higher motion coherences or longer viewing durations led to larger deviations compared to weaker motion coherences or shorter viewing durations (Figure 1.2). They explained their results using a model that integrates the difference in activity between populations of motion detectors representing the

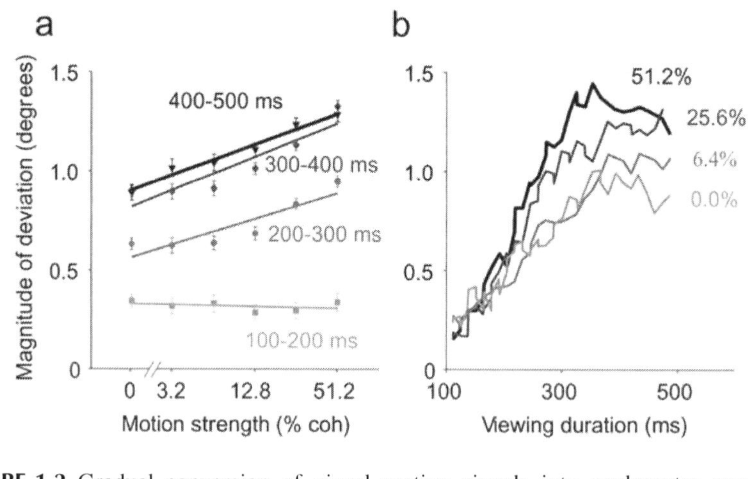

FIGURE 1.2 Gradual conversion of visual motion signals into oculomotor commands revealed by microstimulation of the FEF. Magnitude of evoked-saccade deviation toward the chosen target, plotted against (a) motion strength and (b) viewing duration. (From Gold, J.I. and Shadlen, M.N., Representation of a perceptual decision in developing oculomotor commands, *Nature*, 404, 390, 2000. Reprinted by permission from *Nature* ©2000.)

correct and incorrect directions of motion. Thus, probing the state of the FEF prior to commitment revealed a vector-averaging that depended on the quality and duration of motion information. This finding is strong evidence against the idea the motor elements of the FEF simply receive the final outcome of sensory processing.

It also seems unlikely that motor elements downstream of the FEF, like the SC, receive only the final outcome of the sensory processing guiding target selection. Sommer and Wurtz[38] used antidromic stimulation to show that the FEF sends visual-, memory-, and motor-related signals to the SC. This suggests that the observations made by Gold and Shadlen[37] and Bichot and colleagues[32] are probably not unique to the FEF. In fact, signals related to all stages of the visuomotor transformation have been observed in the SC,[2,39–41] area LIP,[35,42,43] and the dorsolateral prefrontal cortex.[44]

Making decisions requires assessing the relative value of potential actions in addition to analyzing the sensory evidence favoring one action over others (see References 45 and 46). Leon and Shadlen recently showed that FEF neurons are not modulated by reward magnitude, suggesting that signals in the FEF must at least be combined with signals representing the value of potential actions. These value judgments must also be learned, requiring the integration of the outcomes of previous actions over extended periods of time. In the following sections, we review evidence indicating that a number of other cortical areas compute these critical decision variables.

1.3 SUPPLEMENTARY EYE FIELDS

The SEF is located on the dorsomedial aspect of the frontal lobe. Early work indicated that the SEF was involved in eye movements,[48] but Schlag and Schlag-Rey[49] were the first to thoroughly characterize the oculomotor properties of the SEF. They identified the SEF as an area rostral to the supplementary motor area where saccades could be elicited by electrical microstimulation using low currents. Interestingly, a large region of cortex where saccades cannot be elicited by microstimulation separates the SEF from the FEF, and numerous studies have been directed at determining the anatomical and physiological differences between the frontal cortical eye fields.

A role for the supplementary eye fields in the guidance of eye movements is indicated by its anatomical connectivity with a number of cortical and subcortical oculomotor structures. In general, the pattern of connectivity for the SEF overlaps broadly with that of the FEF, with which it is densely interconnected.[23,24,50] The differences in connectivity have been reviewed in detail elsewhere (see Reference 51), but several are worth noting here. Compared to the FEF, the SEF: (1) receives less input from higher-level visual areas, (2) is more extensively interconnected with the prefrontal cortices, and (3) is much more densely interconnected with the anterior and posterior cingulate motor areas.

Both cortical eye fields project directly to the SC. Fibers from the SEF terminate densely in layers I, IV, and VI but not II or III of the SC.[23,52] In contrast, the FEF sends additional projections to the visual layers II and III.[7] One further consistent difference is that tracer injections of similar size yield more tangentially widespread distributions of label from the SEF in comparison to the FEF.[23,52] In addition to influencing eye movements through its direct projections to the SC and to the FEF, the SEF innervates a number of oculomotor areas in the brainstem including the nucleus raphe interpositus, interstitial nucleus of Cajal, nucleus prepositus hypoglossi, and the nucleus reticularis tegmenti pontis.[52] While SEF microstimulation can elicit saccades following lesions of the FEF or SC,[53] the direct brainstem projections do not seem to independently move the eyes since paired lesions of the FEF and the SC virtually eliminate all saccadic eye movements.[1]

Physiological studies using single-unit recording (e.g., see References 49, 54, and 55) and intracortical microstimulation (e.g., see References 49, 57, and 58) have largely focused on the oculomotor function of the SEF. More recently, investigators have begun to explore the response properties of SEF neurons using complex behavioral tasks. These tasks have revealed differences between the FEF and the SEF, and for the remainder of this section, we review work that suggests that the SEF participates in (1) signaling the context and consequences of directed eye movements, (2) the learning and storage of visuomotor associations, and (3) the execution of temporally extended motor programs.

One prerequisite for adaptive, goal-directed behavior is the ability to monitor performance, which requires recognizing the outcomes of prior actions. In a recent set of experiments, Stuphorn, Taylor, and Schall[60] implicated the SEF in performance monitoring. They trained monkeys to perform an oculomotor countermanding task that required a gaze shift to an eccentric target within 500 milliseconds after extinction of a central fixation point. On a fraction of these trials reappearance of the fixation point instructed the monkeys to cancel the planned eye movement and maintain fixation to receive a juice reward. Stuphorn and colleagues identified several types of neuronal activity in the SEF related to performance monitoring. Some neurons showed increased firing rates following a failure to cancel a saccade, regardless of its direction or endpoint. Complementing these error-detecting neurons were cells that showed an increase in firing rate that peaked immediately after the receipt of primary or secondary reinforcement (see also Amador et al.[61]; Mann et al.[62]). Together with the error-detecting neurons, these reward neurons appear to signal the outcomes of an eye movement in the context of an operant task.

A third class of neurons became active whenever the monkeys successfully cancelled a planned eye-movement. Similar responses have been noted in the SEF following correct movement cancellation during go/no-go tasks,[54,62] but Stuphorn and colleagues extended these findings by showing that the magnitude of this activity was correlated with the likelihood that the monkeys failed to cancel a movement. Since the monkeys were making more errors, and presumably receiving fewer rewards, it is possible that these neurons signal overall task difficulty or reward rate. They may also signal the degree of conflict between the desire to initiate a movement and the desire to cancel a movement plan.[60,63] This hypothesis is consistent with increased activation of SEF neurons during anti-saccades,[64] which require the suppression of reflexive or memory-guided pro-saccades. Importantly, the SEF response types identified by Stuphorn and colleagues were not observed in the FEF of monkeys performing the countermanding task,[19] indicating that the SEF and FEF send distinct, but complementary, information to the SC. Together, the findings described above indicate that the SEF is well situated to signal the context and consequences of directed eye movements for the guidance of goal-directed behavior.

The ability to monitor the context and consequences of actions is not only useful for guiding learned behavior, but is necessary for acquiring novel behaviors. The premotor cortex is critical for learning arbitrary associations between visual stimuli and actions.[65] Prompted by physiological work showing changes during visuomotor learning in the premotor cortex,[66] Chen and Wise[67] trained monkeys to associate directions of gaze with arbitrary visual images in order to explore the role of the SEF in learning visuomotor mappings. They found that many neurons in the SEF increased their activity in parallel with behavioral learning. Moreover, the magnitude and time-course of this learning-dependent activity came to resemble responses to familiar stimuli associated with the same direction of gaze. Chen and Wise[68] also observed that over the course of learning the preferred direction of directionally selective neurons could actually reorient to match that of well-learned associations. This lability of tuning is reminiscent of work by Mann, Thau, and Schiller showing that the saccade evoked by microstimulating at a particular cortical location could actually change depending on the current task requirements.[62] The learning-dependent neurons identified by Chen and Wise were much more common in the SEF than the FEF,[69] and they hypothesized that the SEF computes the appropriate goal for a particular context, which it then transmits to the FEF for eye movement programming. This is supported by the recent finding that SEF neurons signal the goal of an impending saccade in a free-choice task over a second before the movement is actually made, nearly 500 milliseconds before similar differential activity is observed in the FEF or area LIP.[70]

The concept of visuomotor association extends naturally to include the dimension of time. That is, the appropriate action may differ depending on the temporal context of the task, and keeping track of time or numerical order is important for performing the task correctly. For instance, stopping at a red light or driving at a green light are clearly arbitrary mappings, but the appropriate action to take at a yellow light depends on when the traffic light transitioned from green to yellow. Many behaviors, such as visually guided motor sequences, can be described as visuomotor mappings that unfold in time.

Work in humans suggests that the SEF is necessary for the correct execution of oculomotor sequences. Both patients with lesions including the SEF[71] as well as normal subjects undergoing transcranial magnetic stimulation of the SEF[72,73] show increases in errors while performing sequences of memory-guided saccades. This agrees with the findings in monkeys that lesions[74] and reversible inactivation[75] of the SEF induce mild deficits in performing sequences of two saccades. Lu, Matsuzawa, and Hikosaka[76] explored the role that single-neurons in the SEF have in oculomotor sequencing. They trained monkeys to perform sequences of five saccades, where each saccade was made from a central fixation point to one of four targets. Only two of the four potential targets were presented prior to each saccade, and the monkeys were free to make a movement to one of the two upon extinction of the fixation target. A critical feature of the task was that a pair of targets could appear more than once in a sequence, so that a leftward saccade might be required for the first movement, but a rightward saccade might be required for the second movement, even though the targets were identical. Lu and colleagues found that nearly three-quarters of the task-related neurons in the SEF responded in a sequence-selective manner while the monkeys performed well-learned sequences. These neurons fired differentially depending on which sequence the monkey was performing. Lu and colleagues hypothesized that different oculomotor sequences might be encoded by separate populations of SEF neurons.

Taken together, these studies reveal that the SEF is functionally and anatomically distinct from the FEF. The SEF appears to be involved with the learning and monitoring of eye movements, or sequences of eye movements, that are directed to accomplish a goal. However, there exist numerous cell classes in the SEF, and further experiments are required to show what types of information the SEF sends to the FEF and the SC.

1.4 PARIETAL CORTEX

The parietal lobe, in particular the inferior parietal lobule, has long been recognized as crucial to normal visuospatial orienting. Patients with damage to parietal cortex often present with a syndrome, known as hemineglect, characterized by an inability or reluctance to orient, and sometimes even acknowledge, visual stimuli contralateral to the lesion.[77–79] Both reversible inactivation and acute lesions of the inferior parietal lobule in monkeys induce saccades that are slowed and hypometric when guided by visible and, in particular, remembered targets contralateral to the lesion.[80,81] Disturbances of visuospatial orienting associated with parietal dysfunction in both humans and monkeys thus suggest that this region plays an important role in the sensory-motor transformations that orient gaze.

This hypothesis is supported by the anatomical connections of parietal cortex. Extrastriate visual areas associated with the dorsal stream,[82] such as area V3, area MT, and area MST, project directly to the inferior parietal lobule, in particular the LIP and area 7a.[26,83,84] Parietal cortex, in particular area LIP, in turn projects directly to both the FEF[22] and the SC.[25] Portions of parietal cortex, especially area LIP, are thus ideally situated to intervene between visual signals and movements of the eyes. Projections back to parietal cortex from FEF[20,22] and SC[85] suggest that signals in these areas can significantly influence ongoing sensory-motor processing in the parietal lobe.

Electrophysiological recordings from single neurons in parietal cortex strongly support the proposal that this brain region participates in the transformation of sensory signals into oculomotor commands. In the 1970s, Mountcastle and colleagues demonstrated that neurons in parietal cortex were activated when monkeys made visually guided saccades to a restricted region of space.[86,87] Following these important early studies, Robinson, Goldberg and colleagues[88] showed that, in addition to firing action potentials prior to saccade onset, many parietal neurons responded following the illumination of a visual stimulus that subjects were instructed to ignore. Moreover, these responses were enhanced when subjects were instructed to attend to the stimulus, either overtly with a gaze shift or covertly in order to detect the dimming of that stimulus.[89,90] Around the same time, Gnadt and Andersen[91] demonstrated that many LIP neurons were also activated during the delay period intervening between the offset of a visual target and the onset of a gaze shift

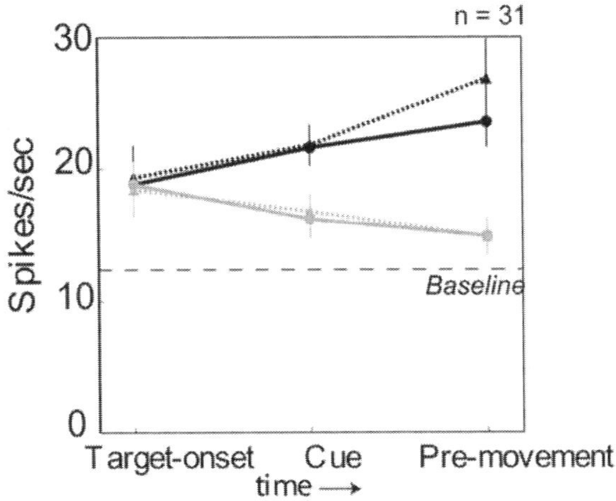

FIGURE 1.3 Neuronal activity in area LIP discriminates saccade targets (black lines) from distractors (grey lines), but not distractor relevance (solid vs. dotted grey lines). (Adapted from Platt, M.L. and Glimcher, P.W., Responses of intraparietal neurons to saccadic targets and visual distractors, *J Neurophysiol*, 78, 1574, 1997. With permission.)

to the remembered location of that target. Subsequent studies extended these observations to saccades guided by acoustic stimuli.[92–94]

These neurophysiological studies revealed that signals in parietal cortex could be linked to sensory encoding, motor preparation, spatial memory, or covert attention. Many attempts were made to unequivocally assign signals in parietal cortex to the representation of a single category of information. In our own work,[95] we attempted to dissociate visual, attentional, and saccade preparatory signals in LIP using a task in which a centrally located color cue instructed a monkey that a gaze shift to one of two diametrically opposed yellow lights (the *target*) would be rewarded (Figure 1.3). Subjects were required to maintain gaze at the central colored light until cued to initiate the required movement. In one block of trials, offset of the central colored light cued the subject to shift gaze rapidly to the target (Figure 1.3: *solid lines*), while in a second block of trials offset of the light to which the subject was not instructed to shift gaze (the *distractor*) served as the saccade initiation cue (Figure 1.3: *dotted lines*).

We probed the activation of LIP neurons under three conditions: (1) when the response field stimulus was a saccade target (Figure 1.3: *black lines)*); (2) when the response field stimulus was a behaviorally irrelevant distractor (Figure 1.3: *grey solid line*); and (3) when the response field stimulus was a behaviorally meaningful distractor (Figure 1.3: *grey dotted line*). We made two primary observations. First, the neuronal activity associated with saccade targets was always greater than the activity associated with either relevant or irrelevant distractors. Second, the neuronal activity associated with distractors was independent of their behavioral relevance. These observations led us to conclude initially that the activation of LIP neurons signals the amplitude and direction of an impending saccade, rather than the location of a visual stimulus or the amount of attention allocated to a particular visual stimulus. However, the observation that neuronal activity associated with distractors did not fall to baseline levels of activity (Figure 1.3: *dashed line*) was difficult to reconcile with this hypothesis. If the activation of LIP neurons signaled an impending saccade, then why should there be any activity in LIP associated with distractors?

In the final analysis, our study failed to unequivocally separate visual, attentional, and motor preparatory signals in LIP. Many other attempts to segregate LIP signals into exclusive categories have also yielded inconclusive results (e.g., see References 90, 92, 96–98), although some researchers have succeeded at classifying subpopulations of parietal neurons as predominately sensory or motor.[99,100] Intriguingly, some of these studies indicated that some parietal neurons were activated by both visual and motor events, even when these events were spatially and temporally segregated.[100] These observations argue against any simplistic categorization of signals carried by parietal neurons as exclusively sensory, motor, or attentional in nature. Instead, parietal cortex may participate in a network that transforms incoming sensory signals into motor plans according to current behavioral goals.

Shadlen and Newsome[42,43] explicitly tested this hypothesis in an experiment requiring monkeys to report, with a gaze shift to one of two response targets, the net direction of motion in a random dot display. The difficulty of the discrimination was varied by systematically changing the proportion of dots moving coherently in any one direction. Intriguingly, neuronal activity in area LIP was higher whenever the motion stimulus was consistent with a movement into the response field compared to motion consistent with a movement out of the response field. Moreover, the rate at which neuronal activity increased from stimulus onset to saccade onset was a direct function of motion strength. The temporal profile of LIP activation in this task is consistent with the temporal integration of motion signals, such as those carried by neurons in the middle temporal area MT,[101] associated with a particular gaze shift. These data indicate that the information carried by LIP neurons reflects the gradual transformation of visual signals into oculomotor commands. Based on the results of this experiment, Shadlen and Newsome concluded that area LIP plays a role in the decision processes linking visual perception with orienting saccades.

While the decision to shift gaze is often guided by immediate perception, target choice, like other types of decisions, can also be influenced by prior experience.[46,102,103] Indeed, a wealth of psychological and economic studies of decision-making has demonstrated that most actions are chosen as a systematic function of their outcomes in the past.[104,105] This perspective suggests that, if parietal cortex does participate in the oculomotor decision process, then parietal neurons will also carry information reflecting the outcome of prior saccades. We tested this proposal in a series of studies in which we systematically varied the amount of reward and the probability of reward associated with a gaze shift of a particular direction and amplitude.

In the first study, we cued monkeys on each trial to shift gaze to one of two peripheral yellow lights, one of which was positioned within the response field of a LIP neuron under study. Across blocks of trials, the volume of fruit juice associated with correct gaze shifts to each peripheral light was systematically varied, while the total amount of fruit juice associated with both lights was held constant. For most LIP neurons, firing rate was systematically modulated by the relative value associated with gaze shifts into the response field (Figure 1.4a). Moreover, the modulation of

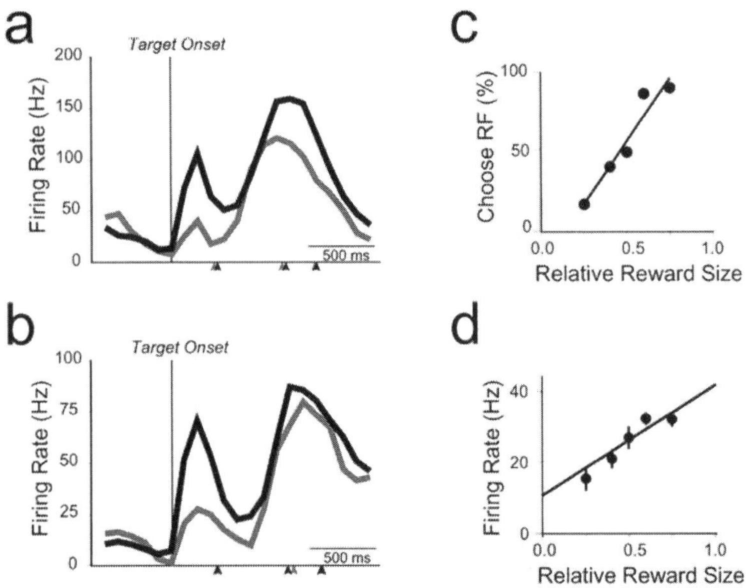

FIGURE 1.4 Representation of saccade value in area LIP. Increases in reward size (a) or probability (b) increase the activity of single LIP neurons on a cued saccade task. Thick black lines, high value; thick grey lines, low value. Arrows indicate average times of saccade direction cue, saccade initiation cue, and movement onset, on high (black) and low (grey) value trials. When monkeys are permitted to freely choose between two differentially reinforced targets, both movement choice (c) and neuronal activity associated with gaze shifts to the response field target (d) are a systematic function of relative movement value. (Adapted from Platt, M.L., Neural correlates of decisions, *Curr Opin Neurobiol*, 12, 141, 2002. With permission.)

firing rate by reward size peaked early in each trial and diminished around the time of the saccade.[102] Similar results were found when we held the amount of reward constant and varied the probability that the subject would be required to shift gaze to a target in the response field of each neuron under study (Figure 1.4b).

These data demonstrate that LIP neurons carry information reflecting the amount and probability of reward associated with a particular saccade, two variables known from behavioral studies to influence decision-making. The results of this study, however, do not address whether parietal neurons carry information correlated with the decisions subjects actually make when choosing between saccades associated with different value. To address this question, we conducted another experiment in which we permitted monkeys to choose freely between two targets, one of which was positioned within the response field of a neuron under study. Across blocks of trials, the volume of fruit juice delivered for shifting gaze to each target was systematically varied, as in the previous experiment. Under these conditions, both the choices subjects made and neuronal activity in parietal cortex were a systematic function of saccade value (Figure 1.4c,d). In fact, reward-related modulations in neuronal activity occurred before the onset of the peripheral targets (see also Coe et al.[70]; Shadlen and Newsome[43]), further reinforcing the conclusion that representations in LIP are not purely sensory.

Taken together, the available evidence suggests that neurons in parietal cortex represent the information, be it sensory data or reward expectations derived from prior experience, favoring the generation of one saccade over another. Adopting a Bayesian approach, Gold and Shadlen[46] proposed that neurons in parietal, as well as prefrontal, cortex compute the logarithm of the likelihood ratio associated with potential saccades by integrating current sensory evidence favoring each particular gaze shift with estimates of its behavioral value. According to these authors, the oculomotor system is then biased to generate the saccade associated with the highest log likelihood ratio. This proposal specifically suggests that parietal neurons integrate information from multiple sources and over time to derive the optimal oculomotor response. While this powerful model succeeds in accounting for many of the observed physiological response properties of LIP neurons, thus suggesting that parietal cortex may indeed serve as a log likelihood estimator for orienting saccades, the role of attention in this process remains to be fully explicated.

1.5 PREFRONTAL CORTEX

The prefrontal cortex (PFC) has been implicated in a wide array of behavioral and cognitive functions, including working memory,[106–108] attention,[65,109] planning,[65] decision-making,[44,110] motivation and emotion,[111,112] and executive control.[109,113,114] The prefrontal cortex itself may be divided into a more ventral and medial aspect principally comprised of Brodmann's areas 11, 12, 13, and 14,[115] and including for the purposes of this chapter the frontopolar cortex area 10,[116] as well as a more dorsal and lateral aspect, principally comprised of areas 9 and 46.[115,117,118] Medial PFC has been postulated by some researchers to be more strongly involved in emotional control and decision-making,[111,112,119–121] whereas lateral PFC has been argued to be more strongly involved in working memory, attention, and planning.[122–124]

The earliest suggestions that prefrontal cortex contributes to higher cognitive functions derived from neurological studies of patients with prefrontal lesions (e.g., see Reference 125), as well as studies of behavioral performance in lesioned monkeys (e.g., see Reference 119). Specifically, these early studies suggested that damage to primarily dorsal and lateral aspects of prefrontal cortex was associated with deficits in working memory, as measured by performance on delayed response tasks, whereas damage to primarily ventral and medial portions of prefrontal cortex was associated with deficits in behavioral inhibition, as measured by performance on discrimination reversal and go/no-go tasks. The foregoing distinctions, however, remain controversial, in part due to a lack of correspondence between humans and monkeys both in terms of the effects of lesions of these subregions on behavior and in terms of the underlying anatomy.[125,126] In fact, subsequent ablation studies in monkeys [122,123,127,128] have demonstrated that deficits in working memory and behavioral inhibition can be induced by damage to either dorsolateral or ventromedial prefrontal cortex, thereby blurring the distinctions drawn between the putative functions of these regions based on lesion studies.

Neurophysiological studies have also been employed in an effort to probe the function of prefrontal cortex. Early studies demonstrated that lateral PFC neurons maintain activation during delay periods on delayed response tasks,[124] consistent with findings from studies of monkeys and humans with prefrontal lesions demonstrating impaired performance on these same tasks. In addition, it has been argued that lateral PFC neurons are segregated anatomically according to the type of information held in working memory. Specifically, neurons in dorsolateral areas 9 and 46 were initially thought to be active during delay periods only when responses were directed to a specific location,[124,129,130] whereas neurons located in the inferior prefrontal convexity were thought to respond during delay periods only when particular objects needed to be remembered independent of their spatial location.[131] Recent work, however, suggests the possibility that subpopulations of prefrontal neurons within each of these areas may process either spatial information, object attributes, or both,[132,133] thus mitigating any strong conclusions regarding segregation of function within the lateral PFC.

Anatomically, dorsolateral PFC is well-situated to participate in oculomotor control. Dorsolateral PFC principally receives inputs from posterior parietal cortex (area 7a), medial parietal cortex (area 7m), posterior cingulate cortex (areas 23 and 31), retrosplenial cortex,[25] and the parvicellular mediodorsal nucleus of the thalamus,[134] areas thought to be important for visuospatial and mnemonic functions based on neurophysiological and neurological data. In turn, the dorsolateral prefrontal cortex projects directly to the intermediate and deep layers of the SC,[6,25,135] the FEF, and posterior parietal cortex,[25,26] as well as to the midline reticular formation of the pons.[6,25] The prefrontal cortex also projects directly to other brain areas known from physiological and neurological data to have functions related to vision and eye movements, including the anterior and posterior cingulate cortex, the caudate nucleus, and the pulvinar.[25,136]

These anatomical data suggest that dorsolateral PFC may participate directly in oculomotor processing. In support of this contention, Goldman-Rakic and colleagues first demonstrated that dorsolateral PFC neurons maintain activation during the delay period when subjects perform memory-guided saccade trials,[108,129,137] in a fashion similar to delay period responses of neurons in area LIP of posterior parietal cortex on the same type of task.[91] Moreover, muscimol inactivation of small sites in dorsolateral PFC results in deficits for contraversive memory-guided saccades with error proportional to delay period length.[138]

More recent neurophysiological studies have demonstrated that, in addition to contributing to memory-guided saccades, dorsolateral PFC neurons carry information correlated with the sensory evidence favoring the generation of a particular eye movement. Specifically, Kim and Shadlen[44] reported that the activation of lateral PFC neurons is correlated with both the level of coherent motion and psychophysical performance in a random-dot motion discrimination task, in which decisions were reported with an eye movement. Similarly, Constantinidis, Goldman-Rakic, and colleagues[139] showed that the activation of lateral PFC neurons, even during a delay period, is correlated with both brightness and psychophysical performance on a luminance discrimination task, in which decisions were also reported with a saccade.

These data are consistent with the proposal that populations of dorsolateral PFC neurons encode information used to compute an oculomotor decision, particularly when that information must be stored in memory during a delay. This supposition is supported by the observation that dorsolateral PFC neurons are also sensitive to learned rules specifying the location of a saccade target.[133] The observation that some dorsolateral PFC neurons preferentially encode visual stimulus attributes such as location and brightness, independent of saccade metrics,[140] while other dorsolateral PFC neurons encode saccade metrics, independent of visual stimulus attributes,[141] suggests that, like posterior parietal cortex, dorsolateral PFC may also be involved in the sensory-motor transformations that link sensory representations with oculomotor outputs.

These data support the inclusion of the dorsolateral PFC within the oculomotor system. The observation of strong interconnections between dorsolateral PFC and ventromedial PFC, which is itself robustly interconnected with the amygdala and temporal visual cortical areas (reviewed in Rolls[111,121]) suggests that dorsolateral PFC may participate in the oculomotor decision process by integrating information across multiple sensory modalities with motivational and emotional information derived from past experience.[46] Indeed, the sensitivity of lateral PFC neurons to reward[47,140] suggests that afferents from ventromedial and orbitofrontal PFC may be specifically dedicated to processing the motivational content of saccades under consideration by the oculomotor decision network, although more global motivational inputs from midbrain dopamine neurons may also contribute to these computations in lateral PFC.[142] The complexity of processing intimated by the wide array of anatomical interconnections shared by PFC suggests that, by virtue of its simplicity, oculomotor control will continue to serve as a useful model for PFC function in general.

1.6 CINGULATE CORTEX

Although the cingulate cortex, within the depths of the cingulate sulcus as well as along the medial wall of the cingulate gyrus, has long been recognized as an important site of visceral, motivational, and emotional information-processing,[143,144] it has also been linked to both visual[145–147] and motor[148–150] processing based on anatomical and physiological evidence. Recently, two areas within cingulate cortex have been implicated specifically in oculomotor processing. A region in anterior cingulate cortex (CGa), located in Brodmann's area 24 in medial frontal cortex, has been shown by both clinical and functional imaging studies to contribute to the control of voluntary saccades.[151–154] A portion of posterior cingulate cortex (CGp), lying along the ventral bank of the cingulate sulcus as well as along the cingulate gyrus and comprised principally of Brodmann's areas 23, 29, 30, and 31, has also been implicated in the control of eye movements and visual attention based on anatomical,[145,147,155–157] electrophysiological,[146,149] and neuroimaging studies.[151,152,158,159]

Anatomically, cingulate cortex does not communicate directly with the SC. Cingulate cortex does, however, make extensive connections with other areas involved in oculomotor control. As noted previously, both the CGa and the CGp are strongly interconnected with the SEF,[23] with limited reciprocal connections with the

FEF.[160–162] In addition, lateral PFC areas 9 and 46 project heavily to both anterior and posterior cingulate cortex.[25,161,163–165]

Moreover, the lateral and medial surfaces of the parietal lobe are strongly interconnected with cingulate cortex. Area 7m, on the medial surface of the parietal lobe, is reciprocally interconnected with CGp, and to a lesser extent, CGa.[22,161,166,167] Area 7m is reciprocally connected with area LIP,[22] and stimulation in area 7m has recently been shown to evoke saccadic eye movements with currents as low as 50 microamperes.[168] In addition, the lateral surface of the inferior parietal lobule (area 7 or PG) is also reciprocally interconnected with CGp,[22,161] and retrosplenial cortex within the cingulate gyrus projects to areas 7a and LIP as well.[22] Finally, autoradiographic experiments suggest that the prefrontal and parietal projections to cingulate cortex may interdigitate in a columnar fashion, suggesting a topographical convergence of information in these areas.[25]

In addition to these oculomotor and visuospatial connections, cingulate cortex makes direct anatomical contact with brain areas involved in motivational processing. Nearly all parts of orbitofrontal cortex, for example, project to both CGa and CGp.[161] Moreover, anterior cingulate cortex receives substantial projections from the amygdala.[155,169,170] In turn, anterior cingulate cortex projects to the laterobasal and central nuclei of the amygdala.[171,172] Finally, anterior and posterior cingulate cortices are also strongly interconnected, thereby permitting extensive communication of motivational, as well as visuospatial and oculomotor, information.[162] These anatomical data suggest the hypothesis that cingulate cortex contributes motivational information to the oculomotor decision process.[103,156,173,174]

Anterior cingulate cortex has been implicated in oculomotor control principally through recent neuroimaging and neurological studies. Specifically, several recent studies have revealed hemodynamic changes in anterior cingulate cortex using PET[151–153] and fMRI[159] imaging in humans asked to shift gaze to visible or remembered targets. The potential importance of anterior cingulate cortex for oculomotor control suggested by neuroimaging studies is buttressed by recent neurological studies demonstrating deficits in oculomotor performance, including visually guided saccades, memory-guided saccades, antisaccades, and memorized sequences of visually guided saccades, in patients with damage to this area.[154,175] While these studies demonstrate that anterior cingulate cortex participates in the cortical control of saccades, they do not address the possibility that this region contributes motivational information to the oculomotor decision process.

The activation patterns of anterior cingulate neurons have just begun to be studied in monkeys performing oculomotor tasks. In a recent review, Schall and colleagues[63] reported that some CGa neurons respond to primary reinforcers, as well as to visual targets, on saccade countermanding tasks. Intriguingly, electrophysiological studies have also demonstrated that neurons in anterior cingulate cortex carry signals related to both expected and experienced reward outcomes associated with manual responses,[176,177] and that these signals can faithfully predict movement choices.[176] Although the preliminary nature of the data reported by Schall and colleagues precludes drawing any strong conclusions regarding the specific role of anterior cingulate cortex in oculomotor control, the results of these studies do suggest the possibility that one role of CGa may be to monitor the motivational outcomes of visually guided saccades. Such signals could serve to update representations of expected saccade value computed during ongoing oculomotor decision processing in parietal and prefrontal cortex.[63,156,178]

Recent neurophysiological and neuroimaging studies have also implicated posterior cingulate cortex in eye movement processing. Olson and Musil[146] first demonstrated that single neurons in posterior cingulate cortex of the cat fire action potentials in response to visual stimulation, and also respond following saccadic eye movements. Olson, Goldberg, and colleagues later demonstrated that CGp neurons are activated just after saccade onset when monkeys shift gaze to contralateral visual targets.[149] Moreover, they reported that the activity of many CGp neurons is modulated by both eye position and saccade direction and amplitude.

Subsequent studies have confirmed the temporal pattern of neuronal activation in CGp reported by Olson and colleagues, but have also suggested that many CGp neurons are activated following the delivery of juice reinforcement on delayed saccade trials as well.[174,179,180] The post-saccadic and post-reward responses of most CGp neurons are also modulated by reward magnitude.[174,180] These neurophysiological reports suggest that CGp may participate in the assignment of motivational value to oculomotor signals, thereby contributing to the neural decision processes that eventually generate movements of the eyes.

In addition to these neurophysiological studies in animals, one recent neuroimaging study has reported hemodynamic changes in posterior cingulate cortex in humans instructed to shift gaze to a visual target.[159] In that study, fMRI scans were made in human subjects while they either fixated a central stimulus, shifted gaze to one of seven visual targets, or smoothly pursued a small target moving at an average speed of 10 degrees/sec along the horizontal meridian. The authors found that many cortical areas, including CGp and CGa, were activated in both the pursuit and saccade conditions. Intriguingly, activation in CGp was greater for smooth pursuit than for saccades, and peak activation was localized within the depths of the cingulate sulcus, just anterior to the marginal ramus. The locus of this activation is consistent with the region in which neurons with oculomotor responses have been recorded in the monkey.[149,179] In agreement with CGp activation revealed in humans using fMRI, Porro and colleagues[181] recently used the 2-deoxyglucose technique to demonstrate increased metabolic activity in the cingulate gyrus and retrosplenial cortex during ketamine-induced oculomotor nystagmus in the rat.

In summary, anatomical data implicate both anterior and posterior cingulate cortex in the cortical control of eye movements. Recent observations of cingulate activation during and after eye movements accord well with this implication. Moreover, anatomical and electrophysiological data intimate that cingulate cortex may play an important role in assigning motivational value to oculomotor signals, and may contribute to the reward-related modulation of saccade-related activity observed in area LIP, the dorsolateral PFC, and the SEF. These data suggest that frontal, parietal, and cingulate cortices may form a network for oculomotor decision-making that contributes to the selection, preparation, and initiation of a gaze shift by downstream oculomotor control structures in the FEF and SC.[178] This hypothesis awaits further testing with neurophysiological techniques in animals and neuroimaging techniques in humans.

1.7 BASAL GANGLIA

In primates, a wide array of brain areas involved in oculomotor processing project to the caudate nucleus of the basal ganglia. Dense inputs to the caudate nucleus originate in a number of oculomotor areas including the FEF, SEF, area LIP, dorsolateral PFC, and the cingulate cortices.[25,182–184] The caudate nucleus, in turn, sends its principal outputs to the substantia nigra pars reticulata (SNr), which together with the globus pallidus serve as the principal output nuclei of the basal ganglia. The SNr is composed largely of GABA-ergic neurons[185], and receives inhibitory inputs from the caudate[186,187] and excitatory inputs from the subthalamic nucleus.[188] The SNr projects back to many of the cortical areas that innervate the caudate, principally via the ventral anterior and mediodorsal nuclei of the thalamus.[189–191] In addition, anatomical studies have clearly established that the SNr sends projections to the intermediate layers of the superior colliculus.[185,192] This nigrotectal pathway forms both a major output of the SNr and a major afferent source for the SC,[193–195] placing the basal ganglia in a position to influence oculomotor signals impinging on the SC.

Hikosaka and colleagues have developed a model of saccadic initiation that synthesizes a large body of anatomical, physiological and clinical work. They propose that cortical signals from areas like the FEF access the caudate and are then conducted as inhibitory signals to the SNr. This inhibition produces a decrease in SNr activity, which releases the SC from tonic inhibition, permitting the initiation of a saccade. This hypothesis is supported by five physiological observations (reviewed in detail by Hikosaka and Wurtz,[196]; Hikosaka et al.[197]). (1) Neurons in the SNr have high baseline firing rates, and show decreases in firing rate during oculomotor tasks.[198–201] (2) Many saccade-related neurons in the SNr can be antidromically activated by microstimulating the intermediate layers of the SC,[202] and SNr microstimulation evokes monosynaptic hyperpolarization in the SC.[203–204] (3) GABA powerfully modulates SC function; injection of the GABA agonist muscimol or the GABA antagonist bicuculline into the SC results in severe deficits in generating saccades and inhibiting saccades, respectively.[205] (4) Injecting muscimol into the SNr increases saccade frequency and substantially reduces the average duration of intersaccadic fixation, similar to the effects seen when bicuculline is injected into the SC.[206,207] (5) Electrical microstimulation of the caudate nucleus can inhibit some SNr neurons,[208] and these cells show saccade-related decreases in activation that coincide temporally with saccade-related increases in caudate activity. Taken together, this body of evidence provides strong support for the role of the SNr in disinhibiting SC neurons.

While decreases in SNr activity are consistent with disinhibition of the SC, recent work has revealed a population of SNr neurons that show saccade-related *increases* in activity.[209–211] Handel and Glimcher[209] identified four classes of SNr activity using a delayed saccade task. Two classes showed decreases in activity similar to that originally observed by Hikosaka and Wurtz. Handel and Glimcher also described two novel cell classes, comprising 41% of the neurons they recorded from, which showed increases in activity during their saccade task. One class, termed bursters, showed increases in activity following the presentation of contralateral saccadic targets and/or before the generation of saccades. The second class, termed pause-bursters, showed increased activation following target presentation in the contralateral hemisphere but decreased activation following target presentation in the ipsilateral hemisphere. All four classes of SNr neurons showed a strong tendency for maximal modulation from a high baseline during contralateral saccades, regardless of whether the modulation was an increase or a decrease from baseline. Quantitative analyses revealed that the movement fields of all SNr neurons were well described as tilted planes. This is distinct from movement fields in the SC, area LIP, and the FEF, which are better characterized as Gaussian,[17,212,213] suggesting that SNr neurons specify the metrics of upcoming saccades in a quantitatively different manner from neurons in the SC.

It is noteworthy that while Hikosaka and colleagues[208] observed decreases in SNr activity following microstimulation of the caudate nucleus, nearly half of their visual- or saccade-related SNr cells showed either excitation or complex combinations of excitation and inhibition after caudate stimulation. Hikosaka and colleagues suggest that increases may be caused by excitatory inputs from the subthalamic nucleus, which is disinhibited by caudate excitation.[197,208] This indirect pathway may also mediate the increases in firing rate seen in bursters and pause-bursters, but it remains unclear what role these neurons have in generating saccades. One possibility is that these neurons act to inhibit competing saccades at the level of the SC, thereby aiding in the selection of a single movement.[210,214] A related idea is that bursters and pause-bursters inhibit fixation neurons in the rostral SC,[215] facilitating gaze shifts from the current locus of fixation. Thus, neurons with increases in activity could complement the direct nigrotectal disinhibition mediated by neurons showing decreases in activity.

Neurons in the SNr are not obligately linked to the production of all saccades, and exhibit differential activity depending on behavioral context (e.g., Hikosaka and Wurtz,[198–200]). Early observations suggested that SNr neurons were never modulated during spontaneous eye movements.[198] One important difference between a spontaneous saccade and a saccade made in an operant task is the reinforcement contingency, which raises the possibility that SNr neurons participate in coding the value of rewards associated with particular movements. This is supported by the fact that most SNr neurons are only modulated during saccades that occur in close temporal proximity to rewards. Handel and Glimcher[216] compared firing rates during fixational saccades that aligned the eyes with a central stimulus to begin a trial with firing rates during terminal saccades that aligned the eyes with an eccentric target at the end of a trial. Although both types of saccade predicted reward, only terminal saccades were immediately followed by reward (Figure 1.5). A quantitative movement field analysis showed that all SNr neurons were modulated during terminal saccades, but were unmodulated or only weakly modulated during fixational movements. These results suggest that SNr neurons carry signals related to the reinforcement that can be expected for a particular eye movement.

Sato and Hikosaka[211] explicitly tested the hypothesis that SNr neurons are modulated by reward contingencies. They trained monkeys to make memory-guided saccades to one of two eccentric targets. One of the targets was randomly cued on each trial, and for blocks of 40 trials, only movements to one target were rewarded.

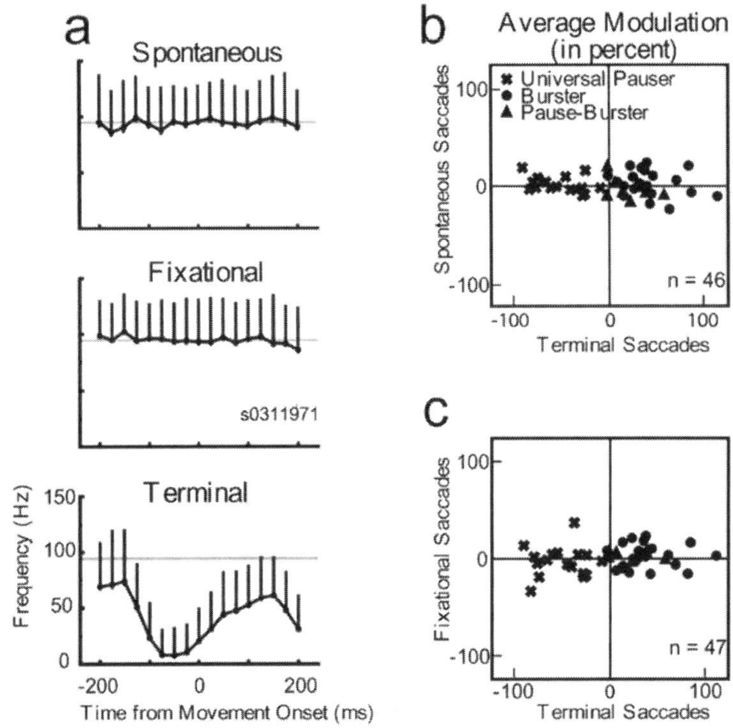

FIGURE 1.5 (a) Comparison of activity for a typical universal pauser during spontaneous, fixational and terminal saccades. Perievent histograms show the mean rate, standard deviations (upward bars), and standard errors (downward bars) in 25 millisecond bins. Gray line indicates baseline activity. Correlations between spontaneous and terminal saccades (b) and fixational and terminal saccades (c). Average modulation is derived from planar fits across all possible movements for each neuron, and is expressed as a percentage relative to baseline activity. (Adapted from Handel, A. and Glimcher, P.W., Contextual modulation of substantia nigra pars reticulata neurons, *J Neurophysiol*, 83, 3042, 2000. With permission.)

Movements to the other target, while unrewarded, were still required. Sato and Hikosaka observed that half their neurons showed decreases in response to a visual cue, and tended to decrease more when the cue indicated an upcoming reward. The remaining neurons showed increases in response to visual cues, and were more heterogeneous in their response properties. Of the cells differentially modulated by reward contingency, roughly half increased their activity more when the cue indicated an upcoming reward, while the other half increased their activity more when the cue indicated no reward. Sato and Hikosaka hypothesize that the larger decreases in activity act to increase the likelihood of a saccade to the rewarded target by selectively disinhibiting the SC. This idea is supported by behavioral work showing that monkeys performing this differential reward task reliably produce faster saccades to the rewarded target.[217]

Projections from the caudate nucleus are a likely source of the reward-related effects observed by Sato and Hikosaka. Caudate neurons show responses related to receipt of reward, expectation of reward, and reward contingency during a variety of oculomotor tasks.[218–221] Kawagoe and colleagues[219] identified one class of caudate neurons that increased their activity when a particular target was rewarded, and a second class that decreased their activity when a particular target was rewarded. Caudate neurons receive reward-related information from a variety of sources including the dopaminergic neurons in the SNc,[222] as well as the orbitofrontal cortex.[182] These caudate neurons, which are thought to project to the SNr, could produce reward-dependent decreases or increases of activity in the SNr.

The dopamine neurons in the SNc are another potential source of the reward-related activity in the SNr. SNc neurons discharge following receipt of rewards or sensory stimuli that predict rewards (reviewed by Schultz[223]), and there is abundant evidence that the two nuclei of the substantia nigra are functionally integrated. Anatomical studies of the nucleus show large apical dendrites extending from the SNc into the SNr, forming an anatomical substrate for this interaction.[224] Dopamine can be released from these dendritic processes,[225] and D1 and D2 dopamine receptors have been localized in the SNr.[226] Finally, dopamine application has been shown to affect both spontaneous activity and striatonigral GABA-ergic inhibition in the SNr.[227,228] While all of these anatomical and physiological data indicate that it is possible for SNc activity to modulate SNr activity, it nonetheless remains the case that very little is known about how these two cell groups jointly process information or participate in the generation of eye movements.

The behavioral contingency of neural activity in the caudate nucleus and the SNr suggest that the oculomotor functions of these areas may be more closely related to the selection, planning and initiation of reinforced movements than simply to the generation of movements per se. The results described above represent part of a growing body of evidence (reviewed in References 229 and 230) that a major function of the basal ganglia is the acquisition of goal-directed behavior.

1.8 SUMMARY

The foregoing review suggests that the SC receives orienting signals originating in areas that appear to be computationally similar, like the FEF, and from areas that appear to be computationally distinct, like the posterior cingulate cortex. All of these areas form a coherent and heavily interconnected network that can select orienting movements from the animal's repertoire for execution based upon sensory and nonsensory signals.

The basal ganglia, for example, appear to carry information important for the execution of saccades that yield rewards. These signals also make their way to the frontal cortices via the thalamus where they can alter activity in those areas and thereby influence saccade generation. In a similar way, parietal areas that participate in saccade generation send strong projections to areas like the FEF, as well as to the SC, thereby apparently influencing saccade generation via more than one route. It is for this reason that it is critical that we begin to think of the SC as a participant, albeit a critical one, in an orienting movement control *network*.

The importance of this observation becomes especially clear when we consider the activity that has already been documented in each of these areas. Current evidence indicates that when a saccade is produced all of these areas work together, generating a common set of signals that leads to the regulation and generation of a single orienting movement. We know very little about how this complex network-level coordination is achieved. We know almost nothing, for example, about whether the SC receives conflicting signals under some conditions or about how such conflicting signals might be resolved either by the SC itself or by the larger network in which it is embedded. Obviously, if we are to truly understand the functional role of the SC we will have to understand the larger network in which it is embedded.

The work reviewed briefly in this chapter represents the first stages of a process aimed at understanding the role of the SC within a larger framework. We are now just beginning to understand what each of these other areas does and how it might influence the SC. The next step will be to begin to ask how these areas interact as a network in order to define more fully the functional role of the SC itself.

ACKNOWLEDGMENTS

The authors would like to thank M. Bendiksby, H.L. Dean, J.D. Roitman, H.M. Bayer, V.M. Ciaramitaro, M.C. Dorris, and M. Yanike for making helpful comments on the manuscript. MLP is supported by the McDonnell-Pew program in Cognitive Neuroscience, the Alfred P. Sloan Foundation, the Whitehall Foundation, the EJLB Foundation, the Klingenstein Foundation, and EY 013496. PLG and BL are supported by the Klingenstein Foundation, the James S. McDonnell Foundation, the McKnight Foundation, and EY 10536. BL is supported by a Department of Defense NDSEG fellowship.

REFERENCES

1. Schiller, P.H., True, S.D., and Conway, J.L., Deficits in eye movements following frontal eye-field and superior colliculus ablations, *J Neurophysiol*, 44, 1175, 1980.
2. Horwitz, G.D. and Newsome, W.T., Separate signals for target selection and movement specification in the superior colliculus, *Science*, 284, 1158, 1999.
3. Ferrier, D., The localization of function in the brain, *Proceedings of the Royal Society of London*, 22, 228, 1874.
4. Robinson, D.A. and Fuchs, A.F., Eye movements evoked by stimulation of frontal eye fields, *J Neurophysiol*, 32, 637, 1969.
5. Bruce, C.J. et al., Primate frontal eye fields. II. Physiological and anatomical correlates of electrically evoked eye movements, *J Neurophysiol*, 54, 714, 1985.
6. Fries, W., Cortical projections to the superior colliculus in the macaque monkey: a retrograde study using horseradish peroxidase, *J Comp Neurol*, 230, 55, 1984.
7. Huerta, M.F., Krubitzer, L.A., and Kaas, J.H., Frontal eye field as defined by intracortical microstimulation in squirrel monkeys, owl monkeys, and macaque monkeys: I. Subcortical connections, *J Comp Neurol*, 253, 415, 1986.
8. Segraves, M.A. and Goldberg, M.E., Functional properties of corticotectal neurons in the monkey's frontal eye field, *J Neurophysiol*, 58, 1387, 1987.
9. Sommer, M.A. and Wurtz, R.H., Composition and topographic organization of signals sent from the frontal eye field to the superior colliculus, *J Neurophysiol*, 83, 1979, 2000.

10. Stanton, G.B., Goldberg, M.E., and Bruce, C.J., Frontal eye field efferents in the macaque monkey: II. Topography of terminal fields in midbrain and pons, *J Comp Neurol*, 271, 493, 1988.
11. Segraves, M.A., Activity of monkey frontal eye field neurons projecting to oculomotor regions of the pons, *J Neurophysiol*, 68, 1967, 1992.
12. Hanes, D.P. and Wurtz, R.H., Interaction of the frontal eye field and superior colliculus for saccade generation, *J Neurophysiol*, 85, 804, 2001.
13. Schall, J.D. and Thompson, K.G., Neural selection and control of visually guided eye movements, *Annu Rev Neurosci*, 22, 241, 1999.
14. Bizzi, E., Discharge of frontal eye field neurons during saccadic and following eye movements in unanesthetized monkeys, *Exp Brain Res*, 6, 69, 1968.
15. Wurtz, R.H. and Mohler, C.W., Enhancement of visual responses in monkey striate cortex and frontal eye fields, *J Neurophysiol*, 39, 766, 1976.
16. Goldberg, M.E. and Bushnell, M.C., Behavioral enhancement of visual responses in monkey cerebral cortex. II. Modulation in frontal eye fields specifically related to saccades, *J Neurophysiol*, 46, 773, 1981.
17. Bruce, C.J. and Goldberg, M.E., Primate frontal eye fields.I. Single neurons discharging before saccades, *J Neurophysiol*, 53, 603, 1985.
18. Gottlieb, J.P., MacAvoy, M.G., and Bruce, C.J., Neural responses related to smooth-pursuit eye movements and their correspondence with electrically elicited smooth eye movements in the primate frontal eye field, *J Neurophysiol*, 72, 1634, 1994.
19. Hanes, D.P., Patterson, W.F., 2nd, and Schall, J.D., Role of frontal eye fields in countermanding saccades: visual, movement, and fixation activity, *J Neurophysiol*, 79, 817, 1998.
20. Stanton, G.B., Bruce, C.J., and Goldberg, M.E., Topography of projections to posterior cortical areas from the macaque frontal eye fields, *J Comp Neurol*, 353, 291, 1995.
21. Schall, J.D. et al., Topography of visual cortex connections with frontal eye field in macaque: convergence and segregation of processing streams, *J Neurosci*, 15, 4464, 1995.
22. Cavada, C. and Goldman-Rakic, P.S., Posterior parietal cortex in rhesus monkey: II. Evidence for segregated corticocortical networks linking sensory and limbic areas with the frontal lobe, *J Comp Neurol*, 287, 422, 1989.
23. Huerta, M.F. and Kaas, J.H., Supplementary eye field as defined by intracortical microstimulation: Connections in macaques, *J Comp Neurol*, 293, 299, 1990.
24. Schall, J.D., Morel, A., and Kaas, J.H., Topography of supplementary eye field afferents to frontal eye field in macaque: implications for mapping between saccade coordinate systems, *Vis Neurosci*, 10, 385, 1993.
25. Selemon, L.D. and Goldman-Rakic, P.S., Common cortical and subcortical targets of the dorsolateral prefrontal and posterior parietal cortices in the rhesus monkey: evidence for a distributed neural network subserving spatially guided behavior, *J Neurosci*, 8, 4049, 1988.
26. Andersen, R.A. et al., Corticocortical connections of anatomically and physiologically defined subdivisions within the inferior parietal lobule, *J Comp Neurol*, 296, 65, 1990.
27. Thompson, K.G. et al., Perceptual and motor processing stages identified in the activity of macaque frontal eye field neurons during visual search, *J Neurophysiol*, 76, 4040, 1996.
28. Bichot, N.P. et al., Reliability of macaque frontal eye field neurons signaling saccade targets during visual search, *J Neurosci*, 21, 713, 2001.

29. Sato, T. et al., Search efficiency but not response interference affects visual selection in frontal eye field, *Neuron*, 30, 583, 2001.
30. Thompson, K.G., Bichot, N.P., and Schall, J.D., Dissociation of visual discrimination from saccade programming in macaque frontal eye field, *J Neurophysiol*, 77, 1046, 1997.
31. Murthy, A., Thompson, K.G., and Schall, J.D., Dynamic dissociation of visual selection from saccade programming in frontal eye field, *J Neurophysiol*, 86, 2634, 2001.
32. Bichot, N.P., Chenchal Rao, S., and Schall, J.D., Continuous processing in macaque frontal cortex during visual search, *Neuropsychologia*, 39, 972, 2001.
33. McPeek, R.M. and Keller, E.L., Short-term priming, concurrent processing, and saccade curvature during a target selection task in the monkey, *Vision Res*, 41, 785, 2001.
34. Hanes, D.P. and Schall, J.D., Neural control of voluntary movement initiation, *Science*, 274, 427, 1996.
35. Roitman, J.D. and Shadlen, M.N., Response of neurons in the lateral intraparietal area during a combined visual discrimination reaction time task, *J Neurosci*, 22, 9475, 2002.
36. McPeek, R.M. and Keller, E.L., Superior colliculus activity related to concurrent processing of saccade goals in a visual search task, *J Neurophysiol*, 87, 1805, 2002.
37. Gold, J.I. and Shadlen, M.N., Representation of a perceptual decision in developing oculomotor commands, *Nature*, 404, 390, 2000.
38. Sommer, M.A. and Wurtz, R.H., Frontal eye field sends delay activity related to movement, memory, and vision to the superior colliculus, *J Neurophysiol*, 85, 1673, 2001.
39. Glimcher, P.W. and Sparks, D.L., Movement selection in advance of action in the superior colliculus, *Nature*, 355, 542, 1992.
40. Kustov, A.A. and Robinson, D.L., Shared neural control of attentional shifts and eye movements, *Nature*, 384, 74, 1996.
41. Dorris, M.C. and Munoz, D.P., Saccadic probability influences motor preparation signals and time to saccadic initiation, *J Neurosci*, 18, 7015, 1998.
42. Shadlen, M.N. and Newsome, W.T., Motion perception: seeing and deciding, *Proc Natl Acad Sci U S A*, 93, 628, 1996.
43. Shadlen, M.N. and Newsome, W.T., Neural basis of a perceptual decision in the parietal cortex (area LIP) of the rhesus monkey, *J Neurophysiol*, 86, 1916, 2001.
44. Kim, J.N. and Shadlen, M.N., Neural correlates of a decision in the dorsolateral prefrontal cortex of the macaque, *Nat Neurosci*, 2, 176, 1999.
45. Glimcher, P.W., Making choices: the neurophysiology of visual-saccadic decision making, *Trends Neurosci*, 24, 654, 2001.
46. Gold, J.I. and Shadlen, M.N., Neural computations that underlie decisions about sensory stimuli, *Trends Cogn Sci*, 5, 10, 2001.
47. Leon, M.I. and Shadlen, M.N., Effect of expected reward magnitude on the response of neurons in the dorsolateral prefrontal cortex of the macaque, *Neuron*, 24, 415, 1999.
48. Woolsey, C.N. et al., Patterns of localization in precentral and "supplementary" motor areas and their relation to the concept of a premotor area., *Res Publ Assoc Res Nerv Ment Dis*, 30, 238, 1952.
49. Schlag, J. and Schlag-Rey, M., Evidence for a supplementary eye field, *J Neurophysiol*, 57, 179, 1987.
50. Tian, J.R. and Lynch, J.C., Corticocortical input to the smooth and saccadic eye movement subregions of the frontal eye field in Cebus monkeys, *J Neurophysiol*, 76, 2754, 1996.

51. Schall, J.D., Visuomotor areas of the frontal lobe, in *Cerebral Cortex: Extrastriate Cortex in Primates (Vol 12)*, Rockland, K.S., Kaas, J.H., and Peters, A., Eds., Plenum Press, New York, 1997, 527.
52. Shook, B.L., Schlag-Rey, M., and Schlag, J., Primate supplementary eye field: I. Comparative aspects of mesencephalic and pontine connections, *J Comp Neurol*, 301, 618, 1990.
53. Tehovnik, E.J., Lee, K., and Schiller, P.H., Stimulation-evoked saccades from the dorsomedial frontal cortex of the rhesus monkey following lesions of the frontal eye fields and superior colliculus, *Exp Brain Res*, 98, 179, 1994.
54. Schall, J.D., Neuronal activity related to visually guided saccadic eye movements in the supplementary motor area of rhesus monkeys, *J Neurophysiol*, 66, 530, 1991.
55. Russo, G.S. and Bruce, C.J., Supplementary eye field: representation of saccades and relationship between neural response fields and elicited eye movements, *J Neurophysiol*, 84, 2605, 2000.
56. Heinen, S.J. and Liu, M., Single-neuron activity in the dorsomedial frontal cortex during smooth-pursuit eye movements to predictable target motion, *Vis Neurosci*, 14, 853, 1997.
57. Schall, J.D., Neuronal activity related to visually guided saccades in the frontal eye fields of rhesus monkeys: comparison with supplementary eye fields, *J Neurophysiol*, 66, 559, 1991.
58. Russo, G.S. and Bruce, C.J., Effect of eye position within the orbit on electrically elicited saccadic eye movements: a comparison of the macaque monkey's frontal and supplementary eye fields, *J Neurophysiol*, 69, 800, 1993.
59. Missal, M. and Heinen, S.J., Facilitation of smooth pursuit initiation by electrical stimulation in the supplementary eye fields, *J Neurophysiol*, 86, 2413, 2001.
60. Stuphorn, V., Taylor, T.L., and Schall, J.D., Performance monitoring by the supplementary eye field, *Nature*, 408, 857, 2000.
61. Amador, N., Schlag-Rey, M., and Schlag, J., Reward-predicting and reward-detecting neuronal activity in the primate supplementary eye field, *J Neurophysiol*, 84, 2166, 2000.
62. Mann, S.E., Thau, R., and Schiller, P.H., Conditional task-related responses in monkey dorsomedial frontal cortex, *Exp Brain Res*, 69, 460, 1988.
63. Schall, J., Stuphorn, V., and Brown, J., Monitoring and control of action by the frontal lobes, *Neuron*, 36, 309, 2002.
64. Schlag-Rey, M. et al., Antisaccade performance predicted by neuronal activity in the supplementary eye field, *Nature*, 390, 398, 1997.
65. Passingham, R.E., *The Frontal Lobes and Voluntary Action*, Oxford University Press, 1993.
66. Mitz, A.R., Godschalk, M., and Wise, S.P., Learning-dependent neuronal activity in the premotor cortex: activity during the acquisition of conditional motor associations, *J Neurosci*, 11, 1855, 1991.
67. Chen, L.L. and Wise, S.P., Neuronal activity in the supplementary eye field during acquisition of conditional oculomotor associations, *J Neurophysiol*, 73, 1101, 1995.
68. Chen, L.L. and Wise, S.P., Evolution of directional preferences in the supplementary eye field during acquisition of conditional oculomotor associations, *J Neurosci*, 16, 3067, 1996.
69. Chen, L.L. and Wise, S.P., Supplementary eye field contrasted with the frontal eye field during acquisition of conditional oculomotor associations, *J Neurophysiol*, 73, 1122, 1995.

70. Coe, B. et al., Visual and anticipatory bias in three cortical eye fields of the monkey during an adaptive decision-making task, *J Neurosci*, 22, 5081, 2002.
71. Gaymard, B., Pierrot-Deseilligny, C., and Rivaud, S., Impairment of sequences of memory-guided saccades after supplementary motor area lesions, *Ann Neurol*, 28, 622, 1990.
72. Muri, R.M. et al., Effects of transcranial magnetic stimulation over the region of the supplementary motor area during sequences of memory-guided saccades, *Exp Brain Res*, 104, 163, 1995.
73. Tobler, P.N. and Muri, R.M., Role of human frontal and supplementary eye fields in double step saccades, *Neuroreport*, 13, 253, 2002.
74. Schiller, P.H. and Chou, I.H., The effects of frontal eye field and dorsomedial frontal cortex lesions on visually guided eye movements, *Nat Neurosci*, 1, 248, 1998.
75. Sommer, M.A. and Tehovnik, E.J., Reversible inactivation of macaque dorsomedial frontal cortex: effects on saccades and fixations, *Exp Brain Res*, 124, 429, 1999.
76. Lu, X., Matsuzawa, M., and Hikosaka, O., A neural correlate of oculomotor sequences in supplementary eye field, *Neuron*, 34, 317, 2002.
77. Balint, R., Seelenlähmung des "Schauens," optische Ataxia, räumliche Störung der Aufmerksamkeit, *Monatsschrift für Psychiatrie und Neurologie*, 25, 51, 1909.
78. Holmes, G., Disturbances of visual orientation, *Br. J. Ophthalmol.*, 2, 449, 1918.
79. Brain, W.R., Visual disorientation with special reference to lesions of the right cerebral hemisphere, *Brain*, 64, 244, 1941.
80. Lynch, J.C. and McLaren, J.W., Deficits of visual attention and saccadic eye movements after lesions of parietooccipital cortex in monkeys, *J Neurophysiol*, 61, 74, 1989.
81. Li, C.S., Mazzoni, P., and Andersen, R.A., Effect of reversible inactivation of macaque lateral intraparietal area on visual and memory saccades, *J Neurophysiol*, 81, 1827, 1999.
82. Ungerleider, L.G. and Mishkin, M., Two cortical visual systems, in *Analysis of Visual Behavior*, Ingle, D.J., Goodale, M.A., and Mansfield, R.J.W., Eds., The MIT Press, Cambridge, MA, 1982.
83. Pandya, D.N. and Seltzer, B., Intrinsic connections and architectonics of posterior parietal cortex in the rhesus monkey, *J Comp Neurol*, 204, 196, 1982.
84. Blatt, G.J., Andersen, R.A., and Stoner, G.R., Visual receptive field organization and cortico-cortical connections of the lateral intraparietal area (area LIP) in the macaque, *J Comp Neurol*, 299, 421, 1990.
85. Clower, D.M. et al., The inferior parietal lobule is the target of output from the superior colliculus, hippocampus, and cerebellum, *J Neurosci*, 21, 6283, 2001.
86. Mountcastle, V.B. et al., Posterior parietal association cortex of the monkey: command functions for operations within extrapersonal space, *J Neurophysiol*, 38, 871, 1975.
87. Lynch, J.C. et al., Parietal lobe mechanisms for directed visual attention, *J Neurophysiol*, 40, 362, 1977.
88. Robinson, D.L., Goldberg, M.E., and Stanton, G.B., Parietal association cortex in the primate: sensory mechanisms and behavioral modulations, *J Neurophysiol*, 41, 910, 1978.
89. Bushnell, M.C., Goldberg, M.E., and Robinson, D.L., Behavioral enhancement of visual responses in monkey cerebral cortex. I. Modulation in posterior parietal cortex related to selective visual attention, *J Neurophysiol*, 46, 755, 1981.
90. Goldberg, M.E., Colby, C.L., and Duhamel, J.R., Representation of visuomotor space in the parietal lobe of the monkey, *Cold Spring Harb Symp Quant Biol*, 55, 729, 1990.

91. Gnadt, J.W. and Andersen, R.A., Memory related motor planning activity in posterior parietal cortex of macaque, *Exp Brain Res*, 70, 216, 1988.
92. Mazzoni, P. et al., Spatially tuned auditory responses in area LIP of macaques performing delayed memory saccades to acoustic targets, *J Neurophysiol*, 75, 1233, 1996.
93. Grunewald, A., Linden, J.F., and Andersen, R.A., Responses to auditory stimuli in macaque lateral intraparietal area. I. Effects of training, *J Neurophysiol*, 82, 330, 1999.
94. Linden, J.F., Grunewald, A., and Andersen, R.A., Responses to auditory stimuli in macaque lateral intraparietal area. II. Behavioral modulation, *J Neurophysiol*, 82, 343, 1999.
95. Platt, M.L. and Glimcher, P.W., Responses of intraparietal neurons to saccadic targets and visual distractors, *J Neurophysiol*, 78, 1574, 1997.
96. Mazzoni, P. et al., Motor intention activity in the macaque's lateral intraparietal area. I. Dissociation of motor plan from sensory memory, *J Neurophysiol*, 76, 1439, 1996.
97. Bracewell, R.M. et al., Motor intention activity in the macaque's lateral intraparietal area. II. Changes of motor plan, *J Neurophysiol*, 76, 1457, 1996.
98. Gottlieb, J.P., Kusunoki, M., and Goldberg, M.E., The representation of visual salience in monkey parietal cortex, *Nature*, 391, 481, 1998.
99. Gottlieb, J. and Goldberg, M.E., Activity of neurons in the lateral intraparietal area of the monkey during an antisaccade task, *Nat Neurosci*, 2, 906, 1999.
100. Zhang, M. and Barash, S., Neuronal switching of sensorimotor transformations for antisaccades, *Nature*, 408, 971, 2000.
101. Newsome, W.T., Britten, K.H., and Movshon, J.A., Neuronal correlates of a perceptual decision, *Nature*, 341, 52, 1989.
102. Platt, M.L. and Glimcher, P.W., Neural correlates of decision variables in parietal cortex, *Nature*, 400, 233, 1999.
103. Platt, M.L., Neural correlates of decisions, *Curr Opin Neurobiol*, 12, 141, 2002.
104. Charnov, E.L., Optimal foraging, the marginal value theorem, *Theor Popul Biol*, 9, 129, 1976.
105. Herrnstein, R.J. and Vaughan, W., Melioration and behavioral allocation, in *Limits to Action: The Allocation of Individual Behavior*, Staddon, J.E.R., Ed., Academic Press, New York, NY, 1980, 143.
106. Mishkin, M., Effects of small frontal lesions on delayed alternation in monkeys, *Journal of Neurophysiology*, 20, 615, 1957.
107. Gross, C.G. and Weiskrantz, L., Evidence for dissociation of impairment on auditory discrimination and delayed response following lateral frontal lesions in monkeys, *Exp. Neurol.*, 5, 453, 1962.
108. Funahashi, S., Bruce, C.J., and Goldman-Rakic, P.S., Neuronal activity related to saccadic eye movements in the monkey's dorsolateral prefrontal cortex, *J Neurophysiol*, 65, 1464, 1991.
109. Shallice, T., Specific impairments of planning, *Philos Trans R Soc Lond B Biol Sci*, 298, 199, 1982.
110. Bechara, A. et al., Deciding advantageously before knowing the advantageous strategy, *Science*, 275, 1293, 1997.
111. Rolls, E.T., *The Brain and Emotion*, Oxford University Press, Oxford, 1999.
112. Damasio, A.R., The somatic marker hypothesis and the possible functions of the prefrontal cortex, in *The Prefrontal Cortex: Executive and Cognitive Functions*, Roberts, A.C., Robbins, T.W., and Weiskrantz, L., Eds., Oxford University Press, New York, 1998.
113. Baddeley, A. and Hitch, G.J., Working memory, in *Recent Advances in Learning and Motivation Vol. VIII*, Bower, G., Ed., Academic Press, New York, NY, 1974.

114. Shallice, T. and Burgess, P.W., Deficits in strategy application following frontal lobe damage in man, *Brain*, 114 (Pt 2), 727, 1991.
115. Petrides, M. and Pandya, D.N., Comparative architectonic analysis of the human and macaque frontal cortex, in *Handbook of Neuropsychology*, Boller, F. and Grafman, J., Eds., Elsevier Science Publishers, 1994, 17.
116. Barbas, H., Complementary roles of prefrontal cortical regions in cognition, memory, and emotion in primates, *Adv Neurol*, 84, 87, 2000.
117. Brodmann, K., Vergleichende Lokalisationslehre der Grosshirnrinde in ihren Prinzipien dargestellt auf Grund des Zellenbaus, Verlag von Johann Ambrosius Barth, Leipzig, 1909.
118. Walker, A.E., A cytoarchitectural study of the prefrontal area of the macaque monkey, *J Comp Neurol*, 73, 59, 1940.
119. Iversen, S.D. and Mishkin, M., Perseverative interference in monkeys following selective lesions of the inferior prefrontal convexity, *Exp Brain Res*, 11, 376, 1970.
120. Bechara, A. et al., Dissociation of working memory from decision making within the human prefrontal cortex, *J Neurosci*, 18, 428, 1998.
121. Rolls, E.T., The orbitofrontal cortex, in *The Prefrontal Cortex: Executive and Cognitive Functions*, Roberts, A.C., Robbins, T.W., and Weiskrantz, L., Eds., Oxford University Press, New York, 1998.
122. Passingham, R., Delayed matching after selective prefrontal lesions in monkeys (*Macaca mulatta*), *Brain Res*, 92, 89, 1975.
123. Mishkin, M. and Manning, F.J., Non-spatial memory after selective prefrontal lesions in monkeys, *Brain Res*, 143, 313, 1978.
124. Funahashi, S., Bruce, C.J., and Goldman-Rakic, P.S., Mnemonic coding of visual space in the monkey's dorsolateral prefrontal cortex, *J Neurophysiol*, 61, 331, 1989.
125. Milner, B., Effects of different brain lesions on card sorting. The role of the frontal lobes, *Arch. Neurol.*, 9, 90, 1963.
126. Mishkin, M., Perseveration of central sets after frontal lesions in monkeys, in *The Frontal Granular Cortex*, Warren, J.M. and Akert, K., Eds., McGraw Hill, New York, 1964, 219.
127. Diamond, A. and Goldman-Rakic, P.S., Comparison of human infants and rhesus monkeys on Piaget's AB task: evidence for dependence on dorsolateral prefrontal cortex, *Exp Brain Res*, 74, 24, 1989.
128. Kowalska, D.M., Bachevalier, J., and Mishkin, M., The role of the inferior prefrontal convexity in performance of delayed nonmatching-to-sample, *Neuropsychologia*, 29, 583, 1991.
129. Funahashi, S., Bruce, C.J., and Goldman-Rakic, P.S., Dorsolateral prefrontal lesions and oculomotor delayed-response performance: evidence for mnemonic "scotomas," *J Neurosci*, 13, 1479, 1993.
130. Funahashi, S., Chafee, M.V., and Goldman-Rakic, P.S., Prefrontal neuronal activity in rhesus monkeys performing a delayed anti-saccade task, *Nature*, 365, 753, 1993.
131. Wilson, F.A., Scalaidhe, S.P., and Goldman-Rakic, P.S., Dissociation of object and spatial processing domains in primate prefrontal cortex, *Science*, 260, 1955, 1993.
132. Rao, S.C., Rainer, G., and Miller, E.K., Integration of what and where in the primate prefrontal cortex, *Science*, 276, 821, 1997.
133. Asaad, W.F., Rainer, G., and Miller, E.K., Neural activity in the primate prefrontal cortex during associative learning, *Neuron*, 21, 1399, 1998.
134. Barbas, H., Henion, T.H., and Dermon, C.R., Diverse thalamic projections to the prefrontal cortex in the rhesus monkey, *J Comp Neurol*, 313, 65, 1991.

135. Hardy, S.G., Projections to the midbrain from the medial versus lateral prefrontal cortices of the rat, *Neurosci Lett*, 63, 159, 1986.
136. Pandya, D.N. and Yeterian, E.H., Comparison of prefrontal architecture and connections, in *The Prefrontal Cortex: Executive and Cognitive Functions*, Roberts, A.C., Robbins, T.W., and Weiskrantz, L., Eds., Oxford University Press, New York, 1998.
137. Funahashi, S., Bruce, C.J., and Goldman-Rakic, P.S., Visuospatial coding in primate prefrontal neurons revealed by oculomotor paradigms, *J Neurophysiol*, 63, 814, 1990.
138. Sawaguchi, T. and Iba, M., Prefrontal cortical representation of visuospatial working memory in monkeys examined by local inactivation with muscimol, *J Neurophysiol*, 86, 2041, 2001.
139. Constantinidis, C., Franowicz, M.N., and Goldman-Rakic, P.S., The sensory nature of mnemonic representation in the primate prefrontal cortex, *Nat Neurosci*, 4, 311, 2001.
140. Takeda, K. and Funahashi, S., Prefrontal task-related activity representing visual cue location or saccade direction in spatial working memory tasks, *J Neurophysiol*, 87, 567, 2002.
141. Kobayashi, S. et al., Influence of reward expectation on visuospatial processing in macaque lateral prefrontal cortex, *J Neurophysiol*, 87, 1488, 2002.
142. Sawaguchi, T., The effects of dopamine and its antagonists on directional delay-period activity of prefrontal neurons in monkeys during an oculomotor delayed-response task, *Neurosci Res*, 41, 115, 2001.
143. Papez, J.W., A proposed mechanism of emotion, *Arch. Neurol. Psychiatr.*, 38, 725, 1937.
144. MacLean, P.D., Psychosomatic disease and the "visceral brain": Recent developments bearing on the Papez theory of emotion, *Psychosom. Med.*, 11, 338, 1949.
145. Musil, S.Y. and Olson, C.R., Organization of cortical and subcortical projections to anterior cingulate cortex in the cat, *J Comp Neurol*, 272, 203, 1988.
146. Olson, C.R. and Musil, S.Y., Posterior cingulate cortex: sensory and oculomotor properties of single neurons in behaving cat, *Cereb Cortex*, 2, 485, 1992.
147. Olson, C.R. and Musil, S.Y., Topographic organization of cortical and subcortical projections to posterior cingulate cortex in the cat: evidence for somatic, ocular, and complex subregions, *J Comp Neurol*, 324, 237, 1992.
148. Shima, K. et al., Two movement-related foci in the primate cingulate cortex observed in signal-triggered and self-paced forelimb movements, *J Neurophysiol*, 65, 188, 1991.
149. Olson, C.R., Musil, S.Y., and Goldberg, M.E., Single neurons in posterior cingulate cortex of behaving macaque: eye movement signals, *J Neurophysiol*, 76, 3285, 1996.
150. Picard, N. and Strick, P.L., Motor areas of the medial wall: a review of their location and functional activation, *Cereb Cortex*, 6, 342, 1996.
151. Paus, T. et al., Role of the human anterior cingulate cortex in the control of oculomotor, manual, and speech responses: a positron emission tomography study, *J Neurophysiol*, 70, 453, 1993.
152. Petit, L. et al., PET study of voluntary saccadic eye movements in humans: basal ganglia-thalamocortical system and cingulate cortex involvement, *J Neurophysiol*, 69, 1009, 1993.
153. Petit, L. et al., Functional anatomy of a prelearned sequence of horizontal saccades in humans, *J Neurosci*, 16, 3714, 1996.
154. Gaymard, B. et al., Effects of anterior cingulate cortex lesions on ocular saccades in humans, *Exp Brain Res*, 120, 173, 1998.

155. Vogt, B.A., Pandya, D.N., and Rosene, D.L., Cingulate cortex of the rhesus monkey: I. Cytoarchitecture and thalamic afferents, *J Comp Neurol*, 262, 256, 1987.
156. Vogt, B.A., Finch, D.M., and Olson, C.R., Functional heterogeneity in cingulate cortex: the anterior executive and posterior evaluative regions, *Cereb Cortex*, 2, 435, 1992.
157. Musil, S.Y. and Olson, C.R., Organization of cortical and subcortical projections to medial prefrontal cortex in the cat, *J Comp Neurol*, 272, 219, 1988.
158. Doricchi, F. et al., Neural control of fast-regular saccades and antisaccades: an investigation using positron emission tomography, *Exp Brain Res*, 116, 50, 1997.
159. Berman, R.A. et al., Cortical networks subserving pursuit and saccadic eye movements in humans: an FMRI study, *Hum Brain Mapp*, 8, 209, 1999.
160. Barbas, H. and Mesulam, M.M., Organization of afferent input to subdivisions of area 8 in the rhesus monkey, *J Comp Neurol*, 200, 407, 1981.
161. Vogt, B.A. and Pandya, D.N., Cingulate cortex of the rhesus monkey: II. Cortical afferents, *J Comp Neurol*, 262, 271, 1987.
162. Van Hoesen, G., Morecraft, R.J., and Vogt, B.A., Connections of the monkey cingulate cortex, in *Neurobiology of Cingulate Cortex and Limbic Thalamus*, Vogt, B.A. and Gabriel, M., Eds., Birkhauser, Boston, 1993, 249.
163. Pandya, D.N. and Kuypers, H.G., Cortico-cortical connections in the rhesus monkey, *Brain Res*, 13, 13, 1969.
164. Goldman, P.S. and Nauta, W.J., Columnar distribution of cortico-cortical fibers in the frontal association, limbic, and motor cortex of the developing rhesus monkey, *Brain Res*, 122, 393, 1977.
165. Barbas, H. and Pandya, D.N., Architecture and intrinsic connections of the prefrontal cortex in the rhesus monkey, *J Comp Neurol*, 286, 353, 1989.
166. Baleydier, C. and Mauguiere, F., The duality of the cingulate gyrus in monkey. Neuroanatomical study and functional hypothesis, *Brain*, 103, 525, 1980.
167. Pandya, D.N., Van Hoesen, G.W., and Mesulam, M.M., Efferent connections of the cingulate gyrus in the rhesus monkey, *Exp Brain Res*, 42, 319, 1981.
168. Thier, P. and Andersen, R.A., Electrical microstimulation distinguishes distinct saccade-related areas in the posterior parietal cortex, *J Neurophysiol*, 80, 1713, 1998.
169. Van Hoesen, G.W., The differential distribution, diversity and sprouting of cortical projections to the amygdala in the rhesus monkey, in *The Amygdaloid Complex*, Ben-Ari, Y., Ed., Elsevier/North-Holland Biomedical Press, New York, 1981, 77.
170. Amaral, D.G. and Price, J.L., Amygdalo-cortical projections in the monkey (*Macaca fascicularis*), *J Comp Neurol*, 230, 465, 1984.
171. Pandya, D.N., Van Hoesen, G.W., and Domesick, V.B., A cingulo-amygdaloid projection in the rhesus monkey, *Brain Res*, 61, 369, 1973.
172. Muller-Preuss, P. and Jurgens, U., Projections from the 'cingular' vocalization area in the squirrel monkey, *Brain Res*, 103, 29, 1976.
173. Gabriel, M. et al., Neuronal activity of cingulate cortex, anteroventral thalamus and hippocampal formation in discriminative conditioning: encoding and extraction of the significance of conditioned stimuli, *Prog Psychobiol Physiol Psychol*, 9, 125, 1980.
174. Crowley, J.C., Haghighian, G., and Platt, M.L. Posterior cingulate neurons encode eye movement value, in *Society for Neuroscience Abstracts*, 2001.
175. Gaymard, B. et al., Cortical control of saccades, *Exp Brain Res*, 123, 159, 1998.
176. Shima, K. and Tanji, J., Role for cingulate motor area cells in voluntary movement selection based on reward, *Science*, 282, 1335, 1998.

177. Shidara, M. and Richmond, B.J., Anterior cingulate: single neuronal signals related to degree of reward expectancy, *Science*, 296, 1709, 2002.
178. Mesulam, M.M., Spatial attention and neglect: parietal, frontal and cingulate contributions to the mental representation and attentional targeting of salient extrapersonal events, *Philos Trans R Soc Lond B Biol Sci*, 354, 1325, 1999.
179. Platt, M.L., McCoy, A.N., and Crowley, J.C. Reference frames for spatial representations in posterior cingulate cortex, in *Society for Neuroscience Abstracts*, 2001.
180. McCoy, A.N. and Platt, M.L. Reward modulation of neuronal activity in posterior cingulate cortex, in *Society for Neuroscience Abstracts*, 2002.
181. Porro, C.A. et al., Neural circuits underlying ketamine-induced oculomotor behavior in the rat: 2-deoxyglucose studies, *Exp Brain Res*, 124, 8, 1999.
182. Selemon, L.D. and Goldman-Rakic, P.S., Longitudinal topography and interdigitation of corticostriatal projections in the rhesus monkey, *J Neurosci*, 5, 776, 1985.
183. Shook, B.L., Schlag-Rey, M., and Schlag, J., Primate supplementary eye field. II. Comparative aspects of connections with the thalamus, corpus striatum, and related forebrain nuclei, *J Comp Neurol*, 307, 562, 1991.
184. Parthasarathy, H.B., Schall, J.D., and Graybiel, A.M., Distributed but convergent ordering of corticostriatal projections: analysis of the frontal eye field and the supplementary eye field in the macaque monkey, *J Neurosci*, 12, 4468, 1992.
185. Ribak, C.E. et al., Immunocytochemical localization of glutamate decarboxylase in rat substantia nigra, *Brain Res*, 116, 287, 1976.
186. Ribak, C.E., Vaughn, J.E., and Roberts, E., The GABA neurons and their axon terminals in rat corpus striatum as demonstrated by GAD immunocytochemistry, *J Comp Neurol*, 187, 261, 1979.
187. Parent, A., Bouchard, C., and Smith, Y., The striatopallidal and striatonigral projections: two distinct fiber systems in primate, *Brain Res*, 303, 385, 1984.
188. Rinvik, E. and Ottersen, O.P., Terminals of subthalamonigral fibres are enriched with glutamate-like immunoreactivity: an electron microscopic, immunogold analysis in the cat, *J Chem Neuroanat*, 6, 19, 1993.
189. Ilinsky, I.A., Jouandet, M.L., and Goldman-Rakic, P.S., Organization of the nigrothalamocortical system in the rhesus monkey, *J Comp Neurol*, 236, 315, 1985.
190. Lynch, J.C., Hoover, J.E., and Strick, P.L., Input to the primate frontal eye field from the substantia nigra, superior colliculus, and dentate nucleus demonstrated by transneuronal transport, *Exp Brain Res*, 100, 181, 1994.
191. Middleton, F.A. and Strick, P.L., Basal-ganglia 'projections' to the prefrontal cortex of the primate, *Cereb Cortex*, 12, 926, 2002.
192. Jayaraman, A., Batton, R.R., 3rd, and Carpenter, M.B., Nigrotectal projections in the monkey: an autoradiographic study, *Brain Res*, 135, 147, 1977.
193. Bentivoglio, M., van der Kooy, D., and Kuypers, H.G., The organization of the efferent projections of the substantia nigra in the rat. A retrograde fluorescent double labeling study, *Brain Res*, 174, 1, 1979.
194. Beckstead, R.M., Long collateral branches of substantia nigra pars reticulata axons to thalamus, superior colliculus and reticular formation in monkey and cat. Multiple retrograde neuronal labeling with fluorescent dyes, *Neuroscience*, 10, 767, 1983.
195. May, P.J. and Hall, W.C., The sources of the nigrotectal pathway, *Neuroscience*, 19, 159, 1986.
196. Hikosaka, O. and Wurtz, R.H., The basal ganglia, in *The Neurobiology of Saccadic Eye Movements: Reviews of Oculomotor Research Volume 3*, Wurtz, R.H. and Goldberg, M.E., Eds., Elsevier Science Publishers, 1989, 257.

197. Hikosaka, O., Takikawa, Y., and Kawagoe, R., Role of the basal ganglia in the control of purposive saccadic eye movements, *Physiol Rev*, 80, 953, 2000.
198. Hikosaka, O. and Wurtz, R.H., Visual and oculomotor functions of monkey substantia nigra pars reticulata. I. Relation of visual and auditory responses to saccades, *J Neurophysiol*, 49, 1230, 1983.
199. Hikosaka, O. and Wurtz, R.H., Visual and oculomotor functions of monkey substantia nigra pars reticulata. II. Visual responses related to fixation of gaze, *J Neurophysiol*, 49, 1254, 1983.
200. Hikosaka, O. and Wurtz, R.H., Visual and oculomotor functions of monkey substantia nigra pars reticulata. III. Memory-contingent visual and saccade responses, *J Neurophysiol*, 49, 1268, 1983.
201. Joseph, J.P. and Boussaoud, D., Role of the cat substantia nigra pars reticulata in eye and head movements. I. Neural activity, *Exp Brain Res*, 57, 286, 1985.
202. Hikosaka, O. and Wurtz, R.H., Visual and oculomotor functions of monkey substantia nigra pars reticulata. IV. Relation of substantia nigra to superior colliculus, *J Neurophysiol*, 49, 1285, 1983.
203. Karabelas, A.B. and Moschovakis, A.K., Nigral inhibitory termination on efferent neurons of the superior colliculus: an intracellular horseradish peroxidase study in the cat, *J Comp Neurol*, 239, 309, 1985.
204. Chevalier, G. et al., The nigro-tectal pathway. An electrophysiological reinvestigation in the rat, *Brain Res*, 213, 253, 1981.
205. Hikosaka, O. and Wurtz, R.H., Modification of saccadic eye movements by GABA-related substances. I. Effect of muscimol and bicuculline in monkey superior colliculus, *J Neurophysiol*, 53, 266, 1985.
206. Hikosaka, O. and Wurtz, R.H., Modification of saccadic eye movements by GABA-related substances. II. Effects of muscimol in monkey substantia nigra pars reticulata, *J Neurophysiol*, 53, 292, 1985.
207. Boussaoud, D. and Joseph, J.P., Role of the cat substantia nigra pars reticulata in eye and head movements. II. Effects of local pharmacological injections, *Exp Brain Res*, 57, 297, 1985.
208. Hikosaka, O., Sakamoto, M., and Miyashita, N., Effects of caudate nucleus stimulation on substantia nigra cell activity in monkey, *Exp Brain Res*, 95, 457, 1993.
209. Handel, A. and Glimcher, P.W., Quantitative analysis of substantia nigra pars reticulata activity during a visually guided saccade task, *J Neurophysiol*, 82, 3458, 1999.
210. Basso, M.A. and Wurtz, R.H., Neuronal activity in substantia nigra pars reticulata during target selection, *J Neurosci*, 22, 1883, 2002.
211. Sato, M. and Hikosaka, O., Role of primate substantia nigra pars reticulata in reward-oriented saccadic eye movement, *J Neurosci*, 22, 2363, 2002.
212. Ottes, F.P., Van Gisbergen, J.A., and Eggermont, J.J., Visuomotor fields of the superior colliculus: a quantitative model, *Vision Res*, 26, 857, 1986.
213. Platt, M.L. and Glimcher, P.W., Response fields of intraparietal neurons quantified with multiple saccadic targets, *Exp Brain Res*, 121, 65, 1998.
214. Mink, J.W., The basal ganglia: focused selection and inhibition of competing motor programs, *Prog Neurobiol*, 50, 381, 1996.
215. Munoz, D.P. and Wurtz, R.H., Fixation cells in monkey superior colliculus. I. Characteristics of cell discharge, *J Neurophysiol*, 70, 559, 1993.
216. Handel, A. and Glimcher, P.W., Contextual modulation of substantia nigra pars reticulata neurons, *J Neurophysiol*, 83, 3042, 2000.
217. Takikawa, Y. et al., Modulation of saccadic eye movements by predicted reward outcome, *Exp Brain Res*, 142, 284, 2002.

218. Hikosaka, O., Sakamoto, M., and Usui, S., Functional properties of monkey caudate neurons. III. Activities related to expectation of target and reward, *J Neurophysiol*, 61, 814, 1989.
219. Kawagoe, R., Takikawa, Y., and Hikosaka, O., Expectation of reward modulates cognitive signals in the basal ganglia, *Nat Neurosci*, 1, 411, 1998.
220. Lauwereyns, J. et al., Feature-based anticipation of cues that predict reward in monkey caudate nucleus, *Neuron*, 33, 463, 2002.
221. Takikawa, Y., Kawagoe, R., and Hikosaka, O., Reward-dependent spatial selectivity of anticipatory activity in monkey caudate neurons, *J Neurophysiol*, 87, 508, 2002.
222. Haber, S.N. and Fudge, J.L., The primate substantia nigra and VTA: integrative circuitry and function, *Crit Rev Neurobiol*, 11, 323, 1997.
223. Schultz, W., Predictive reward signal of dopamine neurons, *J Neurophysiol*, 80, 1, 1998.
224. Francois, C., Yelnik, J., and Percheron, G., Golgi study of the primate substantia nigra. II. Spatial organization of dendritic arborizations in relation to the cytoarchitectonic boundaries and to the striatonigral bundle, *J Comp Neurol*, 265, 473, 1987.
225. Geffen, L.B. et al., Release of dopamine from dendrites in rat substantia nigra, *Nature*, 260, 258, 1976.
226. Boyson, S.J., McGonigle, P., and Molinoff, P.B., Quantitative autoradiographic localization of the D1 and D2 subtypes of dopamine receptors in rat brain, *J Neurosci*, 6, 3177, 1986.
227. Ruffieux, A. and Schultz, W., Dopaminergic activation of reticulata neurones in the substantia nigra, *Nature*, 285, 240, 1980.
228. Waszcak, B.L. and Walters, J.R., Dopamine modulation of the effects of gamma-aminobutyric acid on substantia nigra pars reticulata neurons, *Science*, 220, 218, 1983.
229. Houk, J.C., Adams, J.L., and Barto, A.G., A model of how the basal ganglia generate and use neural signals that predict reinforcement, in *Models of information processing in the Basal Ganglia*, Houk, J.C., Davis, J.L., and Beiser, D.G., Eds., MIT Press, Cambridge, MA, 1995, 248.
230. Schultz, W., Dayan, P., and Montague, P.R., A neural substrate of prediction and reward, *Science*, 275, 1593, 1997.

2 Signal Coding in the Primate Superior Colliculus Revealed through the Use of Artificial Signals

Terrence R. Stanford

CONTENTS

2.1 INTRODUCTION

The superior colliculus (SC), along with its nonmammalian counterpart, the optic tectum, is well known for playing a prominent role in mediating orienting behaviors to exogenous stimuli. A partial list of species for which this capacity of the SC has been examined includes the goldfish, frog, owl, rat, cat, and monkey. In nonhuman primates, study of the oculomotor function of the SC, for all intents and purposes, began in the early 1970s with the pioneering works of David A. Robinson, Robert Wurtz, Peter Schiller, David Sparks and their numerous collaborators and colleagues. Following the lead of Edward Evarts, who combined operant training methods with

0-8493-0097-5/04/$0.00+$1.50

single neuron electrophysiology to study the neural representation of limb movement in monkey motor cortex[1] these investigators began to study the neural basis of visual orienting in alert, behaviorally trained monkeys. Monkeys were a natural choice as a model of visuomotor behavior owing to an extensive oculomotor range, vision as a dominant modality, and the ease with which they could be trained on relatively complex visuomotor tasks. To date, the vast majority of studies of goal-directed orienting in nonhuman primates have focused on saccadic eye movements, rapid changes in eye position designed to place the images of interesting objects on the fovea for detailed viewing. As such, most of what is known about the primate SC concerns saccade-related function, though more recent lines of evidence point toward a more general role in shifting gaze (combined movement of the eyes and head),[2,3] posit involvement in coding for other types of eye movements (e.g., smooth pursuit),[4,5] and even suggest a role in producing arm movements.[6–8]

The SC is, arguably, the best understood of all of the so-called oculomotor regions. Along with a wealth of knowledge of its anatomy, progress in understanding the physiology of the SC has benefited greatly from the power of converging operations wherein results deriving from single-unit electrophysiology, electrical microstimulation, and pharmacological manipulation have provided complementary views of the SC's oculomotor machinery (see References 9 to 13 for detailed discussion). The strength of single-unit recording lies in its spatial and temporal resolution. Clearly, when measuring extracellular action potentials, the single neuron represents the irreducible spatial quantity; equally important is the ability to monitor neural activity with millisecond temporal resolution. This is critical when attempting to relate activity to the occurrence of saccades, which quite often begin and end within the span of 50 milliseconds. Despite these strengths, single neuron recording has some limitations. First, the information represented in neural ensembles can be difficult (and sometimes impossible) to infer from consideration of serially recorded single neurons. Second, the data derived from single neuron electrophysiology is correlative, and while the correlation between a particular neuron's activity and an experimental event (e.g., saccadic eye movement) may be compelling, it falls short of establishing causality.

A complementary approach to studying the relationship between neural activity and its presumed functional consequences is to induce activity by experimental means and measure the impact of the imposed *artificial* signal on behavior. Two such methods, electrical microstimulation and pharmacological inactivation (or activation) have been instrumental to the study of SC function. The advantage of this general approach is clear; within limits, one can be reasonably certain that there is a causal relationship between the experimentally induced activation and behavioral outcome. Thus, to the extent that the properties of the induced signal can be controlled, these methods have the potential to establish input-output functions relating activity to behavior.

This chapter considers the question of what *saccade-related* information is coded within the primate SC. In doing so, the findings of a handful of studies are examined in some detail in order to highlight the way in which artificial signals have contributed to advancing our understanding of this issue. While the contributions of microstimulation and pharmacological manipulation are emphasized, it is not the goal of this chapter to provide a detailed description of these methodologies (see References 14 to 17 for reviews). However, to provide a foundation for considering the interpretations of studies using these methods, the following sections touch briefly on some of their essential strengths and weaknesses.

2.2 ELECTRICAL MICROSTIMULATION

Within any given structure, activity induced by microstimulation is the product of both direct activation of neurons within the vicinity of the electrode and synaptic activation of more remote neurons via intrinsic circuitry. Typically, trains composed of very short (e.g., 0.2 ms), biphasic current pulses are used to recruit relatively large populations of neurons in order to elicit overt behaviors (e.g., saccades). Many factors contribute to the size, shape, and efficacy of the current spread within the neural tissue, and while a complete account is beyond the scope of this review (see Reference 16 for recent review), there are some straightforward principles that can be helpful in considering the results of microstimulation studies. First, the size of the active population, and perhaps the degree to which individual neurons are activated, is positively related to current strength. This falls out of the well-established current-distance relationship, which specifies that the minimum current needed to activate a given neuron increases in proportion to the square of its distance from the electrode tip.[14] Typically, effective current spread is controlled experimentally by varying the amplitude of the elemental current pulse and less frequently, its duration. The remaining parameters apply to the pulse train and include frequency, duration, and their product, the number of pulses. These too are likely to affect the size of the active population; however, their primary effects are likely to be on the firing frequency of the activated neurons (train frequency) and the length of time (train duration) over which this activity is maintained, respectively.

Given the above, the value of microstimulation as a tool for exploring function is readily apparent. With it, the location, extent, response rate, and response duration of a neural population is under experimental control. Thus, as would an engineer testing a circuit, the experimenter can apply a known input (electrical stimulus), measure the resulting output (e.g., evoked saccade), and deduce the nature of the operation that is performed by the activated neural elements (i.e., input-output function). In fact, as will be clear below, this general approach has been used with great success in the study of the SC. However, when interpreting results it is important to consider that the activity evoked by microstimulation does not mimic exactly that which would occur naturally. First, microstimulation cannot be expected to preserve any of the normal subtleties of response timing within the activity of single neurons or across populations of neurons. Second, along with the intended effects of directly activating local neurons and indirectly activating their downstream targets, concurrently stimulated input fibers and passing fibers may produce unintended activity in areas upstream to the stimulated site as well as in regions that may not be directly related to the desired behavior. Evaluating the impact of these factors for interpreting results can be difficult at best.

Despite the aforementioned caveats, microstimulation of the SC has been used in several different ways to provide critical pieces of information relating to SC function. As discussed in the following section, the most straightforward use of the method has been to produce saccades; however, microstimulation has also been used in more subtle ways to modify ongoing saccades (e.g., see Reference 18), and even to influence the probability of initiating a particular saccade.[19]

2.3 PHARMACOLOGICAL MANIPULATION

The infusion of chemical agents to temporarily alter neural activity has largely supplanted ablation as a way of evaluating the contribution of a given brain region to mediating a particular function. Depending on which agent is infused, activity can be either increased or depressed. The two most commonly used deactivating agents are lidocaine and muscimol and each has been used to make *reversible lesions* in the SC. Lidocaine is well known as a local anesthetic that, when infused into the brain, prevents depolarization by blocking sodium channels. Whether assayed by overt changes in behavior or direct measurement of neural activity,[17] lidocaine is relatively fast acting, maximally effective soon after administration, and short-lived (e.g., 20 minutes), making it possible to monitor both the behavioral effect and its recovery multiple times within the same experimental session.

Whereas the short time course of lidocaine's effect can be an asset, lack of specificity is a potential liability. Interpretation of such data must always consider the fact that lidocaine does not discriminate between neurons, presynaptic fibers, and fibers of passage. In contrast, muscimol depresses activity by activating receptors (A-type) normally occupied by the inhibitory transmitter, gamma-amino-butyric acid (GABA). The complement to muscimol is bicuculline, a GABA(A) antagonist, which, by blocking the effects of endogenous transmitter, leads to hyperexcitability. Thus, the effects of muscimol and bicuculline are quite specific and can be attributed to their opposing actions on a specific transmitter system within the structure of interest. However, a potential drawback (though it is sometimes an advantage) is that these effects are relatively slow to develop and long acting thus precluding any opportunity for repetition, or even to observe complete recovery within an experimental session.

The behavioral manifestations of either muscimol or bicuculline are likely the result of simultaneous actions on multiple GABA-ergic substrates comprising intrinsic and extrinsic sources of GABA (see Reference 20 for review). During normal functioning, intrinsic GABAergic neurons may serve to define the locus of activity within the SC (which determines saccade vector — see following section) or perhaps contribute to the process of active fixation (i.e., preventing unsolicited saccades).[21,22] The largest source of GABAergic input to the SC derives from the substantia nigra pars reticulata (SNr), a principal output nucleus of the basal ganglia. A large, well documented, projection from the ipsilateral (to the SC) SNr is known to provide a powerful tonic inhibitory drive to SC saccade-related neurons. Transient release of this inhibition, in concert with descending excitatory influences onto the same SC neurons results in high-frequency saccade-related bursts among SC efferent neurons (see References 23 and 24 for reviews). These bursts, in turn, drive saccades into the opposite hemifield via the SC's projections to downstream premotor neuron pools (see References 9 and 11 to 13 for reviews).

It should be clear from the above that the methodologies of microstimulation and pharmacological manipulation have different strengths and weaknesses and these, in turn, are distinct from those of single-unit electrophysiology. Microstimulation offers relatively good control over the extent, vigor, and temporal profile of the induced activity, but it is relatively nonspecific and cannot be used to suppress activity. Pharmacological inactivation/activation permits suppression or facilitation, but affords less control over the affected population. Both methods, however, impact large neural ensembles and establish causal relationships between neural activity and behavior. As such they represent a powerful complement to the precise, but correlative approach of single-unit electrophysiology. The following sections consider some of the issues that have been resolved by the combined use of these methodologies.

2.4 THE SC MOTOR MAP

It is interesting to note that the earliest evidence for the SC's involvement in generating eye movements derives from the use of stimulation and pharmacological methods in studies that predate even the earliest single-unit recordings. First Adamük,[25] and much later, Hess[26] observed that electrical stimulation of the SC produced conjugate eye movements while at the same time, Apter[27] reported eye movements consequent to application of strychnine (which antagonizes the inhibitory transmitters glycine and GABA) to the cat SC. However, the study of the SC's contribution to oculomotor control began in earnest in the early 1970s, with one of the most influential studies being the classic microstimulation study carried out by David A. Robinson.[28] Robinson systematically explored the relationship between the site of stimulation in the SC and the vector (amplitude and direction) of the evoked eye movements. In doing so, he revealed a two-dimensional *motor map* within the intermediate and deep layers of the SC: amplitude, from small to large, is represented from rostral to caudal while direction, from upward to downward is represented from medial to lateral across the SC (Figure 2.1, left; see legend for details).

In parallel with Robinson's microstimulation results were results from several single-unit recording studies showing that neurons within these same regions of SC responded selectively in association with saccades of varying vector.[29–31] Analogous to sensory receptive fields, these *saccade-related* neurons had movement fields and their systematic topographic organization according to their preferred vector was evidently the basis for Robinson's motor map[32] (Figure 2.1, right; see legend for details). Direct evidence for this correspondence was provided by Schiller and Stryker[30] who, by combining microstimulation and single-unit recording, showed agreement between the preferred vectors of recorded units and the vectors of movements that could be evoked from the same collicular sites. Sparks[33] later reported precise temporal coupling between the presaccadic discharges of these neurons and saccade onset (e.g., see Figure 2.1; bottom right). Ultimately, SC neurons with these physiological properties were shown to project to downstream premotor regions,[12,34] lending further credence to the idea that the activity of these neurons is causal for producing saccades.

The demonstration of what appeared to be a purely topographic code for saccade vector has had critical implications for considering how activity in the SC could be

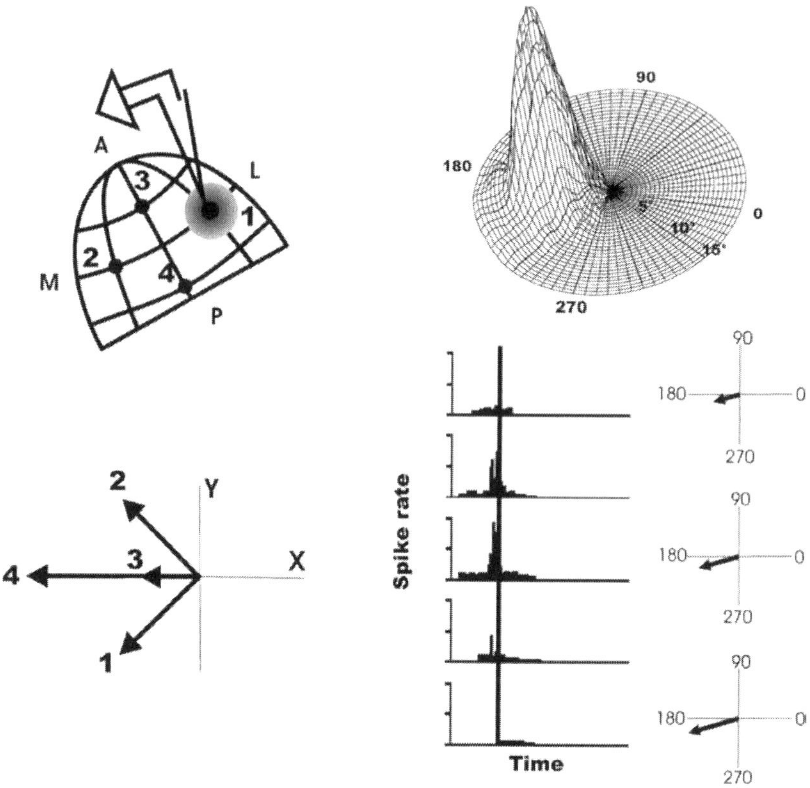

FIGURE 2.1 Two-dimensional SC motor map. Top left: schematic depicting Robinson's[28] microstimulation study of the primate SC. Electrode is shown to Site 1, which is lateral and at an intermediate location along the anterior-posterior axis of the right SC. Three additional sites of stimulation are shown. Site 2: medial location at same A-P location as Site 1. Site 3: anterior location at the midpoint of the medial-lateral axis of the SC. Site 4: posterior location at the midpoint of the medial-lateral axis of the SC. Bottom left: Site-specific vectors of evoked saccades. The pattern of vectors illustrates Robinson's main finding that saccade size and direction are systematically represented along the anterior-posterior and medial-lateral dimensions of the SC, respectively. Saccade amplitude increases from anterior (Site 3) to posterior (Site 4). Proceeding from lateral to medial, saccade direction varies from downward (Site 1) to upward (Site 2). Top right: movement field of a saccade-related neuron recorded in the right SC (Stanford and Sparks, unpublished). Polar plot of discharge rate as a function of saccade amplitude and direction illustrates that the neuron discharges for leftward saccades having an eccentricity between 2° and 10° and a direction (θ) of between 180° and 240°, with maximal firing corresponding to r and θ of approximately 7° and 200°, respectively. Bottom right: saccade-related discharge for movements of varying eccentricity along the best direction (200°). Left: plots of instantaneous firing frequency synchronized on saccade onset (vertical line). From top to bottom are saccade-related bursts associated with saccades of increasing eccentricity. Right: saccade vectors corresponding to top, middle, and bottom firing rate histograms. Schiller and Stryker[30] showed good agreement between movement field centers and the vectors of microstimulation-evoked saccades. Abbreviations: A = anterior; P = posterior; M = medial, L = lateral.

read-out to yield signals appropriate for driving the eyes at saccadic velocities. As noted by Robinson, the SC's topographic code would need to be transformed to yield the rate-coded signals ultimately required by the extraocular muscles (see Reference 11 for review). Extraocular motoneuron pools provide a high-frequency pulse of innervation to the muscles with eye velocity and duration dictated mainly by the frequency and duration of the pulse, respectively. Saccade amplitude, being the product of eye velocity and movement duration thus corresponds to the number of spikes in the pulse (i.e., pulse frequency × duration). It is important to note that these early microstimulation studies emphasized that it is the **site** of stimulation within the SC, **not** the temporal aspects of the artificially induced activity that determined the properties of the evoked movement. Thus, a train of the same frequency, duration and current strength, when applied to different loci in the SC, produced very different saccades. Furthermore, for stimulation at any given site, the vector of an evoked saccade was reportedly impervious to variation in microstimulation parameters (e.g., frequency, duration) despite presumed changes in the activated population's size, strength, and duration. For example, rather than a movement commensurate with the duration of the stimulation train, long stimulation trains produced a sequence of apparently independent saccades of approximately the same amplitude.

The microstimulation results were striking when juxtaposed with those for stimulation of the motoneuron pools directly, whereupon eye movement velocity and duration were observed to correlate with train frequency and duration, respectively.[30] Together, the stimulation and single-unit data suggested that the SC efferent projection carried a relatively high-order motor command that specified a desired change in eye position, in effect a signal of *what* to do, but not *how* to do it. Somehow, the locus of activity within the SC motor map was interpreted by downstream structures to deliver innervations of appropriate strength and duration to each of the six extraocular muscles. Several important insights into the nature of this process came from the pharmacological deactivation studies described in the following section.

2.5 VECTOR AVERAGING

The well-known deactivation study by Lee, Rohrer, and Sparks[35] showed that it is not the site of SC activity *per se* that determines the vector of the saccade, but instead, a weighted average of the preferred vectors of all SC neurons active in association with a given saccade. This conclusion was based on the behavioral effects observed after injecting lidocaine into localized regions of the SC motor map. Lee et al. found that saccades with vectors that were congruent with the movement field centers of neurons in the deactivated region remained accurate despite the absence of activity from what would normally be the most active population (Figure 2.2 — middle row — see legend for details). Inferring from single-unit recording studies (i.e., movement field size) that a saccade is preceded by activity across a large expanse of the SC motor map, it was reasonable to conclude that the appropriate vector was computed by averaging the movement contributions of neurons on the fringe of the deactivated region of the motor map. Also consistent with a vector

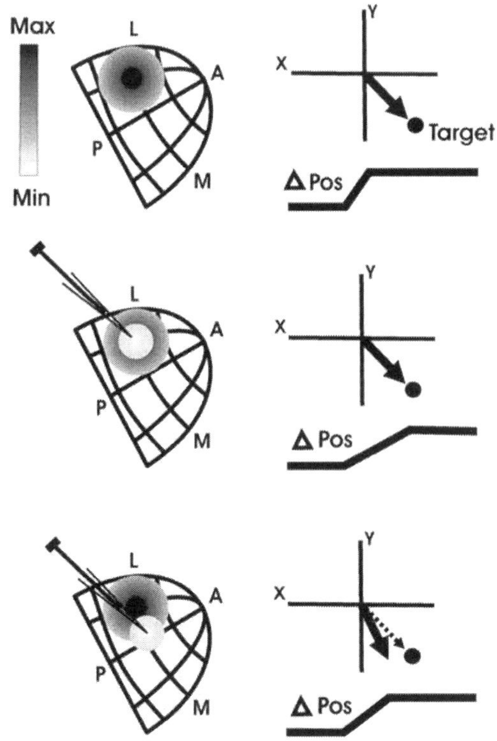

FIGURE 2.2 Vector averaging. Graphical illustration of deactivation results of Lee et al.[35] Top left: control condition prior to injection of lidocaine. Activity in the lateral aspect of the left SC associated with an accurate saccade to a down and right visual target. Top right: spatial (above) and temporal (below) representation of control saccade to the visual target. Vector (arrow) represents the saccade from a central location to the eccentric target. A plot of eye position as a function of time depicts the time course of the saccade. Middle left: lidocaine injection deactivates central region of population. Middle right: saccade to down right target remains accurate, however, reduction in saccade velocity is evident as an increase in saccade duration compared to the control condition. Bottom left: deactivation of neurons medial to the center of the active population. Bottom right: saccade directed to the target is inaccurate and shows an exaggerated downward component due to the deactivation of neurons having movement fields with more upward preferred directions (i.e., medial fringe). Deactivation of neurons at the edge of the population produces a modest reduction in velocity (increase in duration). Abbreviations: A = anterior; P = posterior; M = medial, L = lateral; Pos = eye position.

averaging scheme, saccades to surrounding locations exhibited a pattern of errors that could be accounted for by the loss of the inactivated population's contribution to a vector average computation (Figure 2.2, bottom row; see legend for details).

As suggested by earlier microstimulation studies,[28,30] these findings indicated that the computation of saccade vector was relatively impervious to changes in the overall level of activity within the SC population. However, unlike the microstimulation experiments, these data were not consistent with an exclusively topographic coding scheme within the SC. As had been shown previously by Hikosaka and

Wurtz,[36,37] Lee et al., found deactivation of the SC to produce dramatic decreases in saccadic velocity along with compensatory increases in saccade duration[35] (Figure 2.2, top, middle; see legend for details). Lee et al. also reported large increases in saccadic latency with reduction in the size of the active population. Thus, while the center of mass of SC activity may specify the desired amplitude and direction, it seemed that the overall level of activity contributed to determining when (saccade latency) and how rapidly (saccade speed) the eyes moved.

2.6 MICROSTIMULATION AND THE SC MOTOR MAP REVISITED

The apparent discrepancy between results from deactivation studies[35–37] and earlier microstimulation studies[28,30] was resolved by subsequent microstimulation studies showing that, along with the site of activity, both the level and duration of SC activity contribute to the programming of saccades. An early clue was provided by Van Opstal and colleagues[38] who noted that the vector of a stimulation-evoked saccade was not, in fact, entirely independent of the parameters of stimulation. Analogous to the results from SC deactivation, they found decreases in eye movement velocity with decreases in the size of the active population (i.e., reduced current strength; see Section 2.2). However, in addition, they found that the **size** of evoked saccades decreased monotonically as a function of decreasing current strength. In contrast to the vector averaging scheme supported by the Lee et al. findings, this suggested that in addition to the site of activity, the size of the active population contributes to the computation of saccade amplitude. But, as noted by Van Opstal and colleagues at the time, the effect of current strength on saccade amplitude might also have been the secondary consequence of the reduction in saccade velocity in the absence of a compensatory increase in duration. According to this logic, increases in the duration of the stimulation train, had they been applied, might have compensated for the reductions in velocity and produced saccades of consistent site-specific amplitude.

A relationship between amplitude and train duration was verified in studies by Stanford et al.[39] and Freedman et al.,[2] in which the parameters of microstimulation were varied systematically for many different sites within the SC motor topography. Here the relative influences of the site of stimulation, duration of stimulation, and frequency of stimulation to determining the properties of the evoked saccade[39] or gaze shift[2] were systematically examined. A principal finding of these studies was that the site of stimulation determined the *maximum* amplitude of an evoked movement, however, achievement of this site-specific amplitude was contingent upon adequate duration of the stimulation train (Figure 2.3, see legend for details); insufficient train durations produced saccades that fell short of the asymptotic value with the offset of the movement being clearly linked to the offset of the stimulation train. However, there was not a unique relationship between train duration and movement amplitude, rather this relationship could be systematically modulated by varying the frequency of the stimulation train. Consistent with the observations from deactivation studies[35–37] and with the effects of current level,[38] increasing the frequency of stimulation, and thus presumably, the level of SC activation, was found to increase the

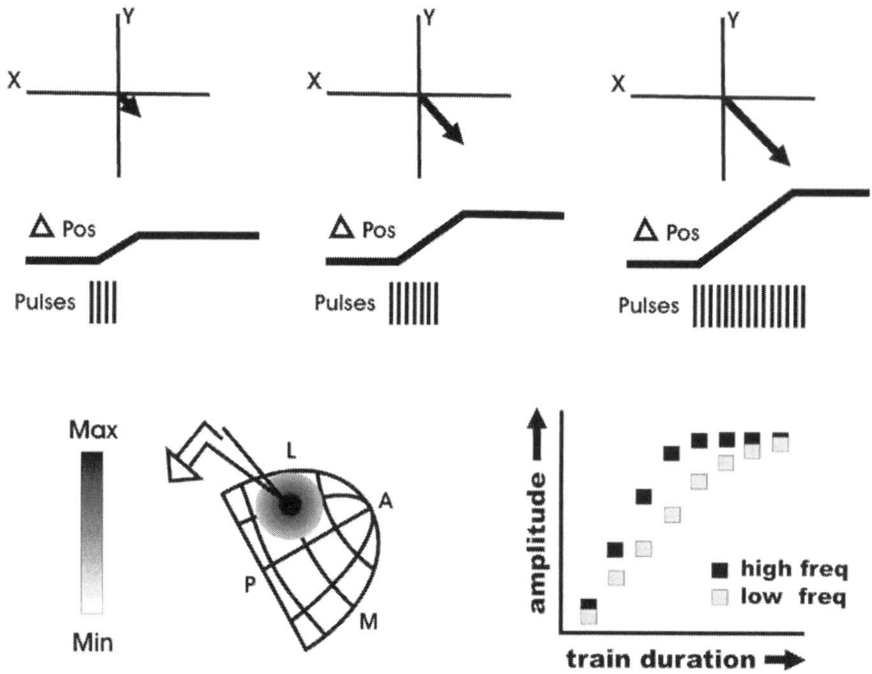

FIGURE 2.3 Effects of microstimulation train duration. Depiction of results obtained by Stanford et al.[39] and Freedman et al.[2] The effect of train duration for stimulation of a single site in the right SC (shown on bottom left). Top row: from left to right the amplitude of the evoked movement increases with increases in train duration. Offset of the movement is linked in time to offset of the stimulation train except on far right where the stimulation train outlasts the movement. In this case, the site-specific amplitude is met and continued stimulation produces no further increment in amplitude. Bottom right: plot summarizing the relationship between train duration and amplitude for two stimulation frequencies (high-frequency — black symbols; low-frequency — gray symbols). Owing to a decrease in velocity with reduced stimulation frequency (see Figure 2.4), longer trains are needed to produce the site-specific saturating amplitude. Abbreviations: A = anterior; P = posterior; M = medial, L = lateral; Pos = eye position.

velocity (and decrease the latency) of the evoked movements (Figure 2.4; see legend for details). This, in turn, reduced the train duration requirement for producing the site-specific maximum amplitude (see Figure 2.3, bottom right).

2.7 DUAL CODING IN THE SC

Although the results from deactivation and later microstimulation studies disagree with those from early microstimulation studies, they are not entirely inconsistent. Rather than contradict the notion of a strictly topographic code, these more recent data simply place constraints on how an anatomical code for saccade amplitude and direction is interpreted downstream. In addition, these more recent findings specify the existence of a second, independent signal derived from the level of

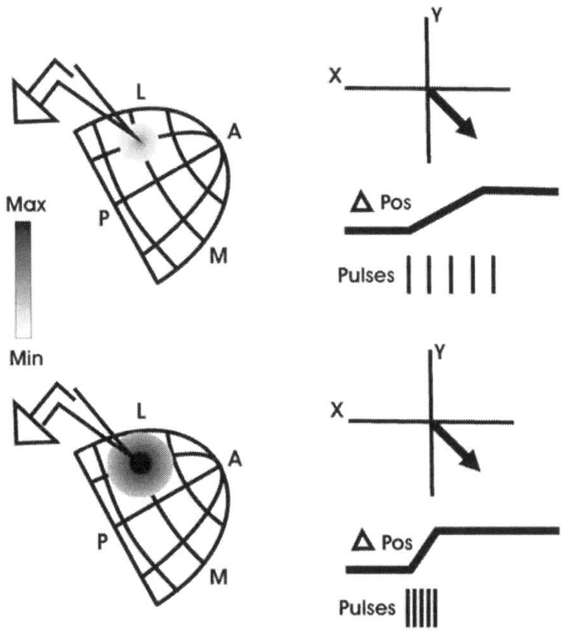

FIGURE 2.4 Effects of train frequency. Depiction of results obtained by Stanford et al.[39] and Freedman et al.[2] Top left: hypothetical extent of SC activity produced by low-frequency stimulation of a lateral site in the right SC. Top right: vector and time course of movement evoked by low-frequency stimulation. Bottom left: extent of activity produced by high-frequency stimulation of same SC site. Bottom right: vector and time course of movement evoked by a high frequency train. For both the high and low-frequency trains, duration is sufficient to produce the site-specific amplitude, however, due to lower peak velocity, low-frequency stimulation requires a longer stimulation train. Abbreviations: A = anterior; P = posterior; M = medial, L = lateral; Pos = eye position.

activity within the motor map, one that influences saccade velocity. According to a traditional view, which incorporates Robinson's model of the downstream circuitry,[40–42] the SC provides a signal of desired eye position or desired eye displacement[43] as input to a downstream local feedback circuit (Figure 2.5, top; see legend for details). This signal, which may reflect the vector average of the active SC population, can be idealized as a step of activity with a frequency (height) proportional to the size of the desired movement. A pulse generator acts on this input, differentiating the step (in the mathematical sense) to produce a high-frequency pulse that drives the eyes at saccadic velocity. In this model, the frequency (height) of the pulse and thus saccade velocity are determined by properties intrinsic to the pulse generator. Saccade duration is not controlled explicitly; a comparator subtracts an ongoing signal of current displacement (or position) from the input signal of desired displacement (or position) and a saccade endures until these two signals are equal and the comparator output (input to the pulse generator) becomes zero. Thus for a given amplitude, saccade duration is constrained to vary as an inverse function of saccade velocity.

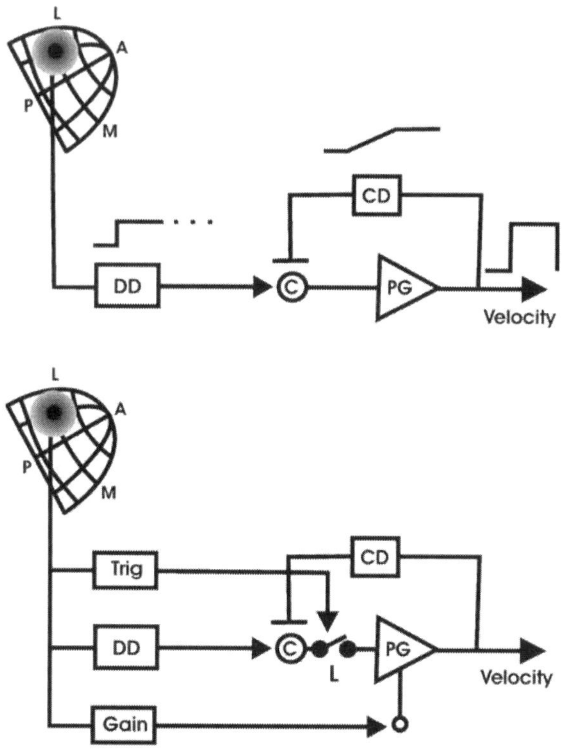

FIGURE 2.5 SC input to downstream pulse generator. Top: essential elements of Robinson's local feedback model as modified by Jurgens et al.[43] In this version, an SC derived signal of desired displacement (i.e., change in eye position) provides input to a downstream comparator. This input signal is compared to a feedback signal of current displacement and this difference drives the pulse generator. The pulse generator transforms this difference signal (also known as motor error) into a velocity command, which is sent to the motor neuron pools that innervate the eye muscles. The velocity command is also sent to a neural integrator that computes current displacement and provides a subtractive feedback signal to the comparator. A saccade endures until the current displacement matches the desired displacement. Bottom: amendments to the model to account for independent effects of the site and level of SC activity. As above, the pulse generator acts on a signal of motor error, however, two additional inputs proportional to the level of SC activity are specified. The first (Gain), influences the gain of the pulse generator and modulates the peak velocity of the saccade. The second (trig) signal contributes to determining saccade onset (symbolically represented by the closing of a latch between the comparator and the pulse generator). See text for details. Abbreviations: DD = desired displacement; CD = current displacement; Trig = trigger signal; C = comparator; PG = pulse generator; L = Latch.

No modification to this classical model of the downstream circuitry is needed to account for a systematic relationship between the duration of the stimulation train and the amplitude (and duration) of stimulation-evoked saccades.[39] Note that the model requires that the step signal persist at the input to the comparator, *at least* until the current displacement (or position) equals desired displacement (or position).

Thus, given the reasonable assumption that electrically evoked SC activation ceases with offset of the microstimulation train (see Section 2.2), a very short stimulation train would lead to premature truncation of this reference input signal, in effect, removing the drive to the pulse generator and bringing the ongoing movement to a halt as was, in fact, found to be the case.

Whereas, the effect of train duration is readily accounted for given the model shown at the top of Figure 2.5, some modification is necessary to explain the effects of deactivation, current level, or stimulation frequency on saccade velocity. All of these results indicate that the level of SC activity can influence the velocity of the movement, but most importantly, that it can do so without altering the computation of saccade amplitude.[35,39] In Robinson's original model[41,42] and its variants (e.g., see Reference 43), saccades having the same amplitude but different velocities could be produced by allowing the intrinsic gain of the pulse generator to vary. Qualitatively, at least, all that is needed to account for the deactivation/microstimulation data is to permit the gain of the pulse generator to be modulated by an *extrinsic* signal (e.g., see Reference 44), with one such signal being proportional to the level of activity within the SC population (Figure 2.5, bottom; see legend for details).

It is also clear from microstimulation and pharmacological manipulation studies that the vigor of SC activity can influence saccade latency. It is generally held that a powerful, tonic inhibitory drive prevents the high-gain pulse generator from transforming neural noise into nonpurposive saccadic eye movements and that a prerequisite to activation of the saccadic pulse generator is the removal of the inhibitory influence (see References 22, 45 to 47 for related reviews). The fact that the latencies of both sensory driven or stimulus-evoked saccades vary inversely with the vigor of SC activation suggests that the SC may also provide at least part of a trigger signal necessary to release the pulse generator from inhibition (Figure 2.5; see legend for details).

The identification of collicular signals relating to both *what* saccade is produced (i.e., amplitude and direction determined by spatial distribution of SC activity) and *how* it is produced (i.e., level of activity influences velocity and latency) is the direct result of observing the consequences of artificially manipulating the activity of large populations of SC neurons. In keeping with this chapter's emphasis on these methods, it is worth noting that neural correlates of these signals are not necessarily evident in the activities of single neurons. Clearly, the well-defined movement fields of saccade-related neurons within the SC are the precursors to the vector-average computation of desired displacement (or eye position), but considered alone, they provide no direct information about how this activity is interpreted downstream. In principle, correlates of saccade velocity could be found in the discharges of single SC neurons, however, in practice, strong evidence for this is not available. These data may be lacking for several reasons. First, deactivation and microstimulation data suggest a role for SC activity in *modulating* the peak velocity attained during a saccade (or gaze shift), not for determining movement velocity on a moment-to-moment basis (but see Reference 48). Given that naturally elicited saccades of comparable amplitude show relatively small variation in peak velocity[49,50] and that a large proportion of SC saccade-related neurons are active in association with each saccade,[51] it is not particularly surprising that strong correlations between saccade velocity and the discharges of *single* SC neurons have not been observed. On the other hand, there is evidence to suggest that the magnitude of relatively low-frequency activity among some SC neurons may reflect the process of motor preparation and thus may be a factor in determining saccadic reaction time.[52] Munoz and colleagues have reported that, given advance knowledge of probable target location, high levels of this preparatory signal predict the occurrence of short-latency, anticipatory saccades (see Reference 53 for a review of these findings). These results are generally consistent with the prolonged latencies observed with deactivation (e.g., see Reference 35) and reductions in stimulation frequency[2,39] and, in addition, agree well with the increases in saccade probability associated with bicuculline (GABA antagonist; see previous section) injections into the SC.[36]

2.8 MULTIPLE CODING FORMATS AS A PRINCIPLE OF SC ORGANIZATION

Whereas the influence of microstimulation parameters on saccade metrics and dynamics was, at the time, novel for the primate SC, similar dependencies had been shown for SC-evoked gaze shifts in the cat,[54,55] head movements in the owl,[56] and head and body movements in rat.[57,58] Analogous results have subsequently been reported from microstimulation studies of the goldfish optic tectum.[59]

Considered across studies and species, there seems to be compelling evidence that the SC uses the orthogonal coding dimensions of location (motor map) and level (firing rate) to independently specify the distance (and direction) and speed (latency, velocity) of orientation behaviors. As such, activity in the SC might provide a convenient substrate for the modulatory influences of external and internal contextual factors such as stimulus salience, prior knowledge, and expected outcome. Indeed, recent studies have begun to examine the SC for neural correlates of cognitive factors and their behavioral consequences. As described below, the use of artificial signals may in fact prove valuable in elucidating what, if anything, the SC contributes to these higher-order aspects of sensorimotor integration.

2.9 SC DECISION-MAKING

The preceding sections have mainly considered those SC signals thought necessary to *generate* saccades or gaze shifts; i.e., the motor commands issued by the SC to engage the downstream pulse generator. From single-unit studies, it is known that saccade-related bursts of primate SC neurons precede eye movement by as little as 20 ms[33] while, for microstimulation, the time from pulse delivery to manifestation as a change in eye velocity is on the order of 10 ms or less.[39,60] Thus, even a generous accounting suggests that, once formulated, the SC motor command is issued and carried out within a 30 to 50 ms time frame. As noted by Carpenter and colleagues[61,62] (see also Reference 63) the addition of afferent (sensory processing; approximately 40 to 100 ms) and efferent (motor processing) delays (30 to 50 ms) predicts saccadic reaction times that are, on average, much shorter than actually observed. That saccades are not initiated as soon as theoretically possible based on known afferent and efferent delays is presumed to reflect the time necessary to make decisions about where to move the eyes in the event of competing alternatives.

The search for neural correlates of putative perceptual-motor decision signals is the focus of an ever-increasing number of studies in many brain regions (see References 64 to 67 for reviews). Signals related to perceptual discrimination and/or response choice are thought to evolve during the period that intervenes between the initial encoding of the sensory stimulus (i.e., stimulus detection) and the issuing of a motor command to generate an appropriate response. Neural correlates of both perceptual discrimination and response selection have been reported for neurons within the primate SC. In one of the earliest studies of this type, Glimcher and Sparks[68] reported that, given two stimuli and thus two possible saccade vectors, the low-frequency (prelude) activity of many SC neurons predicted the monkey's choice. For such a neuron, discharge for a visual stimulus in its receptive/movement field was conditional on the knowledge that the stimulus was, in fact, the intended goal. This activity, which can be interpreted as a signal of motor preparation, reflects the outcome of a prior sensory discrimination between *target* and distracting stimuli made on the basis of some stimulus attribute (a color-location association, in this case). More recently, Horwitz and Newsome,[69–71] reported a neural correlate of this earlier process of perceptual discrimination (direction of motion) in the activity of SC neurons. Rather than discharge only after a target has been selected, these neurons appeared to participate in, or at least reflect, the process of *accumulating* sensory evidence (see Reference 67) for this purpose.

The notion that the SC could participate in higher-order aspects of saccadic programming seems plausible. Consider that neurons in the SC are co-active with those in visuomotor cortical domains during the time period that such decisions are weighed. Whether or not activity in the SC during this time period is essential to or simply reflects the state of computations carried out in cortex is difficult to know. It is in this regard that microstimulation may be of value. In extrastriate area MT, microstimulation has been shown to bias the outcome of a direction of motion judgment. [72,73] In principle, an experiment that combines the right behavioral task, microstimulation parameters, and microstimulation timing might distinguish the SC's respective contributions to sensorimotor decision-making and motor command formation.

2.10 SUMMARY

This chapter has focused on a few studies chosen specifically to illustrate how our current understanding of SC function is, at least in part, the consequence of insights gained through use of artificial signals. While microstimulation and deactivation/activation techniques have contributed to the study of many other brain regions, the SC seems somewhat unique in that these methods have yielded so many essential findings. Clearly, proximity to the final motor output along with a well-organized motor topography are primary reasons that the SC has been so amenable to study with these methodologies.

ACKNOWLEDGMENTS

Supported by NIH/NEI EY12389.

REFERENCES

1. Evarts, E.V., A technique for recording activity of subcortical neurons in moving animals, *Electroencephalogr. Clin. Neurophysiol.*, 24, 83, 1968.
2. Freedman, E.G., Stanford, T.R., and Sparks, D.L., Combined eye-head gaze shifts produced by electrical stimulation of the superior colliculus in rhesus monkeys, *J. Neurophysiol.*, 76, 927, 1996.
3. Freedman, E.G. and Sparks, D.L., Activity of cells in the deeper layers of the superior colliculus of the rhesus monkey: evidence for a gaze displacement command, *J Neurophysiol.*, 78, 1669, 1997.
4. Krauzlis, R.J., Basso, M.A., and Wurtz, R.H., Discharge properties of neurons in the rostral superior colliculus of the monkey during smooth-pursuit eye movements, *J. Neurophysiol.*, 84, 876, 2000.
5. Basso, M.A., Krauzlis, R.J., and Wurtz, R.H., Activation and inactivation of rostral superior colliculus neurons during smooth-pursuit eye movements in monkeys, *J. Neurophysiol.*, 84, 892, 2000.
6. Werner, W., Dannenberg, S., and Hoffmann, K.P., Arm-movement-related neurons in the primate superior colliculus and underlying reticular formation: comparison of neuronal activity with EMGs of muscles of the shoulder, arm and trunk during reaching, *Exp. Brain Res.*, 115, 191, 1997.
7. Stuphorn, V., Bauswein, E., and Hoffmann, K.P., Neurons in the primate superior colliculus coding for arm movements in gaze-related coordinates, *J. Neurophysiol.*, 83, 1283, 2000.
8. Lunenburger, L., et al., A possible role of the superior colliculus in eye-hand coordination, *Prog. Brain Res.*, 134, 109, 2001.
9. Sparks, D.L., Translation of sensory signals into commands for control of saccadic eye movements: role of primate superior colliculus, *Physiol. Rev.*, 66, 118, 1986.
10. Sparks, D.L. and Hartwich-Young, R., The deep layers of the superior colliculus, *Rev. Oculomot. Res.*, 3, 213, 1989.
11. Sparks, D.L. and Mays, L.E., Signal transformations required for the generation of saccadic eye movements, *Annu. Rev. Neurosci.*, 13, 309, 1990.
12. Moschovakis, A.K. and Highstein, S.M., The anatomy and physiology of primate neurons that control rapid eye movements, *Annu. Rev. Neurosci.*, 17, 465, 1994.
13. Moschovakis, A.K., The superior colliculus and eye movement control, *Curr. Opin. Neurobiol.*, 6, 811, 1996.
14. Ranck, J.B., Jr., Which elements are excited in electrical stimulation of mammalian central nervous system: a review, *Brain Res.*, 98, 417, 1975.
15. Sandkuhler, J., Maisch, B., and Zimmermann, M., The use of local anaesthetic microinjections to identify central pathways: a quantitative evaluation of the time course and extent of the neuronal block, *Exp. Brain Res.*, 68, 168, 1987.
16. Tehovnik, E.J., Electrical stimulation of neural tissue to evoke behavioral responses, *J. Neurosci. Methods.*, 65, 1, 1996.
17. Tehovnik, E.J. and Sommer, M.A., Effective spread and timecourse of neural inactivation caused by lidocaine injection in monkey cerebral cortex, *J. Neurosci. Methods.*, 74, 17, 1997.

18. Sparks, D.L. and Mays, L.E., Spatial localization of saccade targets. I. Compensation for stimulation-induced perturbations in eye position, *J. Neurophysiol.*, 49, 45, 1983.
19. Glimcher, P.W. and Sparks, D.L., Effects of low-frequency stimulation of the superior colliculus on spontaneous and visually guided saccades, *J. Neurophysiol.*, 69, 953, 1993.
20. Binns, K.E., The synaptic pharmacology underlying sensory processing in the superior colliculus, *Prog. Neurobiol.*, 59, 129, 1999.
21. Munoz, D.P. and Istvan, P.J., Lateral inhibitory interactions in the intermediate layers of the monkey superior colliculus, *J. Neurophysiol.*, 79, 1193, 1998.
22. Munoz, D.P. and Fecteau, J.H., Vying for dominance: dynamic interactions control visual fixation and saccadic initiation in the superior colliculus, *Prog. Brain Res.*, 140, 3, 2002.
23. Hikosaka, O. and Wurtz, R.H., The basal ganglia, *Rev. Oculomot. Res.*, 3, 257, 1989.
24. Hikosaka, O., Takikawa, Y., and Kawagoe, R., Role of the basal ganglia in the control of purposive saccadic eye movements, *Physiol. Rev.*, 80, 953, 2000.
25. Adamük, E., Uber die Innervation der Augenbewegungen, *Zentralbl. Med. Wiss.*, 9, 65, 1870.
26. Hess, W.R., Bürgi, S., and Bucher, V., Motorische Funktion del Taktalund Tegmentalgtebietes. *Monatsschr. Psychiatr. Neurol.*, 112, 1, 1946.
27. Apter, J.T., Eye movements following strychnization of the superior colliculus of cats. *J. Neurophysiol.*, 9, 73, 1946.
28. Robinson, D.A., Eye movements evoked by collicular stimulation in the alert monkey, *Vision Res.*, 12, 1795, 1972.
29. Schiller, P.H. and Koerner, F., Discharge characteristics of single units in superior colliculus of the alert rhesus monkey, *J. Neurophysiol.*, 34, 920, 1971.
30. Schiller, P.H. and Stryker, M., Single-unit recording and stimulation in superior colliculus of the alert rhesus monkey, *J. Neurophysiol.*, 35, 915, 1972.
31. Wurtz, R.H. and Goldberg, M.E., Activity of superior colliculus in behaving monkey. 3. Cells discharging before eye movements, *J. Neurophysiol.*, 35, 575, 1972.
32. Sparks, D.L., Holland, R., and Guthrie, B.L., Size and distribution of movement fields in the monkey superior colliculus, *Brain Res.*, 113, 21, 1976.
33. Sparks, D.L., Functional properties of neurons in the monkey superior colliculus: coupling of neuronal activity and saccade onset, *Brain Res.*, 156, 1, 1978.
34. Moschovakis, A.K., Karabelas, A.B., and Highstein, S.M., Structure-function relationships in the primate superior colliculus. I. Morphological classification of efferent neurons, *J. Neurophysiol.*, 60, 232, 1988.
35. Lee, C., Rohrer, W.H., and Sparks, D.L., Population coding of saccadic eye movements by neurons in the superior colliculus, *Nature*, 332, 357, 1988.
36. Hikosaka, O. and Wurtz, R.H., Modification of saccadic eye movements by GABA-related substances. I. Effect of muscimol and bicuculline in monkey superior colliculus, *J. Neurophysiol.*, 53, 266, 1985.
37. Hikosaka, O. and Wurtz, R.H., Saccadic eye movements following injection of lidocaine into the superior colliculus, *Exp. Brain Res.*, 61, 531, 1986.
38. Van Opstal, A.J., Van Gisbergen, J.A., and Smit, A.C., Comparison of saccades evoked by visual stimulation and collicular electrical stimulation in the alert monkey, *Exp. Brain Res.*, 79, 299, 1990.
39. Stanford, T.R., Freedman, E.G., and Sparks, D.L., Site and parameters of microstimulation: evidence for independent effects on the properties of saccades evoked from the primate superior colliculus, *J. Neurophysiol.*, 76, 3360, 1996.

40. Van Gisbergen, J.A., Robinson, D.A., and Gielen, S., A quantitative analysis of generation of saccadic eye movements by burst neurons, *J. Neurophysiol.*, 45, 417, 1981.
41. Robinson, D.A., Oculomotor control signals, in *Basic Mechanisms of Ocular Motility and their Clinical Implications*, G. Lennerstrand and P. Bach-y-Rita, Eds., Pergamon, Oxford, 1975, p. 337.
42. Zee, D.S., et al., Slow saccades in spinocerebellar degeneration, *Arch. Neurol.*, 33, 243, 1976.
43. Jürgens, R., Becker, W., and Kornhuber, H.H., Natural and drug-induced variations of velocity and duration of human saccadic eye movements: evidence for a control of the neural pulse generator by local feedback, *Biol. Cybern.*, 39, 87, 1981.
44. Nichols, M.J. and Sparks, D.L., Component stretching during oblique stimulation-evoked saccades: the role of the superior colliculus, *J. Neurophysiol.*, 76, 582, 1996.
45. Hepp, K., et al., Brainstem regions related to saccade generation, *Rev. Oculomot. Res.*, 3, 105, 1989.
46. Scudder, C.A., Kaneko, C.S., and Fuchs, A.F., The brainstem burst generator for saccadic eye movements: a modern synthesis, *Exp. Brain Res.*, 142, 439, 2002.
47. Sparks, D.L., The brainstem control of saccadic eye movements, *Nat. Rev Neurosci.*, 3, 952, 2002.
48. Berthoz, A., Grantyn, A., and Droulez, J., Some collicular efferent neurons code saccadic eye velocity, *Neurosci. Lett.*, 72, 289, 1986.
49. Baloh, R.W., et al., Quantitative measurement of saccade amplitude, duration, and velocity, *Neurology*, 25, 1065, 1975.
50. Bahill, T.A., Clark, M.R., and Stark, L., The main sequence as a tool for studying human eye movements, *Math. Biosci.*, 24, 181, 1975.
51. Van Gisbergen, J.A., Van Opstal, A.J., and Tax, A.A., Collicular ensemble coding of saccades based on vector summation, *Neuroscience*, 21, 541, 1987.
52. Dorris, M.C., Pare, M., and Munoz, D.P., Neuronal activity in monkey superior colliculus related to the initiation of saccadic eye movements, *J. Neurosci.*, 17, 85, 66, 1997.
53. Trappenberg, T.P., et al., A model of saccade initiation based on the competitive integration of exogenous and endogenous signals in the superior colliculus, *J. Cogn. Neurosci.*, 13, 256, 2001.
54. Munoz, D.P., Guitton, D., and Pelisson, D., Control of orienting gaze shifts by the tectoreticulospinal system in the head-free cat. III. Spatiotemporal characteristics of phasic motor discharges, *J. Neurophysiol.*, 66, 1642, 1991.
55. Pare, M., Crommelinck, M., and Guitton, D., Gaze shifts evoked by stimulation of the superior colliculus in the head-free cat conform to the motor map but also depend on stimulus strength and fixation activity, *Exp. Brain Res.*, 101, 123, 1994.
56. du Lac, S. and Knudsen, E.I., Neural maps of head movement vector and speed in the optic tectum of the barn owl, *J. Neurophysiol.*, 63, 131, 1990.
57. King, S.M., Dean, P., and Redgrave, P., Bypassing the saccadic pulse generator: possible control of head movement trajectory by rat superior colliculus, *Eur. J. Neurosci.*, 3, 790, 1991.
58. Tehovnik, E.J., Head and body movements evoked electrically from the caudal superior colliculus of rats: pulse frequency effects, *Behav. Brain Res.*, 34, 71, 1989.
59. Salas, C., et al., Tectal codification of eye movements in goldfish studied by electrical microstimulation, *Neuroscience*, 78, 271, 1997.

60. Miyashita, N. and Hikosaka, O., Minimal synaptic delay in the saccadic output pathway of the superior colliculus studied in awake monkey, *Exp. Brain Res.*, 112, 187, 1996.
61. Reddi, B.A. and Carpenter, R.H., The influence of urgency on decision time, *Nat. Neurosci.*, 3, 827, 2000.
62. Carpenter, R.H. and Williams, M.L., Neural computation of log likelihood in control of saccadic eye movements, *Nature*, 377, 59, 1995.
63. Schall, J.D., Experimental psychology. Race to explain procrastination, *Nature*, 377, 14, 1995.
64. Schall, J.D. and Thompson, K.G., Neural selection and control of visually guided eye movements, *Annu. Rev. Neurosci.*, 22, 241, 1999.
65. Nichols, M.J. and Newsome, W.T., Neurobiology. Monkeys play the odds, *Nature*, 400, 217, 1999.
66. Glimcher, P.W., Making choices: the neurophysiology of visual-saccadic decision making, *Trends Neurosci.*, 24, 654, 2001.
67. Gold, J.I. and Shadlen, M.N., Neural computations that underlie decisions about sensory stimuli, *Trends Cogn. Sci.*, 5, 10, 2001.
68. Glimcher, P.W. and Sparks, D.L., Movement selection in advance of action in the superior colliculus, *Nature*, 355, 542, 1992.
69. Horwitz, G.D. and Newsome, W.T., Separate signals for target selection and movement specification in the superior colliculus, *Science*, 284, 1158, 1999.
70. Horwitz, G.D. and Newsome, W.T., Target selection for saccadic eye movements: direction-selective visual responses in the superior colliculus, *J. Neurophysiol.*, 86, 2527, 2001.
71. Horwitz, G.D. and Newsome, W.T., Target selection for saccadic eye movements: prelude activity in the superior colliculus during a direction-discrimination task, *J. Neurophysiol.*, 86, 2543, 2001.
72. Salzman, C.D., Britten, K.H., and Newsome, W.T., Cortical microstimulation influences perceptual judgements of motion direction, *Nature*, 346, 174, 1990.
73. Seidemann, E., Zohary, E., and Newsome, W.T., Temporal gating of neural signals during performance of a visual discrimination task, *Nature*, 394, 72, 1998.

3 Concurrent, Distributed Control of Saccade Initiation in the Frontal Eye Field and Superior Colliculus

Douglas P. Munoz and Jeffrey D. Schall

CONTENTS

3.1 INTRODUCTION

A central problem in neuroscience is the localization of function. Significant progress on this has been made in the oculomotor system. With progress has come the realization that function is usually not localized within a single structure but rather it is distributed across a network of brain areas. In this review, we contrast two brain areas that play a critical role in the planning and initiation of saccadic eye movements: the superior colliculus (SC) and the frontal eye fields (FEF). We describe the

0-8493-0097-5/04/$0.00+$1.50

role of these structures in the control of visual fixation and initiation of saccadic eye movements by contrasting neural discharges recorded from neurons in each area in a variety of oculomotor tasks.

3.2 THE TIME TO INITIATE A SACCADE

In flight, saccades are incredibly predictable. The velocity and duration of a saccade is mechanistically related to the amplitude of the saccade.[1] However, the time of initiation of the movement is exceedingly unpredictable. Numerous experiments have measured the time that elapses from presentation of a visual stimulus until initiation of a saccade. Each experiment finds that this saccadic reaction time (SRT) ranges from rarely less than 100 ms to as much as 500 ms or more. Moreover, SRT can vary over a wide range across a block of trials even within a single task with constant stimuli and unchanging instructions. The origin of the delay and variability of SRT is a central problem that has received increasing attention with, it is fair to say, notable progress toward its elucidation.

Many models have been developed to explain the stochastic variability of reaction time.[2] Accumulator models have been evaluated in terms of brain function. Accumulator models suppose that in response to a stimulus, some signal grows until it reaches a threshold thereby triggering a movement in response to the stimulus. Models of this sort include three sources for the stochastic variability evident in reaction times (Figure 3.1). One type of accumulator model (Figure 3.1A), assumes that the threshold is constant, but that the average rate of growth of the accumulator is random across trials.[3,4] This architecture can account for a broad range of reaction times measured in a variety of tasks.[5,6] Another type of accumulator model (Figure 3.1B) supposes that the variability in reaction time arises from randomness in the level of the trigger threshold.[7,8] A third scenario (Figure 3.1C) could employ a fixed threshold but a *variable baseline* at target onset.[9] Although this latter model is the mathematical equivalent of variable threshold, it would be implemented very differently in the brain. The alternative models cannot be distinguished on the basis of performance data alone. As a matter of fact, it has been shown that random accumulator and random threshold models generate equivalent predictions.[10]

How are aspects of these models instantiated at the level of the single cell, a single brain area, and the entire saccadic generating circuitry? We first review the saccade generating circuitry including the connectivity between the FEF and the SC. We then review neurophysiological experiments that provide information that seems to differentiate the possibilities illustrated in Figure 3.1.

3.3 OVERVIEW OF THE SACCADE GENERATION NETWORK

A network of cortical and subcortical structures is required for the accurate and timely control of saccadic eye movements (Figure 3.2), including regions within the cerebral cortex, thalamus, basal ganglia, cerebellum, SC, and brainstem reticular formation. Details of this network are reviewed elsewhere.[11–18]

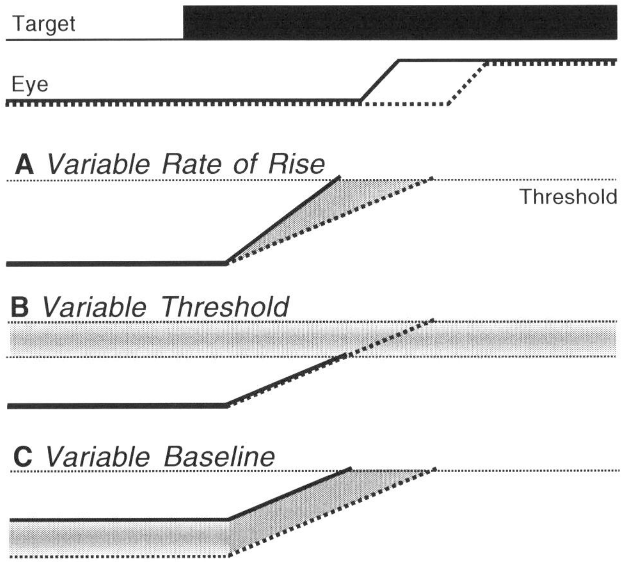

FIGURE 3.1 Accumulator models of saccade initiation. (A) Variability in saccadic reaction time (SRT) could be the result of a variable rate of rise toward the saccadic threshold. (B) Variability in SRT could be the result of a variable threshold. (C) Variability in SRT could be the result of variability in baseline activity before accumulation begins.

3.3.1 Brainstem Saccade Generator

The saccadic burst generator circuit is housed in the brainstem reticular formation.[13,17] Burst neurons (BNs) in the reticular formation innervate the extraocular muscle motoneurons (MNs) to provide the high-frequency burst of spikes necessary to move the eyes. BNs are silent during fixation and discharge action potentials for saccades in a specific direction. Excitatory burst neurons (EBNs) monosynaptically excite the one-direction motoneurons, while inhibitory burst neurons (IBNs), which receive their input from the EBNs, inhibit the antagonist MNs. The EBNs and IBNs for horizontal saccades are located in the pontine and medullary reticular formation, while the BNs for vertical saccades are located in the mesencephalic reticular formation.

Other neurons in the brainstem reticular formation are believed to control the discharge of EBNs and IBNs. The EBNs and IBNs for horizontal and vertical systems are subject to potent monosynaptic inhibition from omnipause neurons (OPNs), also located in the paramedian pontine reticular formation (PPRF), which discharge tonically during all fixations and pause for saccades in all directions. Thus, to generate a saccade, OPNs must first be silenced and then the appropriate pools of EBNs and IBNs are activated to produce the burst in the corresponding MN pools. Following completion of the saccade, OPNs are reactivated to inhibit the EBNs and IBNs.

Neurons with tonic activity proportional to the angle of the eyes in the orbit are also present in the brainstem and they innervate the MNs as well. The activation from these neurons results in the amount of innervation of the MNs to resist the centripetal viscoelastic forces and keep the eyes at an eccentric location in the orbit.

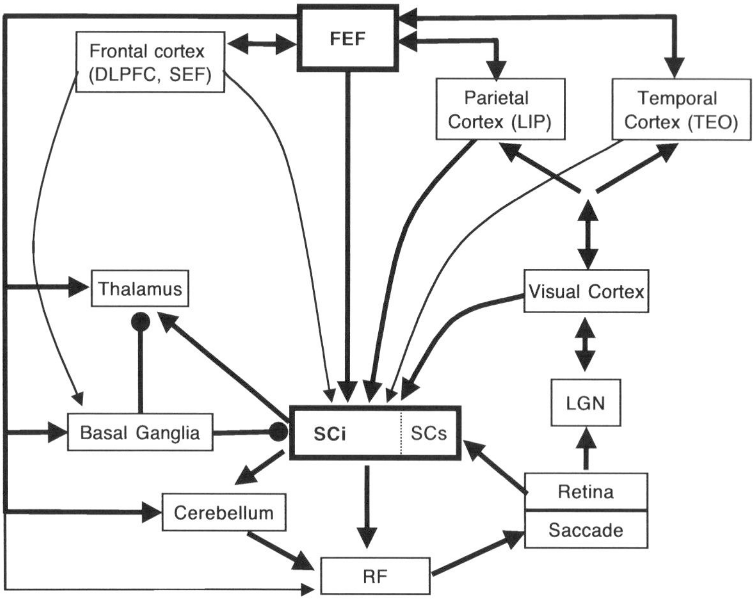

FIGURE 3.2 Schematic of brain areas and pathways involved in saccade generation. See text for additional details. Abbreviations: CN: caudate nucleus; DLPFC: dorsolateral prefrontal cortex; FEF: frontal eye fields; GPe: external segment of globus pallidus; LGN: lateral geniculate nucleus; LIP: lateral intraparietal area; RF: reticular formation; SCi: intermediate layers of superior colliculus; SCs: superficial layers of superior colliculus; SEF: supplementary eye fields; SNr: substantia nigra pars reticulata; STN: subthalamic nucleus.

These tonic neurons, which comprise the neural integrator, are located primarily in the medial vestibular nucleus and the nucleus prepositus hypoglossi for horizontal position control and in the interstitial nucleus of Cajal for vertical control.

Long-lead burst neurons (LLBNs), also located in the brainstem reticular formation, discharge a high frequency burst of action potentials for saccades into the contralateral hemifield. In addition to the burst, these cells also have a low frequency buildup of activity before the burst. It is believed that at least some LLBNs are innervated by descending projections from higher centers such as the SC and FEF and project directly to the EBNs and IBNs to provide the burst input.

3.3.2 Superior Colliculus

The superior colliculus (SC) is a laminated structure in the dorsal mesencephalon. The dorsal-most superficial layers of the SC contain neurons that receive direct retinal inputs as well as inputs from other visual areas.[19] The retinotectal projection arises from no more than 10% of all the ganglion cells.[20] The superficial layers of the SC also receive major afferent inputs from primary visual cortex and many extrastriate visual areas in occipital, parietal, and temporal lobes, as well as areas in the frontal lobe. The superficial layers of the SC project to the dorsal lateral geniculate nucleus, the pregeniculate nucleus, the inferior and lateral pulvinar, and the pretectum. As a result of these retinal and other visual afferents, neurons in the superficial SC have well defined visual receptive fields, and there is an orderly retinotopic map in the superior colliculus with the fovea represented rostrally, and the upper visual field represented medially.[21,22]

The input to and output of the intermediate layers of the SC are diverse. The intermediate layers receive input from a variety of cortical areas including primary visual cortex, extrastriate visual areas, posterior parietal cortex, temporal cortex, the supplementary and frontal eye fields, and prefrontal cortex[23–28]. Subcortical afferents to the intermediate layers include the reticular nucleus and the pregeniculate nucleus; the pretectum; various midbrain structures including in particular the substantia nigra pars reticularis; various pontine and medullary nuclei, including the nucleus reticularis tegmenti pontis; deep cerebellar nuclei and finally the cervical spinal cord (see References 29 and 30 for reviews). The efferents of the intermediate layers of the SC are just as widespread. Ascending projections travel to numerous thalamic nuclei including the ventral anterior, ventral lateral, mediodorsal, central lateral, anterior, medial, and inferior pulvinar, lateral dorsal, reticular thalamic nuclei.[31] Descending projections terminate in the ipsilateral substantia nigra pars reticulata, mesencephalic reticular formation, pons, medulla and spinal cord, and in the contralateral pons, including PPRF, medulla, and spinal cord.[32]

The intermediate layers of the SC contain neurons with discharges that are correlated with saccadic eye movements and visual fixation.[33–49] These neurons are organized into a two-dimensional motor map coding for saccades directed to the contralateral visual field.[50] Neurons increasing their discharges before and during saccades — referred to hereafter as *saccade neurons* — are distributed throughout the extent of these intermediate layers. Each saccade neuron discharges for a range of amplitude and direction saccades that define a movement field.[42,45,46] These saccade neurons can be divided into subclasses based upon prelude[33,51,52] or buildup[42,53,54] activity preceding the saccadic burst. Buildup neurons have low frequency preamble activity before the saccade burst and many of these neurons also have open-ended movement fields.[42] Burst neurons lack the prelude or low frequency buildup activity. There is probably a continuum between these two classes of saccade neurons.[42,55]

Neurons exhibiting tonic discharge during visual fixation and a pause during most saccades — *fixation neurons* — are restricted mainly to the rostro-lateral pole of the motor map beneath the superficial layer's representation of the fovea, forming a continuum with the saccade neurons.[55,56] Fixation neurons have been prescribed a role in the maintenance of active visual fixation;[41,57] however, they may also participate in the execution of microsaccades,[57] smooth pursuit,[58,59] and vergence eye movements.[60]

Local inhibitory connections may help shape the reciprocal activity patterns of saccade and fixation neurons.[18,37,38,55,57] The findings that most saccade and fixation neurons are inhibited at very short latency after microstimulation of remote collicular regions[61] and that injection of GABAergic agonists and antagonists in the rostral[57] versus caudal SC[62] produces reciprocal effects on behavior provides credence to this hypothesis.

The SC can directly influence saccade generation through its direct projection onto LLBNs in the brainstem reticular formation.[63,64] Besides providing a signal that specifies where to make the saccade, the SC also provides a trigger signal through its projection to the OPNs in the brainstem, which are excited monosynaptically but inhibited polysynaptically by stimulation of the intermediate layers of the SC.[64]

Early studies revealed that ablation of the SC impairs the ability to generate saccades, but many of these effects recover with time.[65] However, there remain lasting deficits in saccade initiation that are revealed as increases in saccadic reaction time.[66] More recently, reversible inactivation of focal regions within the SC has revealed that the SC is critical for saccade initiation.[62,67–69]

3.3.3 Cerebral Cortex — Frontal Eye Field

Visual inputs guiding saccades are delivered through the retino-geniculo-cortical pathway as well as through a direct retino-tectal projection. Visual information for the guidance of saccades is processed through multiple extrastriate visual areas. Areas in the temporal lobe represent the visual features of objects and are modulated by the covert allocation of attention.[70,71] Areas in the parietal lobe, such as the lateral intraparietal area (LIP) represent the location and relevance of objects in retinocentric and craniocentric coordinates at the interface of the sensory-motor transformation.[72,73] LIP delivers visual and extraretinal signals to the intermediate layers of the SC[74] as well as to the frontal eye fields (FEF),[75] located in the rostral bank of the arcuate sulcus in macaque monkeys. Broadly considered, the FEF participates in the transformation of visual signals into saccade motor commands.[14]

FEF is innervated in a topographic fashion by areas in both the dorsal and ventral streams of extrastriate visual cortex.[76] As a result of this extensive connectivity with extrastriate visual cortical areas, many neurons in FEF respond to visual stimuli. Physiological recordings in the FEF of monkeys trained to shift gaze to visual targets have found that roughly half of the neurons have visual responses.[77,78] Recent research has demonstrated how these visually responsive neurons in FEF participate in the selection of visual targets for saccades.[79–81]

FEF is also known to play a direct role in producing saccadic eye movements. Low intensity microstimulation of FEF elicits saccades.[82] This direct influence is presumably mediated by a subpopulation of neurons in FEF that discharge specifically before and during saccades.[79,83] These saccade neurons in the FEF appear to be the functional equivalent of the saccade neurons with buildup activity observed in the SC. FEF is also populated by fixation neurons that seem to parallel their counterparts in the SC.[84]

Hanes and Wurtz[69] showed that it was not possible to elicit saccades with microstimulation of the FEF, following reversible deactivation of the SC. Thus, although the FEF and SC have parallel projections to the saccadic premotor circuitry in the brainstem, they may not be weighted equally. Evidence suggests that the serial pathway from FEF to SC to brainstem is dominant.

FEF can influence saccade production through three pathways. One pathway is a major projection to the ipsilateral SC concentrated in the intermediate layers but extending to superficial and deep layers.[23,26,28,85,86] This projection is topographically organized with lateral FEF projecting to rostral SC and medial FEF, to caudal SC.[24] Another major pathway is through the basal ganglia via the ipsilateral striatum and subthalamic nucleus.[86–89] FEF efferents terminate in the region of caudate where neural activity related to saccade production is recorded.[90–92] The terminations in the striatum are topographically organized; the medial aspect of FEF projects to the central part of the head and body of the caudate and dorsomedial putamen while the lateral portion of FEF terminates ventrolaterally in the caudate and ventromedially in the putamen.[89]

The third pathway is a projection to mesencephalic and pontine nuclei.[26,28,93–96] The FEF projects weakly and inconsistently to the ipsilateral nucleus of Darkschewitsch, interstitial nucleus of Cajal, and rostral interstitial nucleus of the medial longitudinal fasciculus. FEF also projects weakly to the paramedian pontine reticular formation and nucleus prepositus hypoglossi and slightly more strongly to the nucleus raphe interpositus. These projections tend to be mainly ipsilateral, but some studies report some contralateral fibers as well. The FEF projection is stronger and clearly bilateral to the nucleus reticularis tegmenti pontis.

Many studies have shown that FEF is reciprocally connected in a topographic manner with a longitudinal zone of thalamic nuclei bordering the internal medullary lamina extending from the ventroanterior nucleus to the medial pulvinar.[26,87,89] The densest connections of FEF are with the lateral part of the mediodorsal nucleus and the medial part of the ventroanterior nucleus. FEF is more weakly connected with the more medial and caudal parts of the mediodorsal nucleus, with area X of the ventrolateral nucleus and with the caudal ventrolateral nucleus and medial pulvinar. Some but not all studies have reported weak FEF connections with the paracentral, centrolateral, and central superior lateral intralaminar nuclei. The FEF connections with the paralaminar nuclei is topographically organized with the dorsomedial part of FEF projecting dorsally and the ventrolateral part of FEF projecting ventrally. The thalamic zones most heavily connected with FEF are themselves innervated by oculomotor afferents from the intermediate and deep layers of the superior colliculus, the substantia nigra pars reticulata, and the dentate nucleus of the cerebellum.[97,98]

Reversible inactivation of FEF impairs monkeys' ability to make saccades.[99–101] This observation complements earlier observations that ablation of FEF causes an initial severe impairment in saccade production that recovers over time.[65,66,102]

Saccade and fixation neurons in the FEF innervate neurons in the intermediate layers of the superior colliculus directly[103–105] and the premotor circuitry in the brainstem reticular formation.[106]

To summarize, the SC and the FEF can influence one another through at least five paths. (1) FEF projects topographically to the intermediate layers of the SC. (2) The intermediate layers of the SC project to the lateral segment of the mediodorsal nucleus that in turn projects to FEF.[107,108] (3) FEF projects to the caudate nucleus, which inhibits the substantia nigra pars reticulata, which inhibits the SC. The substantia nigra pars reticulate also projects to thalamic nuclei that innervate the FEF. (4) The sector of the caudate nucleus receiving FEF afferents also projects to the external segment of the globus pallidus, which in turn projects to the subthalamic nucleus, which influences the substantia nigra pars reticulata. (5) The FEF and SC both project to nucleus reticularis tegmenti pontis, which innervate sectors of the cerebellum that ultimately influence the SC.

3.4 ROLE OF SC AND FEF IN SACCADE INITIATION

The pattern of movement-related activity recorded in the FEF and SC of monkeys performing various saccade tasks has been analyzed to evaluate the alternative models of reaction time (Figure 3.1). Several paradigms have been developed to investigate the neural processes involved in saccade initiation. These tasks range from those requiring reflexive or automatic responses to visual stimuli to those that require less reflexive and more voluntary processing. Another task requires subjects to inhibit the initiation of a partially prepared saccade. We will review several of these tasks here and show how the SC and FEF are involved in the generation of both reflexive (automatic) and voluntary saccadic eye movements.

3.4.1 Reflexive Saccades – The Gap Saccade Task

The gap saccade task[109] requires a subject simply to generate an automatic saccade to a suddenly appearing visual stimulus. Each trial is initiated by the appearance of a central fixation point. After a period of visual fixation, the fixation point disappears leaving the subject momentarily in complete darkness (the gap period) until an eccentric visual target appears. The subject is required to maintain central fixation during the gap period and then initiate a targeting saccade after the detection of the visual target. The disappearance of the fixation point can both release the fixation system and act as a temporal warning signal allowing the subject to prepare for the impending target appearance.[110–116] The introduction of a gap period (e.g., 200 ms) leads to a general reduction in SRT, known as the *gap effect*. It is therefore likely that the excitability levels of various elements of the saccade generating circuitry are altered during the gap period prior to target appearance. Munoz and colleagues[16,42,53,54,114,117,118] have recorded from several different classes of oculomotor neurons in the FEF, the substantia nigra pars reticulata, the SC intermediate layers, and the brainstem paramedian pontine reticular formation while monkeys performed the gap saccade task. Figure 3.3 shows representative examples of the single-cell activity patterns that were correlated to aspects of the task. In these experiments, the gap duration was fixed at 200 ms and the target appeared randomly at a single location in either the right or left hemifield, one location being centered in the response field of the neuron under study. The activation functions for individual neurons shown in Figure 3.3 are aligned on target appearance (left column) and saccade initiation (right column).

Within the FEF, saccade neurons are silent during visual fixation but some of them begin to discharge at a low frequency during the gap period (Figure 3.3A; blue trace; see color insert), starting about 100 ms after the FP disappears,[104,119] while others remain silent until after the target appears (Figure 3.3A, green trace). Saccade neurons then discharge a high frequency visuomotor response after the target appears in their response field and a saccade is initiated (Figure 3.3A; right panel, blue and green traces). FEF fixation neurons (Figure 3.3A; red traces) have a very different pattern of discharge: they are tonically active during visual fixation and there is often

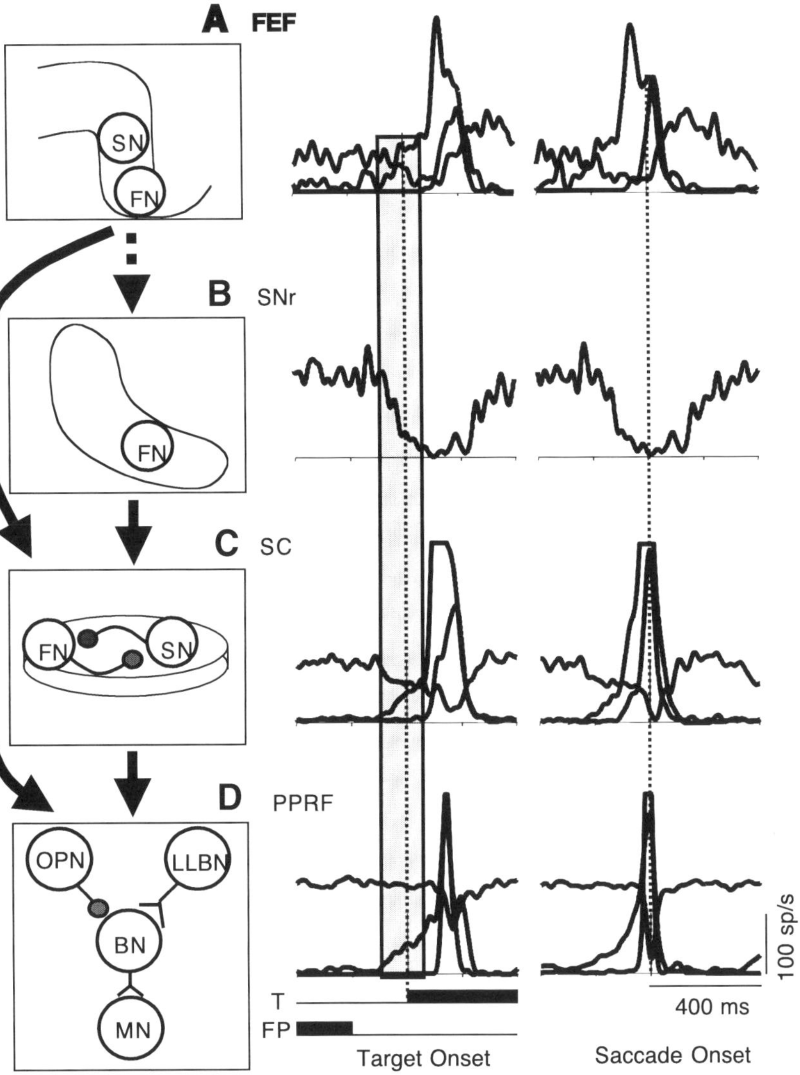

FIGURE 3.3 (See color insert following page 176.) Discharge of neurons in the frontal eye fields (A), substantia nigra pars reticulata (B), intermediate layers of the superior colliculus (C), and paramedian pontine reticular formation (D) during the gap-saccade paradigm. The spike density waveforms are aligned on target appearance (left column) and saccade onset (right column). Cells at all three levels are modulated by the gap period. Saccade neurons are silent during visual fixation of the fixation point and then increase their discharge for saccades. Of these saccade neurons, some become active before target appearance (blue traces), while others remain silent until after target appearance (green traces). Neurons with tonic activity during fixation have a drop in discharge rate during the gap and saccade initiation (red traces). Vertical gray bar denotes the end of gap epoch, highlighting neurons that change their activity during the gap period. Abbreviations: BN: burst neuron; FN: fixation neuron; LLBN: long-lead burst neuron; MN: motoneuron; OPN: omnipause neuron; SN: saccade neuron.

a drop in this activity approximately 100 ms into the gap period. Many FEF fixation neurons also have a pause in their discharge during the initiation of the saccade to the target and a subsequent postsaccadic enhancement in their discharge.[84,103]

Within the substantia nigra pars reticulata, there are also neurons whose discharge is modulated during the gap period of the gap saccade paradigm.[118] Neurons in the SNr that decrease or pause their discharge during saccades, pause neurons,[120,121] can also have a drop in their tonic discharge during the gap period (Figure 3.3B; red traces).

Within the SC, the fixation neurons (Figure 3.3C; red traces) have discharges that are similar to the FEF fixation neurons: they are tonically active during visual fixation, but reduce their activity about 100 ms into the gap period before pausing completely for the ensuing saccades.[114] Reciprocally, the SC saccade-related neurons are silent during visual fixation and discharge a high frequency burst of action potentials for saccades into their response field.[42,44] During the gap period, some of the SC saccade neurons — buildup neurons — additionally display low-frequency pretarget discharges (Figure 3.3C; blue trace), while others — burst neurons — remain silent (Figure 3.3C; green trace).[42,53] Both burst and buildup neurons will then discharge a high frequency burst for the saccade and many of these neurons will also discharge a burst time-locked to target appearance.

In the reticular formation, OPNs continue to discharge at the same tonic rate before and after FP disappearance in the gap task (Figure 3.3 red traces). They do not exhibit the gap-related reduction in activity that was observed among many fixation neurons in the FEF, substantia nigra pars reticulata, or SC.[16,117] In comparison, LLBNs also display low-frequency pretarget activity during the gap period (Figure 3.3D; blue traces) that is similar to that recorded in the SC and FEF. The BNs lack any gap-related activity; they only discharge a high frequency burst for the contraversive saccade (Figure 3.3D; green traces).

Thus fixation and saccade signals interact at several stages of the neuraxis (e.g., FEF, SC, and PPRF) during the gap saccade paradigm. Although we see neurons at several different levels of the circuit being modulated during the gap period, we also see an evolution of processing from FEF, to SC, to the PPRF. One important distinction is that OPNs maintain their tonic discharge during the gap period while fixation neurons in the SC and FEF show a drop in discharge rate. OPN discharge must be maintained to inhibit BNs and avoid premature initiation of saccades. It was hypothesized that a major component of the input to the brainstem OPNs arose from fixation neurons in the superior colliculus.[38,41] If this is true, how do the OPNs maintain the same tonic rate during the gap period? It has been hypothesized that OPNs receive excitatory inputs from both fixation and saccade neurons in the SC.[64,117] Increased input from saccade neurons with buildup activity occurs when input from fixation neurons is reduced. A constant input onto the OPNs ensures that their high tonic discharge continues during the gap period and that, via their potent inhibition of BNs, low-frequency pretarget signals do not trigger a saccade prematurely. Then, when a saccade is triggered, the high frequency burst in the SC is relayed via an inhibitory interneuron to inhibit the OPNs for the duration of the saccade.

3.4.2 Express Saccades: A Special Class of Visually Guided Saccade

Using the gap paradigm often results in a bimodal distribution of SRTs (Figure 3.4A), with a first mode (~100 ms) of *express* saccades and a second mode (~150 ms) of *regular* saccades.[122–124] Express saccades have SRTs that approach the minimal time required for the visual response to travel from the retina, through the brain, to the MN to trigger a motor action.[125] When the spatial location of the saccade target and the probability of its presentation are varied, it can be demonstrated that the disengagement of fixation afforded by the disappearance of the central fixation point leads to a general reduction in SRT, the gap effect.[16,114] However, the introduction of the gap period does not guarantee that express saccades will be generated. Rather, express saccade occurrence is determined by factors such as predictability in the location and timing of appearance of the saccadic target.[116]

Given the special nature of express saccades, an obvious question is do they conform to any of the threshold models of saccadic initiation illustrated in Figure 3.1? Textbooks commonly state that the SC produces reflexive saccades and the FEF produces voluntary saccades. One basis for this seems to be the observation that lesions of the SC but not the FEF prevent express saccade production.[66] It has been supposed, therefore, that FEF is not essential for, nor perhaps even involved in, producing express saccades. However, evidence from recent neurophysiological studies has shown that saccade neurons in the FEF have elevated pretarget activity preceding express saccades,[104] in a manner that parallels what is observed in the SC.[56,126] Thus it remains to be determined whether the FEF activation during express saccades is required to help initiate them, or the result of feedback from the SC.[107,108]

Figure 3.4B and C shows the discharge of individual saccade neurons recorded in the FEF and the SC during the execution of express and regular saccades. There are two important observations to make. First, prior to the execution of express saccades, saccade neurons in the both the FEF and SC have a higher level of pretarget discharge during the gap period, which is consistent with the hypothesis that higher pretarget activity results in shorter SRT.[53,54,104] Second, many saccade neurons have two bursts of action potentials after target appearance for regular saccades, an early, small stimulus-related burst and a later, larger saccade-related burst, but only one burst before express saccades that is equally well aligned on both stimulus presentation and saccade onset.[53,104,126] For express saccades, it is as if the two bursts of activity merged into one *visuo-motor* burst of sufficient magnitude to trigger the saccade with an express latency.

The pretarget activity that appears on saccade neurons during the gap period in the gap paradigm is hypothesized to move the system closer to the threshold for saccade initiation.[53,54,104] When the transient, stimulus-related burst of activity arrives in the SC, it is added to the early pretarget activity. If this early pretarget activity is of sufficient strength, then the visual burst drives the system over the threshold producing an express saccade. If the pretarget activity is low, then the visual burst does not reach the threshold for saccade initiation, and the system must wait for a

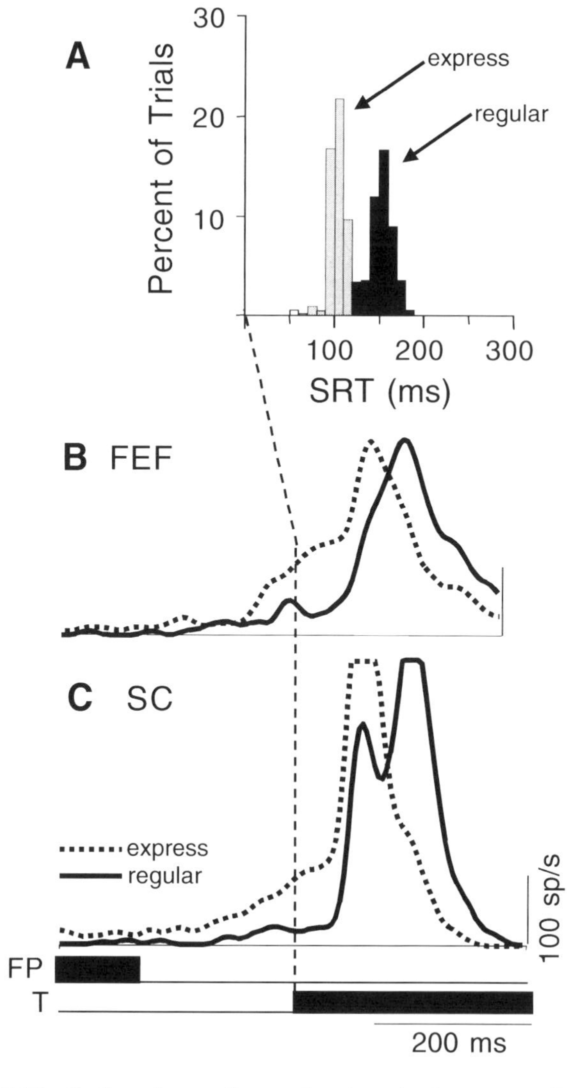

FIGURE 3.4 (A) Distribution of saccadic reaction times in the gap paradigm, highlighting the bimodal distribution of express and regular latency saccades. (B, C) Discharge of an individual FEF and SC neuron, respectively, during generation of express and regular latency saccades.

later signal, which delays saccade initiation and produces a regular-latency saccade. According to this hypothesis, two events are required for express saccades: a high level of pretarget activity and a transient response caused by either the appearance of a new target or motion of an existing target.

3.4.3 Control of Saccade Production — The Countermanding Task

The countermanding task can be used to investigate the control over automatic responses by infrequently presenting an imperative stop signal in a reaction time task.[127,128] The subjects' task is to cancel the planned movement if the stop signal is presented. In the saccade version,[129] monkeys were trained to make a saccade to a peripheral target that appeared when the fixation spot disappeared (Figure 3.5A). On a fraction of trials a *stop signal* was presented (Figure 3.5B); in these experiments the stop signal was the reappearance of the fixation spot. In response to the stop signal, the monkeys were to withhold the planned movement, which they did with variable success. In other words, monkeys could either cancel the planned movement (in which case they earned reinforcement) or fail to cancel the movement (in which case the trial was aborted).

Logan and Cowan[127] showed that performance on this task can be accounted for by a race between a process that generates the movement and a process that cancels the movement. This race model provides an estimate of the *stop signal reaction time*, which is the time needed to cancel the planned movement. The stop signal reaction time corresponds to estimates of the time needed to reprogram a saccade in double-step saccade tasks.[130,131] Oculomotor stop signal reaction times with a foveal spot signal average around 80 to 100 ms in monkeys[129] and are slightly longer in humans.[132–135] This may not be surprising in view of the common observation that saccade latencies in the monkey are somewhat shorter than in humans. Human studies have also examined the influence of nonfoveal and nonvisual stop signals on saccadic countermanding.[133–135]

To collect data in the countermanding task, it is necessary to have no stop signal in most trials. These data afford a kind of analysis that can indicate how the activity of single neurons is related to the initiation of saccades (Figure 3.6A). By grouping trials according to saccade latency, it becomes clear that saccades are initiated when the activity of saccade neurons in the FEF reach a particular level.[83] This relationship has been observed for SC neurons[53,54] and also holds true for movements of the limbs.[136] The variability in reaction time can be accounted for mainly by the time taken by the activity of movement-related neurons to reach the threshold. The origin of such variability in the growth of movement-related activity is not known but may include the state of neuromodulatory systems.[137] The relation of movement-related neural activity to reaction time corresponds to an accumulator architecture with variable growth to a fixed threshold (Figure 3.1A) and directly contradicts an architecture with a fixed growth process and random threshold (Figure 3.1B).

Usually, brain structures are attributed a function in motor control if it can be shown that they play a role in producing movements. The countermanding paradigm permits the investigation of another facet of control, the cancellation of a planned

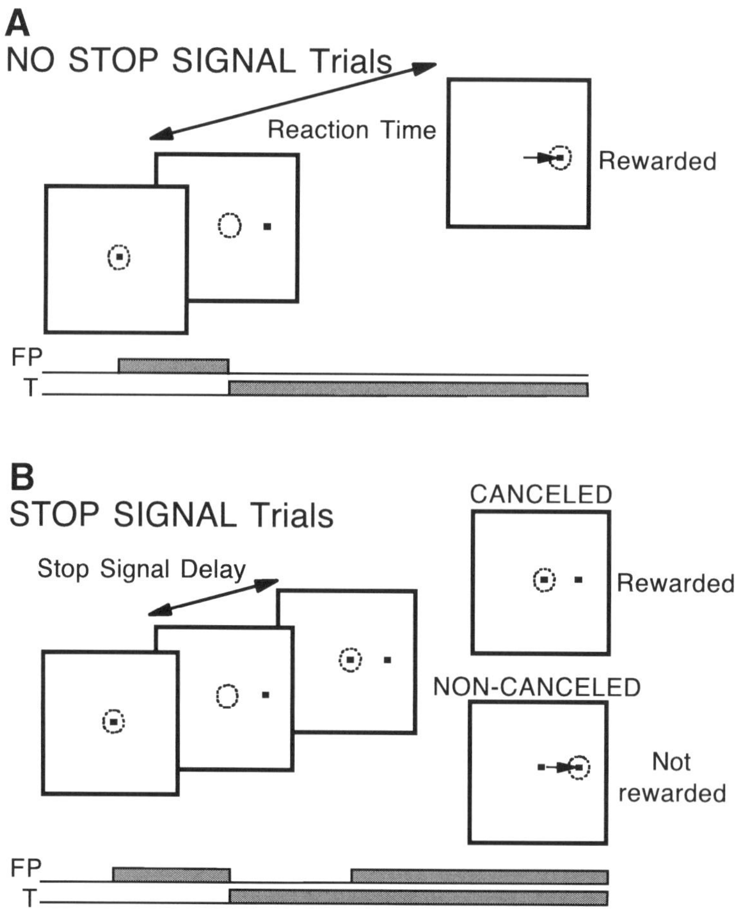

FIGURE 3.5 The countermanding paradigm, consisting of NO STOP SIGNAL trials (75%) and STOP SIGNAL trials (25%). On NO STOP trials, the monkey is reward for initially fixating the central fixation point (FP) and then making a saccadic eye movement to the eccentric target (T). On STOP SIGNAL trials, the monkey is rewarded for canceling the saccade to T after the stop signal (reappearance of FP) is presented.

movement. The chief virtue of the countermanding paradigm is that one can determine whether single neurons generate signals that are logically sufficient not only to initiate movements but also to prevent the production of movements. The logic of the countermanding paradigm establishes two criteria a neuron must meet to play a direct role in the control of movement. First, the neuron must discharge differently when a saccade is initiated versus when a saccade is withheld. Second, this difference in activity must occur by the time that the movement is canceled, i.e., within the stop-signal reaction time.

Neural activity in FEF and SC has been described when monkeys performing the countermanding task cancel partially prepared saccades.[84,138] The activity of

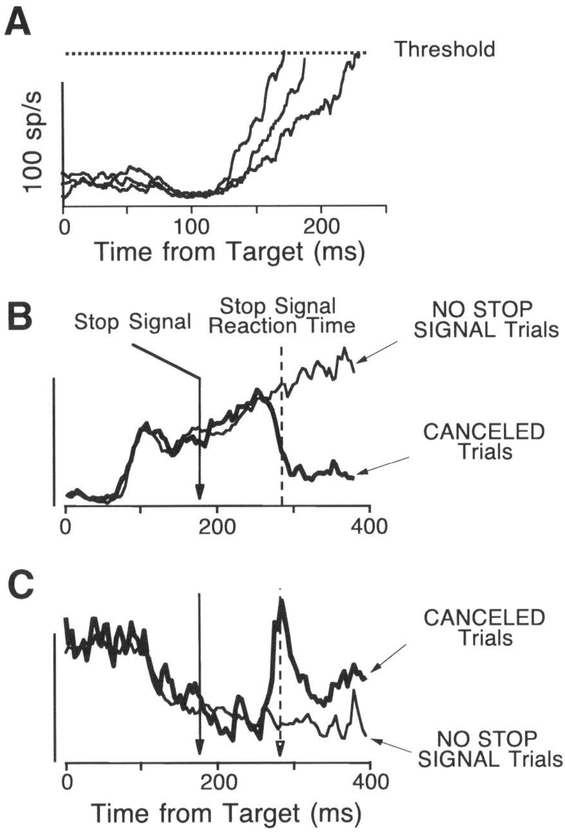

FIGURE 3.6 Discharge of FEF saccade (blues traces) and fixation (red traces) neurons in the oculomotor countermanding task. (A) Discharge of an individual saccade neuron for saccades of different saccadic reaction times (SRT). The slope of the buildup or accumulation of activity was correlated to SRT: the fastest rate of rise preceded the fastest SRTs, the slowest rate of rise preceded the slowest SRTs. (B, C) Discharge of FEF saccade (B) and fixation (C) neurons for NO STOP trials (thin lines) and correctly canceled STOP SIGNAL trials (thick lines). The STOP SIGNAL was presented 180 ms after the target appeared. Note that both saccade and fixation neurons altered their discharge accordingly to halt the initiation of a saccade on STOP SIGNAL trials, before the calculated stop signal reaction time.

saccade-related neurons, which began to grow toward the trigger threshold, failed to reach the threshold activation level when movements were canceled (Figure 3.6B). Instead, when planned saccades were canceled, the activity of saccade neurons decreased rapidly after the stop signal was presented (thick trace in Figure 3.6B). When the saccades were produced in trials with no stop signal, the activity of saccade-related neurons continued to accumulate toward the threshold (thin trace in Figure 3.6B). The same growth of activity to the threshold was observed when the saccade was produced in spite of the stop signal. Moreover, the activity associated with canceling as compared to executing the saccade became different before the stop signal reaction time had elapsed. Similar patterns of activation have been reported for saccade-related neurons in the SC. Therefore, the activity of saccade neurons in the FEF and SC is sufficient to specify whether or not a saccade will be produced.

A reciprocal pattern of neural activity was observed in fixation neurons in FEF[84] (Figure 3.6C) and the SC.[138] If partially prepared eye movements were canceled, fixation neurons that had decreased firing generated a rapid burst of activity before the stop signal reaction time (thick trace in Figure 3.6C). This rapid increase in fixation neuron activity was not observed in trials with no stop signal (thin trace in Figure 3.6C) or before noncanceled saccades.

This modulation before the stop signal reaction time was never observed in neurons with only visual responses and no saccade-related modulation of discharges. The visually responsive cells that had no saccade-related activation were either not modulated when planned saccades were canceled or the modulation occurred well after the stop signal reaction time. Thus, not every neuron in FEF or SC plays a direct role in the control of saccade production. Rather this function may be limited to only saccade-related and fixation neurons.

The findings from SC and FEF using the countermanding paradigm indicate that the preparation of a movement can be a controlled process; it can be canceled if the growth of the activation toward the trigger threshold is sufficiently slow. These data strongly support the rise to threshold model illustrated in Figure 3.1 and reinforce the parallel nature of the signals observed in the SC and FEF.

3.4.4 Voluntary Saccade Production — The Antisaccade Task

Many tasks have also been devised to investigate the role of various brain regions in the generation of voluntary, nonautomatic responses. Intention can be investigated by manipulating the type of eye movement response required when a stimulus is presented. In the prosaccade task, an automatic visuomotor response is required, whereas in an antisaccade task,[139] this automatic response must be inhibited and instead a voluntary response is required to the stimulus mirror position. Contrasting pro- and antisaccade responses recorded in the same experiment can reveal and dissociate aspects of sensory, intention, and premotor processing. Everling and coworkers[104,140–142] recorded the activity of saccade-related neurons in the SC and FEF while monkeys performed a task with randomly interleaved prosaccade (saccade toward the stimulus) and antisaccade (saccade away from the stimulus to the opposite side) trials. Within a block of trials, the color of the initial fixation point conveys the instruction to generate either a prosaccade or antisaccade. On prosaccade trials, the same saccade neurons activated by the stimulus are required to drive the saccade. However, on antisaccade trials, the cells activated by the appearance of the target stimulus must be suppressed so that saccade neurons on the opposite side of the brain can drive the correct antisaccade away from the stimulus.

Figure 3.7 shows the discharge of saccade neurons recorded in the FEF and SC for correct prosaccade (thin traces) and antisaccade (thick traces) tasks when the target stimulus appeared in the response field of the neuron (Figure 3.7A and D), or the saccade was required into the response field of the neuron (Figure 3.7B and E). There are several important observations to note. First, saccade neurons in both the SC and FEF discharged at a lower frequency during fixation of the central fixation point on antisaccade trials (thick traces below thin traces in gray bar in Figure 3.7A, B, D, and E). Second, following appearance of the target stimulus in the saccade neuron's response field (Figure 3.7A and D), a visual response was elicited on both pro- and antisaccade trials. Third, when the saccade was initiated into the response field of the saccade neurons being recorded (Figure 3.7B and E), there was often a drop in activity associated with the appearance of the target stimulus that is the result of inhibitory input resulting from target appearance elsewhere in the visual field. Fourth, following the stimulus related decrease in activity (Figure 3.7B and E), there was an increase in activity of saccade neurons that is aligned with initiation of the correct antisaccade away from the target stimulus. Thus, consistent with SRT distributions,[142] the saccade burst in the FEF and SC is delayed on antisaccade trials compared to prosaccade trials. Finally, on antisaccade trials, the peak of the saccade burst is less for antisaccades than prosaccades.[104,141] In fact, the magnitude of the antisaccade response of many saccade neurons in the SC and FEF is less than the visual response created by the appearance of the visual stimulus. Therefore, presumably other structures or cells must contribute activity toward the threshold for the initiation of a correct antisaccade. One such structure could be the SEF, which has neurons that are more active for antisaccades compared to prosaccades.[143]

What is the nature of the signal used to reduce saccade neuron excitability prior to target appearance in the antisaccade task? Fixation neurons in the FEF and the SC have a discharge pattern on pro- and antisaccade trials that is opposite to saccade neurons. Consistent with their role in saccade suppression, the fixation neurons are more active during the instructed fixation period preceding target appearance on antisaccade trials (thick traces above thin traces in gray bar in Figure 3.7C and F). Thus, to perform the antisaccade task correctly, fixation activity is enhanced prior to target appearance and this acts as a suppression signal in both the FEF and SC to reduce excitability of saccade neurons prior to target appearance. Thus, it is not the case that the SC is silent for antisaccade generation as earlier hypothesized.[144] Rather, fixation neurons in the FEF and SC appear to work in concert to inhibit saccade neurons at the time of target appearance to suppress the initiation of the reflexive prosaccade; saccade neurons in the FEF and SC are then activated to initiate the correct antisaccade.[104,140,141]

The antisaccade task also provides a means to test the threshold models of saccade initiation (Figure 3.1). If there is indeed a fixed threshold for saccade initiation, there should be predictable changes in the discharge of saccade neurons in the FEF and SC for correct and incorrect antisaccades. Figure 3.8A shows the distribution of latencies among correct and error trials. Note that direction errors (reflexive saccades triggered toward the target) were initiated with shorter SRT than correct antisaccades and most of these errors have SRTs in the range of express saccades. Does a high level of pretarget activity in the SC, which shortens the SRT of visually triggered saccades in the gap task (Figure 3.4), increase the occurrence of direction errors in the antisaccade task, by allowing the stimulus-related burst to trigger a reflexive saccade? Neurophysiological recordings from the FEF and SC

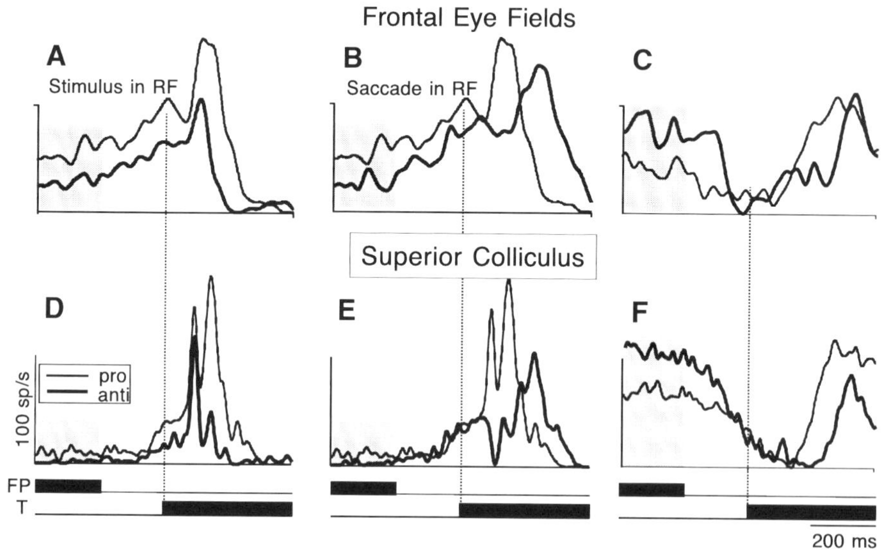

FIGURE 3.7 Discharge of saccade (A, B, D, E) and fixation (C, F) neurons in the FEF (top row) and SC (bottom row) the combined antisaccade (thick traces) and prosaccade (thin traces) task. (A, D) Stimulus appears in the saccade neuron's response field and the saccade is either in (prosaccade) or away from (antisaccade) the response field. (B, E) Saccade is into the saccade neuron's response field and stimulus is either in (prosaccade) or out of (antisaccade) response field. (C, F) Discharge of fixation neuron for pro- and antisaccade trials. Note that during the instructed fixation period (gray bar), the saccade neurons are more active on prosaccade trials, while the fixation neurons are more active on antisaccade trials.

confirm this prediction. Saccade neurons in both the FEF and SC have greater pretarget discharge prior to error trials.[104,140] When this low frequency pretarget activity combines with the visual transient response produced by the appearance of the visual stimulus, a direction error is triggered at express latency. On correct trials, the pretarget discharge is reduced and the transient visual response cannot surpass the saccade threshold. The stimulus was presented in the neuron's response field so that the opposite FEF and SC are required to drive the correct antisaccade. The appearance of the visual stimulus triggers a phasic response in the neurons. On error trials, excessive pretarget activity during the gap allows the stimulus-related response to exceed a threshold and trigger an erroneous prosaccade. Most importantly, note that almost all of the erroneous prosaccades are triggered at express saccade latencies.[142] On correct trials, the pretarget activity is reduced prior to stimulus appearance so that the transient stimulus-related burst does not exceed the threshold activation. Thus, once again the level of pretarget activity among saccade neurons predicts, not only saccade latency, but also the selection of the saccade: too much pretarget activity leads to the initiation of a visually triggered reflexive saccade. This is consistent with the variable baseline model illustrated in Figure 3.1C. To initiate a voluntary antisaccade, pretarget activity within the FEF and SC must be reduced. The reduction of excitability of saccade neurons in the FEF and SC during antisaccade trials may

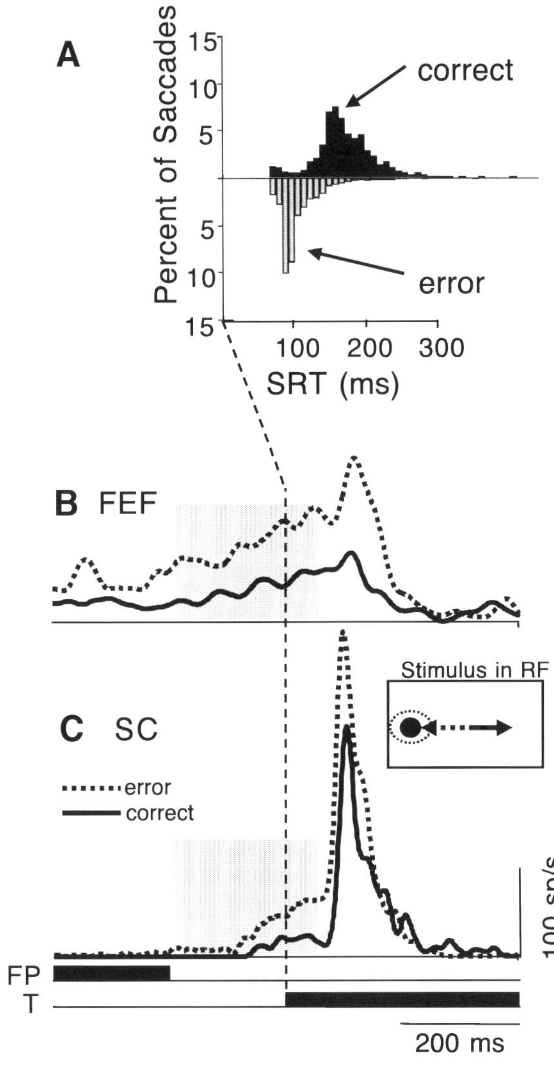

FIGURE 3.8 (A) Distribution of saccadic reaction times in the gap antisaccade paradigm, highlighting the bimodal distribution of correct (black bars) and direction errors (gray bars). (B, C) Discharge of saccade neurons in the FEF and SC associated with correct (solid traces) and erroneous (dotted traces) antisaccades when the stimulus appeared in the neuron's response field. Note the high level of pretarget activity associated with error trials.

be imposed by the supplementary eye field in which neurons are *more* active before and during antisaccades.[143] Another possible source of this control signal is the dorsolateral prefrontal cortex.[145,146]

3.5 CONCLUSIONS

Numerous lines of evidence lead to the conclusion that the FEF and SC function as two critical nodes in a network of saccade generation. Function is distributed across this network, and neurons with common discharge properties can be identified at multiple levels in the circuit, from the cortex, to the basal ganglia, to the SC, to the brainstem reticular formation. The connectivity between areas contributes somewhat to specificity, but there is substantial overlap in function. Although the FEF and SC together are essential for saccade generation, specific function is distributed across the network.

The rise to threshold architecture (Figure 3.1) can account for many of the observations of individual neurons recorded while animals perform very different saccade tasks and can therefore account for much of the behavior. We have described some of the important pre- and posttarget factors that contribute to either altering the baseline or the rate of rise of accumulation toward the threshold. In addition, the threshold level may vary between tasks. All of these mechanisms afford different levels of control over the saccade generating circuitry.

The results we have reviewed present an increasingly clear account of the neural basis of saccade generation. However, a number of outstanding questions remain. For example, how is the initiation of a saccade controlled? The ultimate switch is the balance of activation in the push-pull network of BNs and OPNs in the brainstem.[17] Therefore, it will be necessary to record single unit activity in these structures during tasks like countermanding. Furthermore, a paradox confronts us. Thousands of neurons are necessary to produce a saccade, but the averaged signal from single neurons in SC and FEF are sufficient to specify whether and when gaze will shift. This paradox can be addressed through simultaneous recording of presaccadic activity in multiple neurons throughout the saccade-generating circuit to clarify the mechanisms by which activity is coordinated.

How is activation coordinated across structures? As described above, FEF and SC can influence each other through several anatomical pathways. Most directly, FEF projects to SC, which projects to the thalamus, which then projects back to the FEF. The transmission time of these pathways is 2 to 3 msec.[103,104,107] Even allowing for 10 to 20 ms synaptic integration times, these delays allow enough time for SC and FEF to mutually influence the state of the other in the 100 ms interval during which saccade preparation usually occurs.

Another key issue involves how saccade latency is adjusted to perform tasks. A given visual stimulus can elicit a saccade with a latency less than 100 ms in the context of the gap paradigm or a latency exceeding 400 ms in the context of the countermanding task. How is the context of the task sensed and then how does it influence performance? One avenue of research indicates that areas in the medial frontal lobe such as SEF and parts of the anterior cingulate cortex are sensitive to the consequences of saccades, which is prerequisite for adaptive behavior.[147]

ACKNOWLEDGMENTS

DPM supported by the Canadian Institutes for Health Research and a Canada Research Chair in Neuroscience. JDS supported by NIH R01-EY08890, R01-MH55806, P30-EY08126, the McDonnell-Pew Program in Cognitive Neuroscience and the McKnight Endowment Fund for Neuroscience.

REFERENCES

1. Robinson, D.A., Oculomotor control signals, in *Basic Mechanisms of Ocular Motility and Their Clinical Implications,* Bach-y-Rita, P. and Lennerstrand, G., Eds., Pergamon, Oxford, 1975, pp. 337–374.
2. Luce, R.D., *Response Times: Their Role in Inferring Elementary Mental Organization*, Oxford University Press, Oxford, 1986.
3. Ratcliff, R., A theory of memory retrieval, *Psych. Rev.* 85, 59, 1978.
4. Carpenter, R.H.S., *Movements of the Eyes*, Pion, London, 1988.
5. Carpenter, R.H. and Williams, M.L., Neural computation of log likelihood in control of saccadic eye movements, *Nature*, 377, 59, 1995.
6. Ratcliff, R. and Rouder, J.N., Modeling response times for two-choice decisions, *Psych. Sci.* 9, 347, 1998.
7. Grice, G.R., Nullmeyer, R., and Spiker, V.A., Human reaction time: toward a general theory. *J. Exp. Psych., Gen.* 111, 135, 1982.
8. Nazir, T.A. and Jacobs, A.M., The effects of target discriminability and retinal eccentricity on saccade latencies: An analysis in terms of variable-criterion theory, *Psychol. Res.*, 53, 281, 1991.
9. Trappenberg, T.P., Dorris, M.C., Munoz, D.P., and Klein, R.M., A model of saccade initiation based on the competitive integration of exogenous and endogenous signals in the superior colliculus, *J. Cog. Neurosci.*, 15, 256, 2001.
10. Dzhafarov, E.N., Grice-representability of response time distribution families, *Psychometrika,* 58, 281, 1993.
11. Wurtz, R.H. and Goldberg, M.E., *The Neurobiology of Saccaic Eye Movements*, Elsevier, New York, 1989.
12. Leigh, R.J. and Zee, D.S., *The Neurology of Eye Movements.* 3rd ed., F.A. Davis, Philadelphia, 1999.
13. Moschovakis, A.K., Scudder, C.A., and Highstein, S.M., The microscopic anatomy and physiology of the mammalian saccadic system, *Progr. Neurobiol.*, 50, 133, 1996.
14. Schall, J.D., Visuomotor areas of the frontal lobe, in *Extrastriate Cortex of Primates, Volume 12 of Cerebral Cortex*, Rockland, K., Peters, A., and Kaas, J.H., Eds., Plenum Press, New York, 1997, pp. 527–638.
15. Hikosaka, O., Takikawa, Y., and Kawagoe, R., Role of the basal ganglia in the control of purposive saccadic eye movements, *Physiol. Rev.*, 80, 953, 2000.
16. Munoz, D.P., Dorris, M.C., Paré, M., and Everling, S., On your mark, get set: brainstem circuitry underlying saccadic initiation, *Can. J. Physiol. Pharmacol.*, 78, 934, 2000.
17. Scudder, C.A., Kaneko, C.R.S., and Fuchs, A.F., The brainstem burst generator for saccadic eye movements, *Exp. Brain Res.*, 142, 439, 2002.
18. Munoz, D.P. and Fecteau, J.H., Vying for dominance: dynamic interactions control visual fixation and saccadic initiation in the superior colliculus, *Prog. Brain Res.*, 140, 3, 2002.

19. Robinson, D.L. and McClurkin, J.W., The visual superior colliculus and pulvinar, *Rev. Oculomot. Res.*, 3, 337, 1989.
20. Perry, V.H. and Cowey, A., Retinal ganglion cells that project to the superior colliculus and pretectum in the macaque monkey, *Neuroscience*, 12, 1125, 1984.
21. Cynader, M. and Berman, N., Receptive-field organization of monkey superior colliculus, *J. Neurophysiol.*, 35, 187, 1972.
22. Goldberg, M.E. and Wurtz, R.H., Activity of superior colliculus in behaving monkey, IV. Effects of lesions on eye movements, *J. Neurophysiol.*, 35, 587, 1972.
23. Fries, W., Cortical projections to the superior colliculus in the macaque monkey: A retrograde study using horseradish peroxidase, *J. Comp. Neurol.*, 230, 55, 1984.
24. Komatsu, H. and Suzuki, H., Projections from the functional subdivisions of the frontal eye field to the superior colliculus in the monkey, *Brain Res.*, 327, 324, 1985.
25. Lynch, J.C., Graybiel, A.M., and Lobeck, L.J., The differential projection of two cytoarchitectural subregions of the inferior parietal lobule of macaque upon the deep layers of the superior colliculus, *J. Comp. Neurol.*, 235, 241, 1985.
26. Huerta, M.F., Krubitzer, L.A., and Kaas, J.H., Frontal eye field as defined by intracortical microstimulation on squirrel monkeys, owl monkeys and macaque monkeys, I. Subcortical connections, *J. Comp. Neurol.*, 253, 415, 1986.
27. Huerta, M.F. and Kaas, J.H., Connections of the physiologically defined supplementary eye field, *Soc. Neurosci. Abstr.*, 14, 159, 1988.
28. Stanton, G.B., Goldberg, M.E., and Bruce, C.J., Frontal eye field efferents in the macaque monkey, II. Topography of terminal fields in midbrain and pons, *J. Comp. Neurol.*, 271, 493, 1988.
29. Huerta, M.F. and Harting, J.K., Connectional organization of the superior colliculus, *Trend. Neurosci.*, 7, 286, 1984.
30. Huerta, M.F. and Harting, J.K., The mammalian superior colliculus: Studies of its morphology and connections, in *Comparative Neurology of the Optic Tectum*, Vanegas, H., Ed., Plenum Press, New York, 1984, pp. 687–773.
31. Harting, J.K., Huerta, M.F., Frankfurter, A.L., Strominger, N.L., and Royce, G.J., Ascending pathways from monkey superior colliculus. An autoradiographic analysis, *J. Comp. Neurol.* 192, 853, 1980.
32. Harting, J.K., Descending pathways from the superior colliculus. An autoradiographic analysis in the rhesus monkey (*Macaca mulatta*), *J. Comp. Neurol.*, 173, 583, 1977.
33. Glimcher, P.W. and Sparks, D.L., Movement selection in advance of action in the superior colliculus, *Nature*, 355, 542, 1992.
34. Mays, L.E. and Sparks, D.L., Dissociation of visual and saccade-related responses in superior colliculus neurons, *J. Neurophysiol.*, 43, 207, 1980.
35. Mohler, C.W. and Wurtz, R.H., Organization of monkey superior colliculus: intermediate layer cells discharging before eye movements, *J. Neurophysiol.*, 39, 722, 1976.
36. Moschovakis, A.K., Karabelas, A.B., and Highstein, S.M., Structure-function relationships in the primate superior colliculus, II. Morphological identity of presaccadic neurons, *J. Neurophysiol.*, 60, 263, 1988.
37. Munoz, D.P. and Guitton, D., Fixation and orientation control by the tecto-reticulo-spinal system in the cat whose head is unrestrained, *Rev. Neurol. (Paris)*, 145, 567, 1989.
38. Munoz, D.P. and Guitton, D., Control of orienting gaze shifts by the tectoreticulospinal system in the head-free cat. II. Sustained discharges during motor preparation and fixation, *J. Neurophysiol.*, 66, 1624, 1991.
39. Munoz, D.P., Pelisson, D., and Guitton D., Movement of neural activity on the superior colliculus motor map during gaze shifts, *Science*, 251, 1358, 1991.

40. Munoz, D.P., Guitton, D., and Pelisson, D., Control of orienting gaze shifts by the tectoreticulospinal system in the head-free cat, III. Spatiotemporal characteristics of phasic motor discharges, *J. Neurophysiol.*, 66, 1642, 1991.
41. Munoz, D.P. and Wurtz, R.H., Fixation cells in monkey superior colliculus I. Characteristics of cell discharge, *J. Neurophysiol.*, 70, 559, 1993.
42. Munoz, D.P. and Wurtz, R.H., Saccade-related activity in monkey superior colliculus I. Characteristics of burst and buildup cells, *J. Neurophysiol.*, 73, 2313, 1995.
43. Schiller, P.H. and Koerner, F., Discharge characteristics of single units in superior colliculus of the alert rhesus monkey, *J. Neurophysiol.*, 34, 920, 1971.
44. Sparks, D.L., Functional properties of neurons in the monkey superior colliculus: coupling of neuronal activity and saccade onset, *Brain Res.*, 156, 1, 1978.
45. Sparks, D.L. and Mays, L.E., Movement fields of saccade-related burst neurons in the monkey superior colliculus, *Brain Res.*, 190, 39, 1980.
46. Sparks, D.L., Holland, R., and Guthrie, B.L., Size and distribution of movement fields in the monkey superior colliculus, *Brain Res.*, 113, 21, 1976.
47. Waitzman, D.M., Ma, T.P., Optican, L.M., and Wurtz, R.H., Superior colliculus neurons mediate the dynamic characteristics of saccades, *J. Neurophysiol.*, 66, 1716, 1991.
48. Wurtz, R.H. and Goldberg, M.E., Superior colliculus cell responses related to eye movements in awake monkeys, *Science*, 171, 82, 1971.
49. Wurtz, R.H. and Goldberg, M.E., Activity of superior colliculus in behaving monkey, III. Cells discharging before eye movements, *J. Neurophysiol.*, 35, 575, 1972.
50. Robinson, D.A., Eye movements evoked by collicular stimulation in the alert monkey, *Vis. Res.*, 12, 1795, 1972.
51. Basso, M.A. and Wurtz, R.H., Modulation of neuronal activity by target uncertainty, *Nature*, 389, 66, 1997.
52. Scudder, C.A., Moschovakis, A.K., Karabelas, A.B., and Highstein, S.M., Anatomy and physiology of saccadic long-lead burst neurons recorded in the alert squirrel monkey, I. Descending projections from the mesencephalon, *J. Neurophysiol.*, 76, 332, 1996.
53. Dorris, M.C., Paré, M., and Munoz, D.P., Neural activity in monkey superior colliculus related to the initiation of saccadic eye movements, *J. Neurosci.*, 17, 8566, 1997.
54. Dorris, M.C. and Munoz, D.P., Saccadic probability influences motor preparation signals and time to saccadic initiation, *J. Neurosci.*, 18, 7015, 1998.
55. Munoz, D.P. and Wurtz, R.H., Saccade-related activity in monkey superior colliculus, II. Spread of activity during saccades, *J. Neurophysiol.*, 73, 2334, 1995.
56. Gandhi, N.J. and Keller, E.L., Comparison of saccades perturbed by stimulation of the rostral superior colliculus, the caudal superior colliculus, and the omnipause neuron region, *J. Neurophysiol.*, 82, 3236, 1999.
57. Munoz, D.P. and Wurtz, R.H., Fixation cells in monkey superior colliculus, II. Reversible activation and deactivation, *J. Neurophysiol.*, 70, 576, 1993.
58. Krauzlis, R.J., Basso, M.A., and Wurtz, R.H., Shared motor error for multiple eye movements, *Science*, 276, 1693, 1997.
59. Krauzlis, R.J., Basso, M.A., and Wurtz, R.H., Discharge properties of neurons in the rostral superior colliculus of the monkey during smooth-pursuit eye movements, *J. Neurophysiol.*, 84, 876, 2000.
60. Chaturvedi, V. and Van Gisbergen, J.A., Stimulation in the rostral pole of monkey superior colliculus: effects on vergence eye movements, *Exp. Brain Res.*, 132, 72, 2000.
61. Munoz, D.P. and Istvan, P.J., Lateral inhibitory interactions in the intermediate layers of the monkey superior colliculus, *J. Neurophysiol.*, 79, 1193, 1998.

62. Hikosaka, O. and Wurtz, R.H., Modification of saccadic eye movements by GABA-related substances. I. Effect of muscimol and bicuculline in monkey superior colliculus, *J. Neurophysiol.*, 53, 266, 1985.
63. Buttner, U., Hepp, K., and Henn, V., Neurons in the rostral mesencephalic and paramedian pontine reticular formation generating fast eye movements, in *Control of Gaze by Brain Stem Neurons*, Baker, R. and Berthoz A., Eds., Elsevier/North-Holland, New York, 1977.
64. Raybourn, M.S. and Keller, E.L., Colliculoreticular organization in primate oculomotor system, *J. Neurophysiol.*, 40, 861, 1977.
65. Schiller, P.H., True, S.D., and Conway, J.L., Deficits in eye movements following frontal eye-field and superior colliculus ablations, *J. Neurophysiol.*, 44, 1175, 1980.
66. Schiller, P.H., Sandell, J.H., and Maunsell, J.H., The effect of frontal eye field and superior colliculus lesions on saccadic latencies in the rhesus monkey, *J. Neurophysiol.*, 57, 1033, 1987.
67. Hikosaka, O. and Wurtz, R.H., Saccadic eye movements following injection of lidocaine into the superior colliculus, *Exp. Brain Res.*, 61, 531, 1986.
68. Lee, C., Rohrer, W.H., and Sparks, D.L., Population coding of saccadic eye movements by neurons in the superior colliculus, *Nature*, 332, 357, 1988.
69. Hanes, D.P. and Wurtz, R.H., Interaction of the frontal eye field and superior colliculus for saccade generation, *J. Neurophysiol.*, 85, 804, 2001.
70. McAdams, C.J. and Maunsell, J.H., Effects of attention on the reliability of individual neurons in monkey visual cortex, *Neuron*, 23, 765, 1999.
71. Chelazzi, L., Miller, E.K., Duncan, J., and Desimone, R., Responses of neurons in macaque area V4 during memory-guided visual search, *Cereb. Cortex*, 11, 761, 2001.
72. Colby, C.L. and Goldberg, M.E., Space and attention in parietal cortex, *Annu. Rev. Neurosci.*, 22, 319, 1999.
73. Andersen, R.A., Snyder, L.H., Bradley, D.C., and Xing, J., Multimodal representation of space in the posterior parietal cortex and its use in planning movements, *Annu. Rev. Neurosci.*, 20, 303, 1997.
74. Paré, M. and Wurtz, R.H., Progression in neuronal processing for saccadic eye movements from parietal cortex area lip to superior colliculus, *J. Neurophysiol.*, 85, 2545, 2001.
75. Ferraina, S., Pare, M., and Wurtz, R.H., Comparison of cortico-cortical and cortico-collicular signals for the generation of saccadic eye movements, *J. Neurophysiol.*, 87, 845, 2002.
76. Schall, J.D., Morel, A., King, D.J., Bullier, J., Topography of visual cortical afferents to frontal eye field in macaque: Convergence and segregation of processing streams, *J. Neurosci.*, 15, 4464, 1995.
77. Mohler, C.W., Goldberg, M.E., and Wurtz, R.H., Visual receptive fields of frontal eye field neurons, *Brain Res.*, 61, 385, 1973.
78. Bruce, C.J. and Goldberg, M.E., Primate frontal eye fields, I. Single neurons discharging before saccades, *J. Neurophysiol.*, 53, 603, 1985.
79. Schall, J.D. and Thompson, K.G., Neural selection and control of visually guided eye movements, *Annu. Rev. Neurosci.*, 22, 241, 1999.
80. Schall, J.D., Neural selection and control of saccades by frontal eye field, *Philosophical Transactions of the Royal Society of London*: Series B Biological Sciences, 357, 1073–1082, 2002.
81. Schall, J.D., Thompson, K.G., Bichot, N.P., Murthy, A., and Sato, T.R., Visual processing in the frontal eye field, in *The Primate Visual System*, Kaas, J. and Collins, C., Eds., CRC Press, Boca Raton, FL, 2003.

82. Bruce, C.J., Goldberg, M.E., Bushnell, M.C., and Stanton, G.B., Primate frontal eye fields. II. Physiological and anatomical correlates of electrically evoked eye movements, *J. Neurophysiol.*, 54, 714, 1985.
83. Hanes, D.P. and Schall, J.D., Neural control of voluntary movement initiation, *Science*, 274, 427, 1996.
84. Hanes, D.P., Patterson, W.F., and Schall, J.D., Role of frontal eye fields in countermanding saccades: visual, movement and fixation activity, *J. Neurophysiol.*, 79, 817, 1998.
85. Leichnetz, G.R., Spencer, R.F., Hardy, S.G.P., and Astruc, J., The prefrontal corticotectal projection in the monkey: An anterograde and retrograde horseradish peroxidase study, *Neuroscience*, 6, 1023, 1981.
86. Shook, B.L., Schlag-Ray, M., and Schlag, J., Primate supplementary eye field, II. Comparative aspects of connections with the thalamus, corpus striatum and related forebrain nuclei, *J. Comp. Neurol.*, 307, 562, 1991.
87. Selemon, L.D. and Goldman-Rakic, P.S., Longitudinal topography and interdigitation of corticostriatal projections in the rhesus monkey, *J. Neurosci.*, 5, 776, 1985.
88. Stanton, G.B., Goldberg, M.E., and Bruce, C.J., Frontal eye field efferents in the macaque monkey, I. Subcortical pathways and topography of striatal and thalamic terminal fields, *J. Comp. Neuro.*, 271, 473, 1988.
89. Parthasarathy, H.B., Schall, J.D., and Graybiel, A.M., Distributed but convergent ordering of corticostriatal projections: analysis of the frontal eye field and the supplementary eye field in the macaque monkey, *J. Neurosci.*, 12, 4468, 1992.
90. Hikosaka, O., Sakamoto, M., and Usui, S., Functional properties of monkey caudate neurons. I. Activities related to saccadic eye movements, *J. Neurophysiol.*, 61, 780, 1989.
91. Hikosaka, O., Sakamoto, M., and Usui, S., Functional properties of monkey caudate neurons, II. Visual and auditory responses, *J. Neurophysiol.*, 61, 799, 1989.
92. Hikosaka, O., Sakamoto, M., and Usui, S., Functional properties of monkey caudate neurons, III. Activities related to expectation of target and reward, *J. Neurophysiol.*, 61, 814, 1989.
93. Leichnetz, G.R., Spencer, R.F., and Smith, D.J., Cortical projections to nuclei adjacent to the oculomotor complex in the medial dien-mesencephalic tegmentum in the monkey, *J. Comp. Neurol.*, 228, 359, 1984.
94. Leichnetz, G.R., Smith, D.J., and Spencer, R.F., Cortical projections to the paramedian tegmental and basilar pons in the monkey, *J. Comp. Neurol.*, 228, 388, 1984.
95. Schnyder, H., Reisine, H., Hepp, K., and Henn, V., Frontal eye field projection to the paramedian pontine reticular formation traced with wheat germ agglutinin in the monkey, *Brain Res.*, 329, 151, 1985.
96. Shook, B.L., Schlag-Ray, M., and Schlag, J., Primate supplementary eye field, I. Comparative aspects of mesencephalic and pontine connections, *J. Comp. Neurol.*, 301, 618, 1990.
97. Ilinsky, I.A., Jouandet, M.L., and Goldman-Rakic, P.S., Organization of the nigrothalamocortical system in the rhesus monkey, *J. Comp. Neurol.*, 236, 315, 1985.
98. Lynch, J.C., Hoover, J.E., and Strick, P.L., Input to the primate frontal eye field from the substantia nigra, superior colliculus and dentate nucleus demonstrated by transneuronal transport, *Exp. Brain Res.*, 100, 181, 1994.
99. Dias, E.C., Kiesau, M., and Segraves, M.A., Acute activation and inactivation of macaque frontal eye field with GABA-related drugs, *J Neurophysiol.* 74, 2744, 1995.
100. Sommer, M.A. and Tehovnik, E.J., Reversible inactivation of macaque frontal eye field, *Exp. Brain Res.*, 116, 229, 1997.

101. Dias, E.C. and Segraves, M.A., Muscimol-induced inactivation of monkey frontal eye field: effects on visually and memory-guided saccades, *J. Neurophysiol.* 81, 2191, 1999.
102. Schiller, P.H. and Chou, I.H., The effects of frontal eye field and dorsomedial frontal cortex lesions on visually guided eye movements, *Nat. Neurosci.*, 1, 248, 1998.
103. Segraves, M.A. and Goldberg, M.E., Functional properties of corticotectal neurons in the monkey's frontal eye field, *J. Neurophysiol.*, 58, 1387, 1987.
104. Everling, S. and Munoz, D.P., Neuronal correlates for preparatory set associated with saccade generation in the primate frontal eye field, *J. Neurosci.*, 20, 387, 2000.
105. Sommer, M.A. and Wurtz, R.H., Frontal eye field sends delay activity related to movement, memory, and vision to the superior colliculus, *J. Neurophysiol.*, 85, 1673, 2001.
106. Segraves, M.A., Activity of monkey frontal eye field neurons projecting to oculomotor regions of the pons. *J. Neurophysiol.*, 68, 1967, 1992.
107. Sommer, M.A. and Wurtz, R.H., Frontal eye field neurons orthodromically activated from the superior colliculus, *J. Neurophysiol.*, 80, 3331, 1998.
108. Sommer, M.A. and Wurtz, R.H., A pathway in primate brain for internal monitoring of movements, *Science*, 296, 1480, 2002.
109. Saslow, M.G., Effects of components of displacement-step stimuli upon latency of saccadic eye movements, *J. Opt. Soc. Am.*, 57, 1024, 1967.
110. Ross, L.E. and Ross, S.M., Saccade latency and warning signals: stimulus onset, offset, and change as warning events, *Percept. Psychophys.*, 27, 251, 1980.
111. Reuter-Lorenz, P.A., Hughes, H.C., and Fendrich, R., The reduction of saccadic latency by prior offset of the fixation point: an analysis of the gap effect, *Percept. Psychophys.*, 49, 167, 1991.
112. Jüttner, M. and Wolf, W., Occurrence of human express saccades depends on stimulus uncertainty and stimulus sequence, *Exp. Brain Res.*, 89, 678, 1992.
113. Kingstone, A. and Klein, R.M., Visual offsets facilitate saccadic latency: does pre-disengagement of visuospatial attention mediate the gap effect? *J. Exp. Psychol. Hum. Percept. Perform.*, 19, 1251, 1993.
114. Dorris, M.C. and Munoz, D.P., A neural correlate for the gap effect on saccadic reaction times in monkey, *J. Neurophysiol.*, 73, 2558, 1995.
115. Munoz, D.P. and Corneil, B.D., Evidence for interactions between target selection and visual fixation for saccade generation in humans, *Exp. Brain Res.*, 103, 168, 1995.
116. Paré, M, and Munoz, D.P., Saccadic reaction time in the monkey: advanced preparation of oculomotor programs is primarily responsible for express saccade occurrence, *J. Neurophysiol.*, 76, 3666, 1996.
117. Everling, S., Paré, M., Dorris, M.C., and Munoz, D.P., Comparison of the discharge characteristics of brainstem omnipause neurons and superior colliculus fixation neurons in monkey: implications for control of fixation and saccade behavior, *J. Neurophysiol.*, 79, 511, 1998.
118. Gore, J.L., Everling, S., Kobayashi, Y., Dorris, M.C., and Munoz, D.P., Saccadic preparatory signals in the FEF, SNr, and SC, *Soc. Neurosci. Abstr.*, 28, 464.2, 2002.
119. Dias, E.C. and Bruce, C.J., Physiological correlate of fixation disengagement in the primate's frontal eye field, *J. Neurophysiol.*, 72, 2532, 1994.
120. Hikosaka, O. and Wurtz, R.H., Visual and oculomotor functions of monkey substantia nigra pars reticulata, I. Relation of visual and auditory responses to saccades, *J. Neurophysiol.*, 49, 1230, 1983.
121. Handel, A. and Glimcher, P.W., Quantitative analysis of substantia nigra pars reticulata activity during a visually guided saccade task, *J. Neurophysiol.*, 82, 3458, 1999.

122. Fischer, B. and Boch, R., Saccadic eye movements after extremely short reaction times in the monkey, *Brain Res.*, 260, 21, 1983.
123. Fischer, B. and Ramsperger, E., Human express saccades: extremely short reaction times of goal directed eye movements, *Exp. Brain Res.*, 57, 191, 1984.
124. Fischer, B. and Weber, H., Express saccades and visual attention, *Behav. Brain Sci.*, 16, 533, 1993.
125. Carpenter, R.H.S., Oculomotor procrastination, in *Eye Movements: Cognition and Visual Perception*, Fischer, D.F. and Montey, R.A., Eds., Erlbaum, Hillsdale, NJ, 1981, pp. 237–246.
126. Edelman, J.A. and Keller, E.L., Activity of visuomotor burst neurons in the superior colliculus accompanying express saccades, *J. Neurophysiol.*, 76, 1, 1996.
127. Logan, G.D. and Cowan, W.B., On the ability to inhibit thought and action: A theory of an act of control, *Psycholog. Rev.*, 91, 295, 1984.
128. Logan, G.D., On the ability to inhibit thought and action: A users' guide to the stop signal paradigm, in *Inhibitory Processes in Attention, Memory and Language*, Dagenbach, D.E. and Carr, T.H., Eds., Academic Press, San Diego, 1994, pp. 189–239.
129. Hanes, D.P. and Schall, J.D., Countermanding saccades in macaque, *Vis. Neurosci.*, 12, 929, 1995.
130. Lisberger, S.G., Fuchs, A.F., King, W.M., and Evinger, L.C., Effect of mean reaction time on saccadic responses to two-step stimuli with horizontal and vertical components, *Vision Res.,* 15, 1021, 1975.
131. Becker, W. and Jurgens, R., An analysis of the saccadic system by means of double step stimuli, *Vision Res.*, 19, 976, 1979.
132. Hanes, D.P. and Carpenter, R.H., Countermanding saccades in humans, *Vision Res.*, 39, 2777, 1999.
133. Cabel, D.W., Armstrong, I.T., Reingold, E., and Munoz, D.P., Control of saccade initiation in a countermanding task using visual and auditory stop signals, *Exp. Brain Res.*, 133, 431, 2000.
134. Asrress, K.N. and Carpenter, R.H., Saccadic countermanding: a comparison of central and peripheral stop signals, *Vision Res.*, 41, 2645, 2001.
135. Colonius, H., Ozyurt, J., and Arndt, P.A., Countermanding saccades with auditory stop signals: testing the race model, *Vision Res.*, 41, 1951, 2001.
136. Lecas, J.C., Requin, J., Anger, C., and Vitton, N., Changes in neuronal activity of the monkey precentral cortex during preparation for movement, *J. Neurophysiol.*, 56, 1680, 1986.
137. Aston-Jones, G., Rajkowski, J., Kubiak, P., and Alexinsky, T., Locus coeruleus neurons in monkey are selectively activated by attended cues in a vigilance task, *J. Neurosci.*, 14, 4467, 1994.
138. Hanes, D.P. and Paré, M., Neural control of saccade production studied with the countermanding paradigm, superior colliculus, *Soc. Neuro. Abstr.*, 24, 418, 1998.
139. Hallett, P.E., Primary and secondary saccades to goals defined by instructions, *Vision Res.*, 18, 455, 1978.
140. Everling, S., Dorris, M.C., and Munoz, D.P., Reflex suppression in the anti-saccade task is dependent on prestimulus neural processes, *J. Neurophysiol.*, 80, 1584, 1998.
141. Everling, S., Dorris, M.C., Klein, R.M., and Munoz, D.P., Role of primate superior colliculus in preparation and execution of anti- and pro-saccades, *J. Neurosci.*, 19, 2740, 1999.
142. Bell, A.H., Everling, S., and Munoz, D.P., Influence of stimulus eccentricity and direction on characteristics of pro- and antisaccades in non-human primates, *J. Neurophysiol.*, 84, 2595, 2000.

143. Schlag-Rey, M., Amador, N., Sanchez, H., and Schlag, J., Antisaccade performance predicted by neuronal activity in the supplementary eye field, *Nature*, 390, 398, 1997.
144. Forbes, K. and Klein, R.M., The magnitude of the fixation offset effect with endogenously and exogenously controlled saccade, *J. Cog. Neurosci.*, 8, 344, 1996
145. Guitton, D., Buchtel, H.A., and Douglas R.M., Frontal lobe lesions in man cause difficulties in suppressing reflexive glances and in generating goal-directed saccades, *Exp. Brain Res.*, 58, 455, 1985.
146. Pierrot-Deseilligny, C., Rivaud, S., Gaymard, B., and Agid, Y., Cortical control of reflexive visually-guided saccades, *Brain*, 114, 1473, 1991.
147. Schall, J.D., Stuphorn, V., and Brown, J., Monitoring and control of gaze by the frontal lobes, *Neuron,* 36, 309, 2002.

4 Puffs and Patches: A Brief Chronological Review

John K. Harting

CONTENTS

4.1 INTRODUCTION

The mammalian superior colliculus (SC) is an important site of sensorimotor integration and both sensory and motor components lie in a compact vertical stack spanning no more than 2 mm.[1] One of the most interesting findings related to the circuitry underlying these sensorimotor transformations is its *periodicity*. That is, numerous studies of many different afferent and efferent systems, in a wide variety of mammals, have demonstrated an inconceivable assortment of functionally diverse collicular afferents that exhibit discontinuous distributions, especially within the intermediate and deep layers of the SC. This brief review begins with the first widely cited description of SC patches (interestingly in the superficial layers) and concludes with a discussion of novel data published in the early 2000s. In the end, possible future studies are suggested.

0-8493-0097-5/04/$0.00+$1.50

4.2 ORIGINAL OBSERVATIONS OF PATCHES: RETINO- AND NIGROTECTAL PROJECTIONS

> Vertical organization in the SC "would seem to deal with the integration of multiple inputs of markedly different information content."[2]

The history of clusters, patches, puffs, cylinders, lattices, modules, and domains within the SC began with the observations of Ann Graybiel[3,4] in 1975 and 1976 that two major inputs to the cat SC terminate in a patchy, noncontinuous manner. The first of these pathways to be described was the patchy ipsilateral retinotectal projection (Figure 4.1),[5] which Graybiel[4] found to correspond to regions of weak/sparse input from the contralateral eye.

The second periodic projection observed by Graybiel,[3,6] which previewed the unveiling of a bewildering array of such inputs to the expansive intermediate gray layer of the SC, the stratum griseum intermediale (SGI), was the projection from the substantia nigra pars reticularis. The observations of patchy ipsilateral retino- and nigrotectal pathways led Graybiel[6] to emphasize that the *periodic vertical segmentation* in the SC differs functionally from that seen in the cerebral cortex, since such vertical organization in the SC "would seem to deal with the integration of multiple inputs of markedly different information content." Around this same time, reports of discontinuous, patchy, corticotectal projections to the SGI were reported in the monkey[7] and rat[2] respectively and they gave support and momentum to the nascent concept that the SGI consists of vertically organized functional units.

In the early to mid 1980s, three more afferents were observed distributing in a discontinuous fashion across the SGI. Thus, projections from the spinal trigeminal nucleus,[8,9,17] the nucleus of the posterior commissure,[10] the deep cerebellar nuclei,[11] and the hypothalamus were found to distribute in patches at different depths within the SGI. When combined with the earlier findings of Graybiel, these data suggested that functionally related afferents might be *stacked* in distinct repeating modules (called *afferent modules* by Huerta and Harting[8,12]). These findings are summarized in Figures 4.2 and 4.3.

It should be kept in mind that only the sublaminar tiers of the SGI targeted by the various projections were known in 1984, not which inputs were in the same stacked column or overlapped. Moreover, the patchy projections from the cerebellum were inadvertently left out of this review figure (Figure 4.2). However, Kawamura and colleagues[11] had observed that patchy cerebellotectal fibers from the medial nucleus target the dorsal SGI, while the posterior interposed nucleus innervates the ventral SGI, also in a patchy, discontinuous manner.

4.3 INITIAL OBSERVATIONS OF PATCHY EFFERENTS

> The extensive dendritic trees of SGI neurons,[13] which accounts for their large receptive fields[14] will "allow a non-fragmented sensory representation to be present across the horizontal expanse of the deep layers, even if the incoming sensory axons are not continuously distributed in this plane."[12]

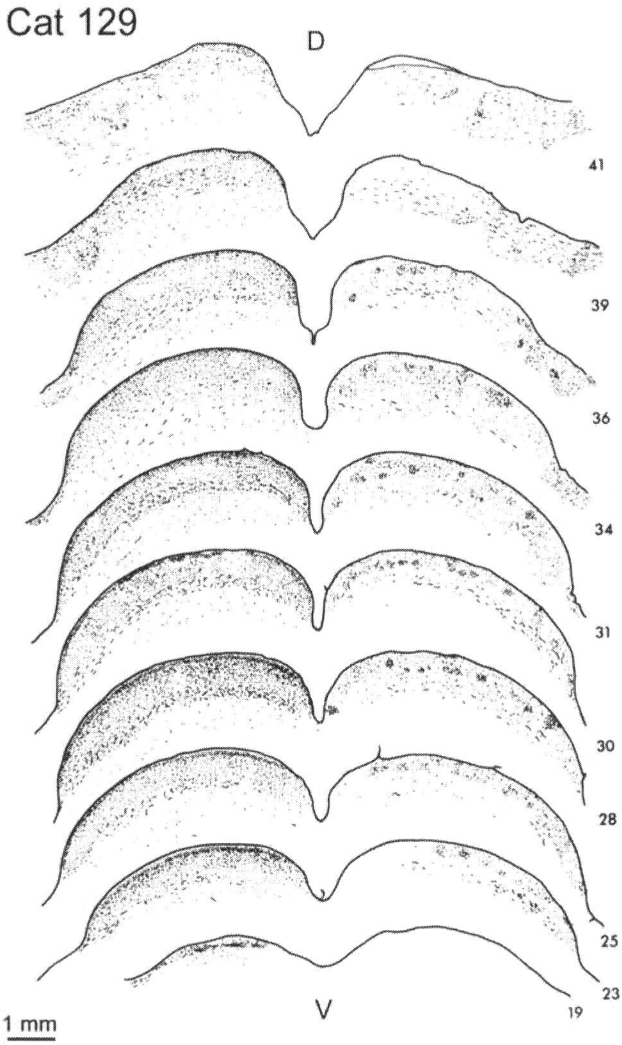

FIGURE 4.1 The distribution of anterogradely transported protein within the SC following an injection of tritiated amino acids into the eye. Note the patches of label/silver grains in the ipsilateral (right side) of the SC. (From Harting, J.K. and Guillery, R.W., Organization of retinocollicular pathways in the cat, *J. Comp. Neurol.*, 166, 133, 1976. With permission.)

Huerta and Harting[8,12] suggested that a stratification of afferents within a single dorsoventral *line* or *column* through the SGI would allow many different afferents to share a single representation in a single spatial map of contralateral space.[14] Around this same time, the vertical organization of outflow channels within the SGI was beginning to be explored, but in a very preliminary fashion. Hashikawa and Kawamura[15], and Kawamura and Hashikawa[16] had earlier noted clumps or patches of tectoreticular cells. Then in 1981, Huerta and colleagues[17] noted retrogradely labeled tecto-trigeminal neurons lying within patches of trigeminotectal axons/ter-

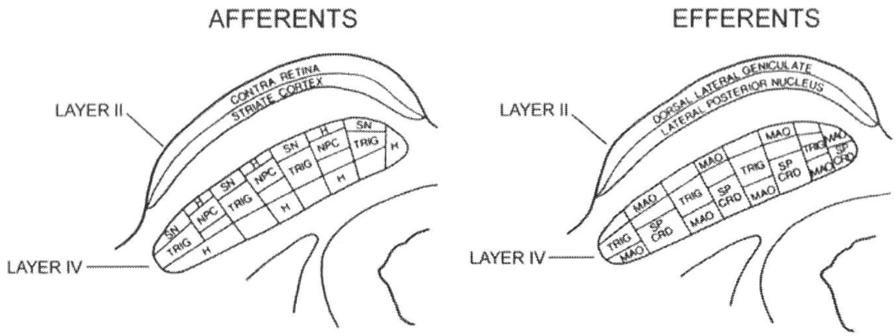

FIGURE 4.2 Modules formed by incoming axons from the hypothalamus (H), nucleus of the posterior commissure (NPC), the substantia nigra (SN) and the spinal trigeminal nucleus (TRIG) are indicated as are modules formed by colliculo-olivary (MAO), colliculo-spinal (SP CRD) and colliculo-trigeminal (TRIG) cells. (From Huerta, M.F. and Harting, J.K., Connectional organization of the superior colliculus, *Trends Neurosci.*, 7, 286, 1984. With permission.)

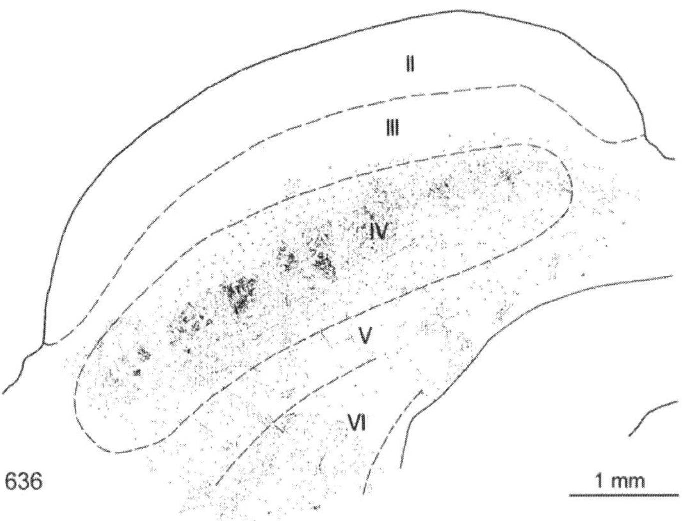

FIGURE 4.3 Distribution of silver grains in the SGI following an injection of a mixture of tritiated proline and leucine into the nucleus of the posterior commissure in the cat. II, stratum griseum superficiae; III, stratum opticum; IV, stratum griseum intermediale; V, stratum album intermediale; VI, stratum griseum profundum. (From Huerta, M.F. and Harting, J.K., The projection from the nucleus of the posterior commissure to the superior colliculus of the cat: patch-like endings within the intermediate and deep grey layers, *Brain Res.*, 238, 426, 1982. With permission.)

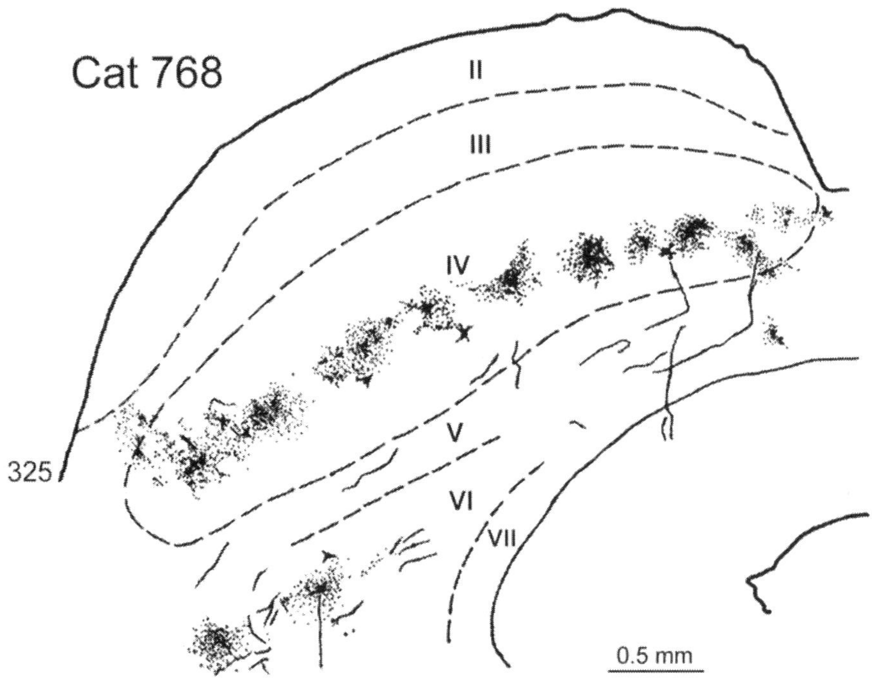

FIGURE 4.4 The pattern of anterogradely transported WGA-HRP (fine dots) and the location of retrogradely labeled cells in the SGI contralateral to an injection into the spinal trigeminal nucleus. VII, stratum album profundum. Other abbreviations as in Figure 4.3. (From Huerta et al., The trigeminocollicular projection in the cat: patch-like endings within the intermediate grey, *Brain Res.*, 211, 1, 1981. With permission.)

minals following injections of wheat germ agglutinin conjugated to horseradish peroxidase (WGA-HRP) into the spinal trigeminal nucleus (Figure 4.4).

This suggested that the organization of tectofugal outputs within the SGI might also be arranged in vertically organized functional units. Findings of patchy tiers of tectofugal neurons that project to the medial superior olive and spinal cord gray gave further support to the hypothesis that connectionally defined modules provide linkage between particular afferents and efferents. This type of organization would "facilitate the separation of information channels within individual collicular layers, thus allowing different types of incoming information to distribute to a select set of collicular targets." [8]

4.4 AChE-RICH AND POOR DOMAINS: A CONCEPTUAL FRAMEWORK

> The intermediate and deep layers of the superior colliculus may have a functional mosaic organization that allows control of eye or axial body movements to be exerted by different combinations of triggering input.[18]

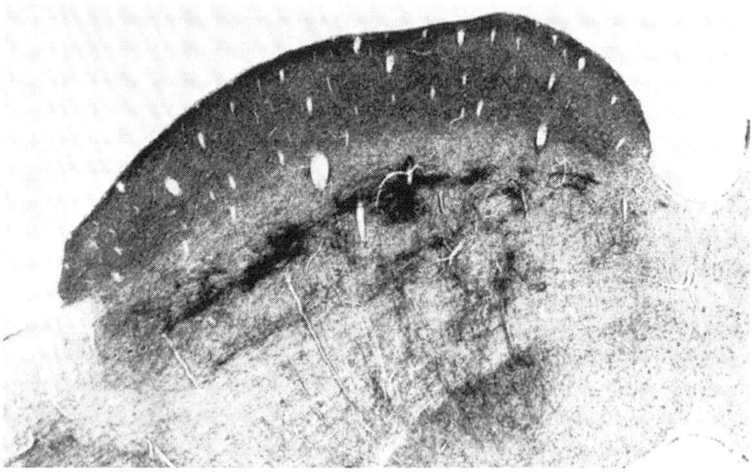

FIGURE 4.5 Shows the staining pattern of AChE in the cat superior colliculus. (From Illing, R.-B. and Graybiel, A.M., Convergence of afferents from frontal cortex and substantia nigra onto acetylcholinesterase-rich patches of the cat's superior colliculus, *Neuroscience*, 14, 455, 1985. With permission.)

Our understanding of SGI patches and puffs gained tremendous momentum with the publication of back-to-back papers by Illing and Graybiel in the mid 1980s.[18,19] They compared the locations of patches of SC afferents with the periodic concentrations of the enzyme acetyl cholinesterase (AChE). Horizontal *tiers* of AChE-rich patches were observed within the dorsal and ventral regions of the SGI and these tiers were interconnected by *streamers* or *bridges* of AChE-rich staining that partitioned the middle region of the SGI into holes or voids (Figure 4.5).

Illing and Graybiel[18] found that medial frontotectal fibers target all AChE-rich patches of the SGI (dorsal and ventral tiers, as well as middle tier *streamers*) while the lateral frontotectal and nigrotectal axons are restricted to the dorsal tier of AChE-rich patches. The overlap of nigrotectal and lateral and medial frontotectal projections within the AChE-rich patches of the dorsal tier showed for the first time that cortical and brain stem regions sharing a role in mechanisms underlying saccadic eye movements project to the same modular zones of the SGI.[18,20,–23] This suggested, "there may be certain attributes of collicular function sharing a dependency on the frontal eye fields and related cortex and on the strio-nigrotectal system."[18] Due to the oculomotor related inputs reaching the AChE rich patches, Illing and Graybiel[18] called the constellation of AChE rich patches and their inputs the *motor* domain of the SGI.

Illing and Graybiel's 1986 paper[19] continued in the same vein as their 1985[18] study, but instead, focused upon pathways ending within the AChE-poor zones. They showed that corticotectal projections from the dorsal bank of the anterior ectosylvian sulcus (AES; most likely the fourth somatosensory area or SIV),[24] the posteromedial lateral suprasylvian cortex (PMLS; extrastriate visual cortex) and the bulbospinal junction (most likely the spinal trigeminal nucleus) terminate in AChE-poor regions of the SGI. These AChE-poor regions were called the *sensory* domain of the SGI by Illing and Graybiel.[19]

The combined findings of the two papers suggested to Illing and Graybiel[19] that "the combination of restricted depth-levels of innervation and local clustering within such levels serves to focus different inputs into highly delimited regions." They went on to say that it is "reasonable to think that much of the complexity of collicular functioning may derive from the modes in which these afferent-fiber grids are sampled." Moreover, "the intermediate and deep layers of the superior colliculus may have a functional mosaic organization that allows control of eye or axial body movements to be exerted by different combinations of triggering input."

4.5 GLOBAL NIGROTECTAL PROJECTIONS

> "Such a major pathway between the substantia nigra and the superior colliculus might consist of a number of different channels."[25]

Illing and Graybiel's[18,19] findings have been expanded and modified in several ways, but their hypothesis regarding motor and sensory domains has been an important organizing theme for all subsequent studies. The first major extension of their model was the finding that nigrotectal projections terminate widely throughout the SGI. That is, in addition to the single dorsal tier of input reported by Illing and Graybiel,[18] middle and ventral tiers of nigrotectal axons/terminals were observed by Harting and colleagues. [25] Just as important, the three tiers of nigral axons/terminals were found to arise from different regions/cells of the SNpr, suggesting that perhaps the three tiers of nigral inputs carried out slightly different functions in orienting behaviors (Figures 4.6 and 4.7).[25]

While we[25] did not compare the locations of the two additional tiers of nigrotectal axons/terminals with the locations of AChE-rich staining, we hypothesized that such overlap occurs. Thus, it was suggested that the dorsal tier of AChE patches consists of nigrotectal and medial and lateral frontotectal axons. The middle tier was considered to consist of nigrotectal axons that overlap with medial frontotectal inputs, while the ventral tier was hypothesized to be the site of commingling nigro- and medial frontotectal fibers (Figure 4.8).[25]

Since the medial and lateral frontotectal projections differ functionally,[20,21] it was suggested that the commingling of different eye/head movement related pathways (three nigrotectal, medial, and lateral frontotectal) within the three different tiers provides the anatomical substrates for each tier of circuitry to participate in slightly different aspects of oculomotor/orienting processing.[25] In support of this idea are the findings of Joseph and Boussaoud,[23] who reported that SNpr cells vary in their responsiveness to saccades, sensory stimuli (visual, auditory, and somatosensory), and head movements. The finding of a very extensive/global nigrotectal projection suggested that this projection is potentially a component of much of the functional mosaic involving different "combinations of triggering inputs" involved in the control of various eye, head, and body movements.[19]

The findings of Illing and Graybiel[18,19] regarding the spatial relationships between AChE and certain SC inputs led Wallace and Fredens[27] to compare the distribution of several afferents with the enzyme NADPH-diaphorase (contained

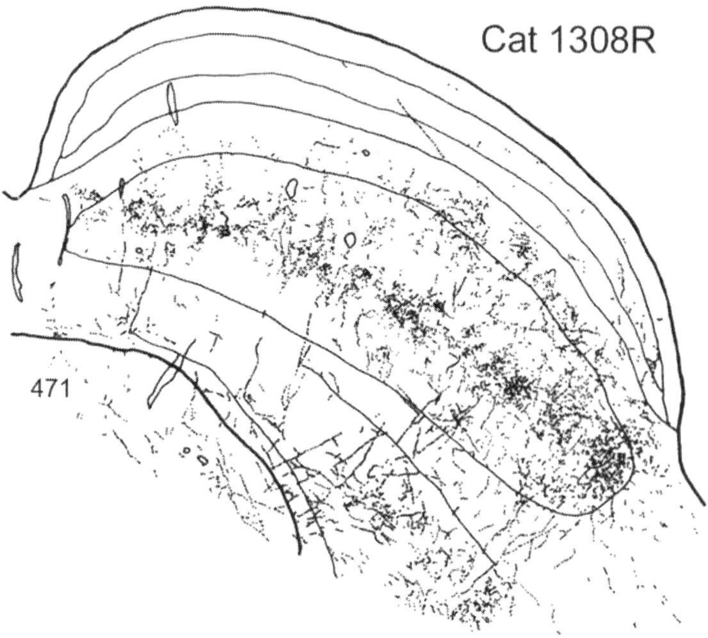

FIGURE 4.6 Shows the location of transported protein in the SGI following an injection of a mixture of tritiated amino acids into the ipsilateral SNpr. In several places the three tiers of nigrotectal terminals/axons can be seen. See also Figure 4.7. (From Harting, J.K. et al., Neuroanatomical studies of the nigrotectal projection in the cat, *J. Comp. Neurol.*, 278, 615, 1988. With permission.)

within cholinergic neurons). They found that cells in the dorsolateral central grey and precommissural nucleus project to NADPH-diaphorase-rich areas of the SC, where they are thought to modulate nociceptive cells. On the other hand, axons/terminals from the nucleus of the brachium of the inferior colliculus were found to distribute to NADPH-diaphorase-poor regions, and are thought to convey auditory information. Interestingly, auditory responsive units in the mouse SC are often arranged in clusters (Figure 4.9).[28]

4.6 POTENTIAL CHOLINERGIC INPUTS

> The pedunculopontine tegmental nucleus might not give rise to all of the cholinergic input to the SGI.[12]

While AChE had been used as a marker for comparing the spatial distribution of afferents within the SGI, relatively little was known regarding the specific sources of cholinergic input to the three tiers. The SGI contains a few intrinsic cholinergic neurons, but the great majority of the AChE and choline acetyltransferase (ChAT) staining is comprised of processes of pedunculopontine tegmental (PTN) and lat-

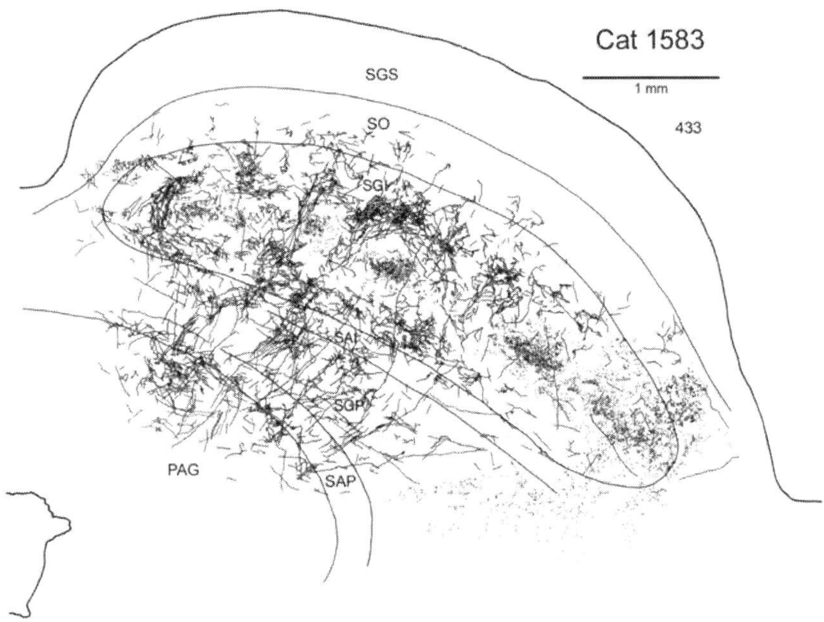

FIGURE 4.7 The distribution of anterogradely transported WGA-HRP in the dorsal and ventral tiers of the SGI following an injection into the ipsilateral SNpr (only the middle tier of nigrotectal axons is labeled in Figure 4.6). Transported protein can also be seen in the middle tier subsequent to an injection of a mixture of tritiated amino acids into the contralateral spinal trigeminal nucleus. SGS, stratum griseum superficiale; SO, stratum opticum; SGI, stratum griseum intermediale; SAI, stratum album profundum; SGP, stratum griseum profundum; stratum album profundum. (From Harting, J.K. and Van Lieshout, D.P., Spatial relationships of axons arising from the substantia nigra, spinal trigeminal nucleus and pedunculopontine tegmental nucleus within the intermediate gray of the cat superior colliculus, *J. Comp. Neurol.*, 305, 543, 1991. With permission.)

erodorsal tegmental (LDTN) neurons.[29,30] Interestingly, anterograde transport studies of the PTN revealed a relatively dense, patchy, projection across only the middle tier of the SGI.[31] Since the injections (a mixture of two tritiated amino acids) invaded relatively small portions of the PTN, areas of the PTN not included in the injections, and LDTN, have been hypothesized to project to the dorsal and ventral tiers (Figure 4.10).

It appears that different regions of the PTN (which potentially innervate different tiers of the SGI) exhibit varying patterns of connectivity and/or functions. For example, it has been reported [32] that cells within the entopeduncular nucleus (equivalent to the primate medial pallidal segment) project upon the medial PTN, while those within the globus pallidus (equivalent to the primate lateral pallidal segment) target the lateral PTN. It remains to be seen if the PTN-tectal projections are connectionally, and possibly functionally, heterogeneous, but to date, no specific PTN pathway(s) has/have been described as ending in the ventral and dorsal tiers of the SGI. Just recently it has been shown that cells in the primate PTN play a role

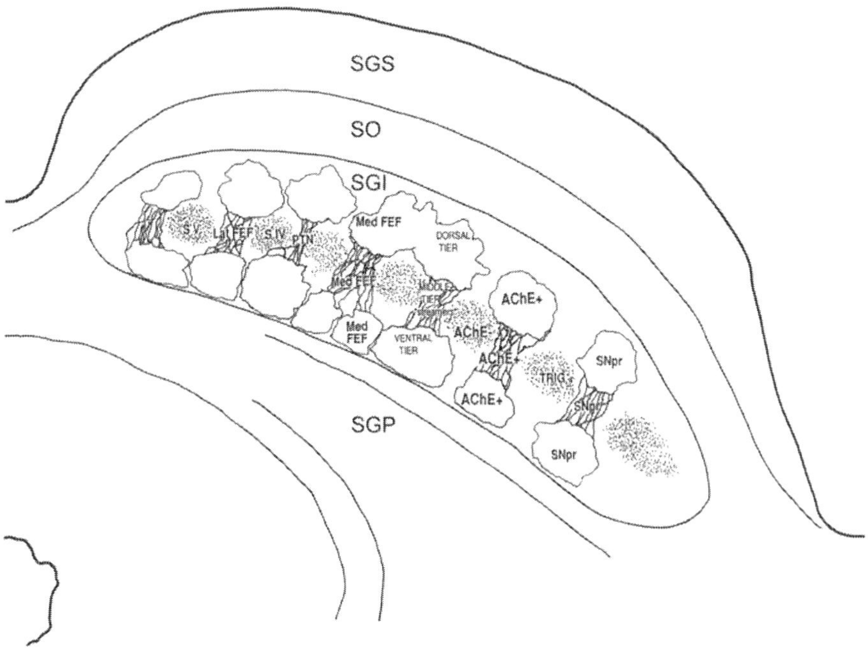

FIGURE 4.8 Schematic showing the spatial relationships between medial (Med FEF) and lateral frontotectal (Lat FEF), nigrotectal (SNpr), trigeminotectal (TRIG) and fourth somatosensory cortex (SIV) projections with AChE-rich and poor zones. Other abbreviations as in Figure 4.7. (From Harting, J.K. and Van Lieshout, D.P., Spatial relationships of axons arising from the substantia nigra, spinal trigeminal nucleus and pedunculopontine tegmental nucleus within the intermediate gray of the cat superior colliculus, *J. Comp. Neurol.*, 305, 543, 1991. With permission.)

in the initiation and execution of saccades. Thus, the PTN receives saccade-related signals from the SGI, basal ganglia, and cerebral cortex, and conveys such information back to these same areas.[33]

4.7 SOMAS AND DENDRITES ARE RELATED TO DIFFERENT DOMAINS

> Both descending and ascending collicular efferents originate from perikarya situated in characteristic positions relative to the collicular compartments defined by elevated acetylcholinesterase activity.[34]

As mentioned earlier, relatively little was known regarding the presence or absence of any modular organization of tectofugal neurons. However, in the early 1990s investigators began to study the spatial relationships of efferent cells to the AChE-rich and poor zones. In particular, Illing[34] noted that the somas of tecto-bulbospinal and tectopontine neurons in the rat occupy AChE-poor regions of the SGI, while the dendrites of such cells extend into the AChE-rich patches. In addition to sug-

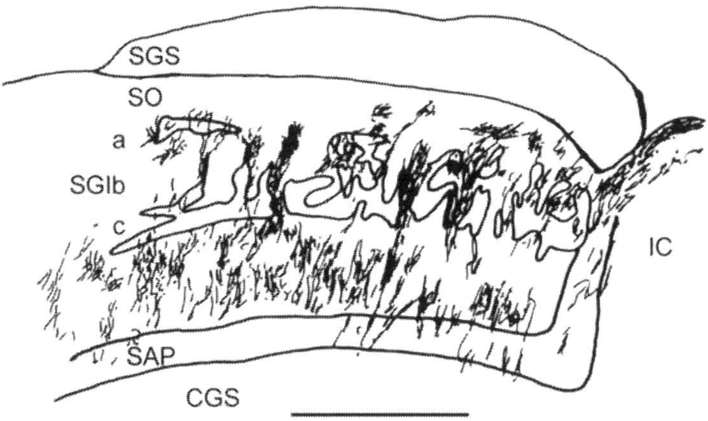

FIGURE 4.9 A parasaggital section showing the distribution of anterogradely labeled axons/fibers (vertical slashes) in the intermediate and deep SC following an injection of leucoagglutinin extracted from Phaseolus vulgaris (PHA-L) into the nucleus of the brachium of the inferior colliculus in a mouse. The areas outlined by a solid line represent NADPH-diaphorase-rich zones. CGS, central grey substance; IC, inferior colliculus. All other abbreviations as in Figure 7. (From Wallace, M.N. and Fredens, K., Relationship of afferent inputs to the lattice of high NADPH-diaphorase activity in the mouse superior colliculus, *Brain Res.*, 78, 435, 1989. With permission.)

gesting that "the compartmental (and apparently fractal) architecture of the multimodal collicular layers provides a structural frame for the specific interlocking of its afferents and efferents,"[34] these data suggested that sensory related inputs to the somas of tectal efferent neurons could shunt, override, or modulate the motor input reaching the dendrites of tectofugal neurons (Figure 4.11).

Jeon and Mize[30] also studied the spatial relationships of tectofugal neurons to the AChE lattice. They confirmed Huerta and Harting's[8,12] observations that SGI cells projecting to the inferior olive arise from the dorsal and ventral tiers of the SGI. More important, however, they observed that cells projecting to the ipsilateral cuneiform nucleus lie within ChAT-positive patches of the dorsal tier. They emphasized that the patches of cuneiform-projecting neurons could be saccade-related. Thus, Peck[35] has observed patches of saccade-related cells in the SGI of the cat. Moreover, saccade related SGI cells in the primate project to homologue of the cuneiform nucleus, the central mesencephalic reticular formation (cMRF; Cohen and Buttner-Ennever[36]) and stimulation of the cMRF induces saccades. Interestingly, Ma and colleagues[37] compared the locations of saccade related neurons to AChE patches in the rhesus monkey and found that such cells lay both inside and outside these zones.

4.8 CORTICOTECTAL PROJECTIONS: PATCHY AND ABUNDANT

Twenty-four different cortical areas can be seen to distribute to SGI. While few areas of the brain exhibit the connectional diversity of SGI afferents, their distinct

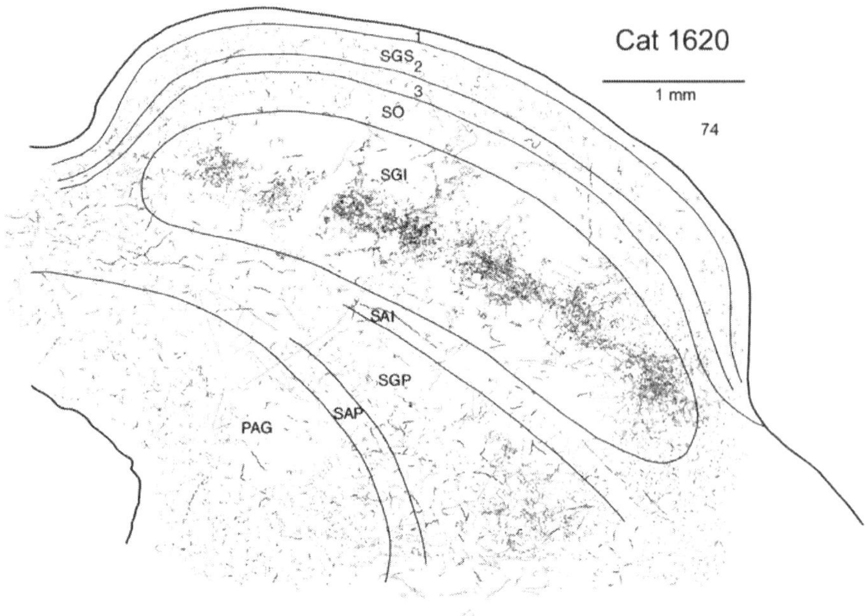

FIGURE 4.10 The light grey dots and dashes indicate the distribution of PHA-L labeled fibers from a relatively small injection into the contralateral PTN. Note that this projection is limited for the most part to the middle tier of the SGI. The blacker dots and dashes represent transported protein from the ipsilateral SNpr. Also, note that only the middle tier or nigrotectal axons have been labeled in this experiment. Abbreviation as in Figure 4.7. (From Harting, J.K. and Van Lieshout, D.P., Spatial relationships of axons arising from the substantia nigra, spinal trigeminal nucleus and pedunculopontine tegmental nucleus within the intermediate gray of the cat superior colliculus, *J. Comp. Neurol.*, 305, 543, 1991. With permission.)

> distributions and modes of terminations reveal an orderliness which most likely has some functional significance.[38]

Wide areas of the cerebral cortex were known to project to the SC[39] and anterograde transport methods were used to explore the potential periodic projections of the majority of these pathways.[38] An attempt was made to relate the distribution of each projection to specific tiers of the SGI and to the evolving concept of sensory and motor domains. One interesting finding was that several extrastriate visual cortical areas (i.e., 19, 20a, 20b, 21a, 21b, posterior suprasylvian [PS], posterolateral lateral suprasylvian area [PLLS] and PMLS) project densely and uniformly to the superficial layers, especially to the SGS. Most important, these same areas also target the dorsal tier of the SGI in a patchy fashion. Since PMLS was shown to project to AChE-poor zones of this dorsal tier,[19] it might be that the other cortical areas that project in a dual fashion to the SGS and dorsal SGI also have this cholinesterase affiliation. Such findings would suggest a functional association between the SGS and the AChE-poor modules of the dorsal SGI. Weyand and

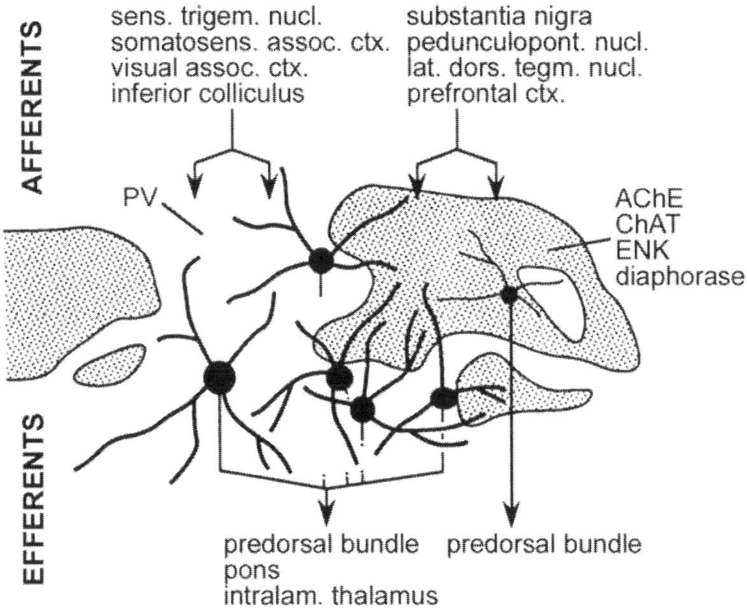

FIGURE 4.11 Summary of spatial relationship between the somas and dendrites of tecto-bulbospinal neurons in the rat. Note that the soma lies in the AChE-poor zone while the dendrites lie in the AChE-rich zone. (From Illing, R.- B., Association of efferent neurons to the compartmental architecture of the superior colliculus, *Proc. Natl. Acad. Sci., USA*, 89, 10900, 1992. With permission.)

Gafka[1] have recently shown that corticotectal cells in both area 17 and PMLS play a role in "enhancing the features of the visual array…." Thus, the dorsal tier of AChE-poor modules and the overlying SGS appear to receive similar functional information from the two pathways (17 and PMLS) and might be sites where signals from several higher cortical visual areas commingle in order to provide a pool of information regarding possible targets of SC modulated eye movements (Figure 4.12).

In contrast to the extrastriate cortical areas that target the SGS and dorsal SGI, other areas such as the dorsal lateral suprasylvian area (DLS) and the anterolateral lateral suprasylvian area (ALLS) were found to distribute to both the dorsal and ventral tiers of SGI, but not to the SGS. It might be that these projections are more concerned with conveying premotor or attentional signals to the SGI. Saccade-related signals have not been observed on corticotectal cells in the PMLS[1] but have been noted on cells in the lateral suprasylvian area.[40]

The comprehensive analysis of corticotectal projections also revealed that the lateral frontotectal projection targets the middle tier of the SGI. This stands in contrast with the earlier findings of Illing and Graybiel,[18] who described this pathway as distributing to AChE-rich patches of the dorsal tier. This finding suggests that the AChE-rich patches of the middle tier are comprised of overlapping nigrotectal, PTN, and medial and lateral frontotectal axons/terminals.

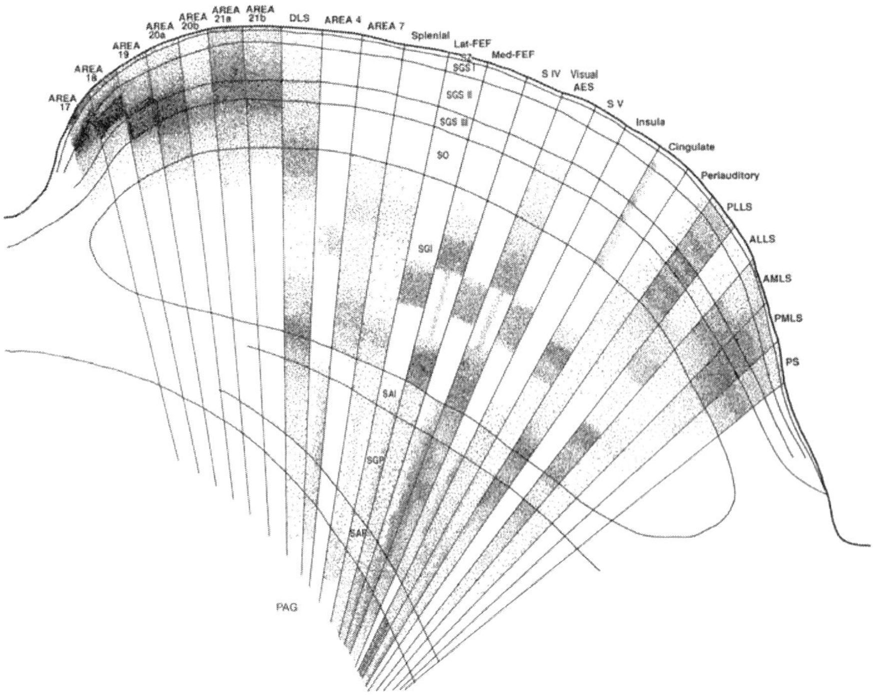

FIGURE 4.12 Summary of the laminar distribution of corticotectal projections in the cat. AES, anterior ectosylvian sulcus area; ALLS, anterolateral lateral suprasylvian area; AMLS, anteromedial lateral suprasylvian area; DLS, dorsolateral suprasylvian area; PAG, periaqueductal grey; PLLS, posterolateral lateral suprasylvian area; PMLS, posteromedial lateral suprasylvian area; PS, posterior suprasylvian area; S IV, fourth somatosensory area; S V, fifth somatosensory area; SZ, stratum zonale. Other abbreviations as in Figure 4.7. (From Harting, J.K., Huerta, M.F., Updyke, B.V., and Van Lieshout, D.P., Corticotectal projections in the cat: anterograde transport studies of twenty-five cortical areas, *J. Comp. Neurol.,* 324, 379, 1992. With permission.)

4.9 CORTICAL AND BRAIN STEM INPUTS TO THE SAME PATCH: DIFFERENT FUNCTIONS?

> The morphology and postsynaptic relationships of ... two spatially and functionally related inputs to the SGI differ[41]

In 1997, we analyzed some of the circuitry in the AChE free zones/patches of the middle tier.[41] A double-label anterograde transport method revealed that trigeminotectal and SIV corticotectal axons/terminals commingle within the same patches in the middle tier of the SGI. As mentioned earlier, it had been suggested that these inputs distribute to the AChE-poor zones.[19] More important, the ultrastructural studies revealed that the trigeminotectal terminals in the patches are relatively small and

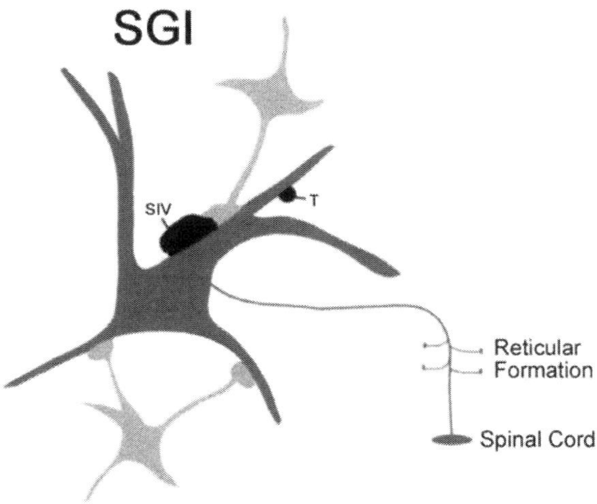

FIGURE 4.13 Shows the spatial relationship between trigeminotectal and S-IV corticotectal terminals on SGI neurons. These projections terminate in AChE-poor zones. (Modified from Harting, J.K., Feig, S., and Van Lieshout, D.P., Cortical somatosensory and trigeminal inputs to the cat superior colliculus: light and electron microscopic analyses, *J. Comp. Neurol.*, 388, 313, 1997. With permission.)

make asymmetrical synapses on distal dendrites of SGI neurons. In contrast, SIV cortical terminals in the same patches are relatively large and make asymmetrical synaptic contacts on proximal dendrites of SGI neurons. These ultrastructural data suggest that the two inputs could play different roles in the overall function of the cell, with one being involved in determining the receptive field of the SGI cell and the other playing a modulatory role (Figure 4.13).[41]

4.10 AUDITORY AND OTHER PATCHY INPUTS

> The present findings strengthen the concept that the cat SGI is functionally dependent upon connections that exhibit a modular organization[42]

The distribution of the projections from the rostral pole of the inferior colliculus (rpIC) to the SGI was compared with those from the spinal trigeminal nucleus. This rpIC-tectal input reaches an extensive part of the SGI (see also Meredith and Clemo[43] for auditory cortical input), but is densest and most patchy within the ventral tier. Interestingly, *streamers* separating the trigeminotectal axons/terminals can be seen in the middle tier, suggesting (if there are terminals in the streamers) that the AChE-rich zones in this tier might be targeted by the auditory input (Figure 4.14).

Several patchy inputs to the SGI of the cat have as yet not been discussed. In particular, patchy projections from the perihypoglossal nucleus[44] distribute primarily across the dorsal tier (Figure 4.15, left side) and (zona) incertotectal fibers (Figure 4.15, right side) have been noted in the ventral tier.[45]

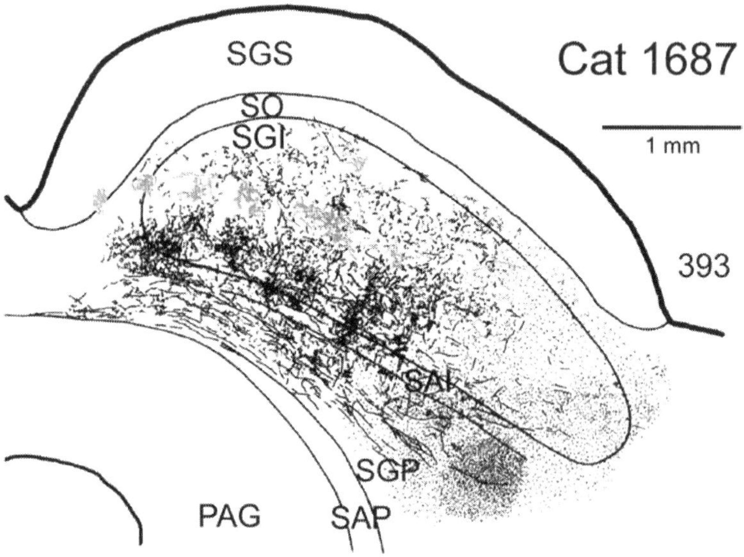

FIGURE 4.14 Black dots and dashes show the distribution of transported protein in the SC following an injection of tritiated amino acids into the ipsilateral rostral pole of the inferior colliculus. The light dots and dashes (across the middle tier) represent biocytin-labeled axons/terminals resulting from an injection into the contralateral spinal trigeminal nucleus. Abbreviations as in Figure 4.7. (From Harting, J.K. and Van Lieshout, D.P., Projections from the rostral pole of the inferior colliculus to the cat superior colliculus, *Brain Res.*, 881, 244, 2000. With permission.)

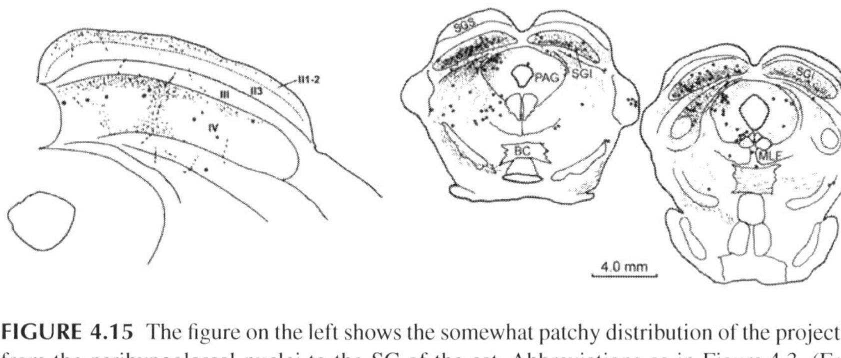

FIGURE 4.15 The figure on the left shows the somewhat patchy distribution of the projection from the perihypoglossal nuclei to the SC of the cat. Abbreviations as in Figure 4.3. (From Higo, S. et al., Differential projections to the superior collicular layers from the perihypoglossal nuclei in the cat, *Brain Res.*, 599, 19, 1992. With permission.) The two drawings on the right show the distribution of fibers (small dots) in the SC arising from cells in the ipsilateral zona incerta. SC-incertal cells are also illustrated by large dots. BC, brachium conjunctivum; MLF, medial longitudinal fasciculus. All other abbreviations as in Figure 4.7. (From May, P.J., Sun, W., and Hall, W.C., Reciprocal connections between the zona incerta and the pretectrum and superior colliculus of the cat, *Neuroscience*, 77, 1091, 1997. With permission.)

4.11 A NEW LOOK AT PATCHES: CYLINDERS, FRAMES, AND WALLS

> Since physiologically the intermediate layers appear to be a high-resolution matrix for the generation of directed orienting movements, there is a possibility that the AChE lattice reflects this organization. In keeping with that hypothesis, one may put forward the proposal that 'AChE compartments' function as orienting modules.[26]

Data from recent studies in the rat by Chevalier and Mana[1,26,46] have reintroduced Graybiel's[6] proposal that there are functional units in the intermediate and deep layers that consist of functionally related *stacked* afferents and tectofugal neurons. These new data clearly elevate the field by showing that the AChE lattice is comprised of approximately 100 vertically oriented compartments or cylinders, each consisting of a dense concentration of AChE called the *frame* and a more diffuse region of AChE immediately adjacent (inside) the frame called the *wall.* [47] The hollow center of each cylinder is AChE free. These cylindrical modules encompass not only the SGI but also the stratum album intermediale (SAI). (While both of these layers most likely correspond to the SGI of the cat, further studies are needed to explore this question). The cylinders are especially apparent in tangentially sectioned tissue and the AChE patches seen in frontal sections are actually bits and pieces of the vertically oriented frames/walls. Most important, this cylindrical organization is not unique to the rat but is also present in the mouse, hamster, cat, and monkey (Figure 4.16).[47]

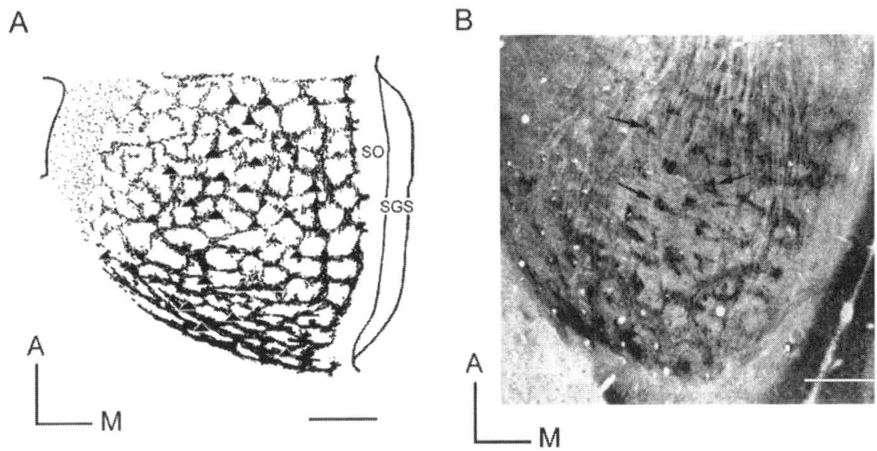

FIGURE 4.16 AChE staining in the intermediate and deep layers of the rat SC sectioned tangentially. Note the honeycomb pattern. A, anterior; M, medial. (From Chevalier, G. and Mana, S., Honeycomb-like structure of the intermediate layers of the rat superior colliculus, with additional observations in several mammals: AChE patterning, *J. Comp. Neurol.*, 419, 137, 2000. With permission.)

This cylindrical/modular AChE lattice has some functional logic, since various combinations of modality related afferents congregate at different depths in each cylinder, where they are associated with the frames and walls of AChE.[26,46] Since tectofugal neurons are also positioned at different depths in these cylinders, Mana and Chevalier[26,46] posit that the cylindrical units function as *orienting modules*. Accordingly, each cylinder "would be an orienting unit, which receives its triggering information from a given location in sensory space and would contain the premotor neurons for the generation of an orienting movement to that location."[46]

In such a model, sensory afferents reaching each cylindrical compartment convey information regarding a single receptive field.[46] This hypothesis is partially based upon anterograde transport data experiments in which a tracer injection into several somatosensory cortical *barrels* (each with their own receptive field) results in the same number of labeled SC modules/cylinders. Interestingly, single trigeminotectal fibers carry signals regarding several whiskers and target a complete row of mediolateral patches at the SGI/SAI interface (see Reference 48). These findings are difficult to reconcile with the idea that individual cylinders receive input limited to single receptive fields (Figure 4.17).

The recent data in the rat give considerable credence to the concept of a *global* nigrotectal system introduced earlier in the cat.[25] They have revealed three nigrotectal channels, somatosensory, visual, and auditory, that distribute predominantly to non-overlapping depths in the cylinders.[46] These three nigrotectal channels are inhibited by functionally corresponding regions of the striatum, and overlap, for the most part, with cortical and brain stem inputs carrying the same modality. This depth correspondence between nigral, cortical and brain stem sensory inputs is greatest for somatosensory-related afferents (nigral, cortical and trigeminal). Tecto-bulbospinal neurons (which project into the contralateral predorsal bundle) lie at the same cylindrical depth as the somatosensory-related pathways, and this spatial association is reflected in the fact that the somatic nigrotectal channel "releases reactivity of the descending crossed cells to somatosensory inputs and… promotes tactilely guided head and body movements."[46]

The depth correspondence between the visual nigrotectal channel and other visual afferents does not appear to be as corresponding as the somatosensory inputs. Documented overlap occurs only at the depth of the SGI/SAI interface, but the projections from only one extrastriate cortical area (Oc2) have been determined. It might be that projections from other visual cortical areas overlap with the visual nigrotectal fibers at other depths. In the case of auditory pathways, little is known regarding the distribution of cortical and brain stem inputs, but the related nigrotectal projection is quite restricted in depth. The findings of Wallace and Fredens[27] in the mouse suggest that projections from the brachium of the inferior colliculus lie ventral to the zone of auditory-related nigral input.

Mana and Chevalier[26] have also shown that tectopontine cells lie dorsal to TBS cells and in particular lie throughout the SGI and the upper/dorsal SAI. Such cells in the hamster have auditory and visual receptive fields[49] and Mana and Chevalier[26] have stressed that they are positioned to be modulated by auditory and visual nigrotectal channels.

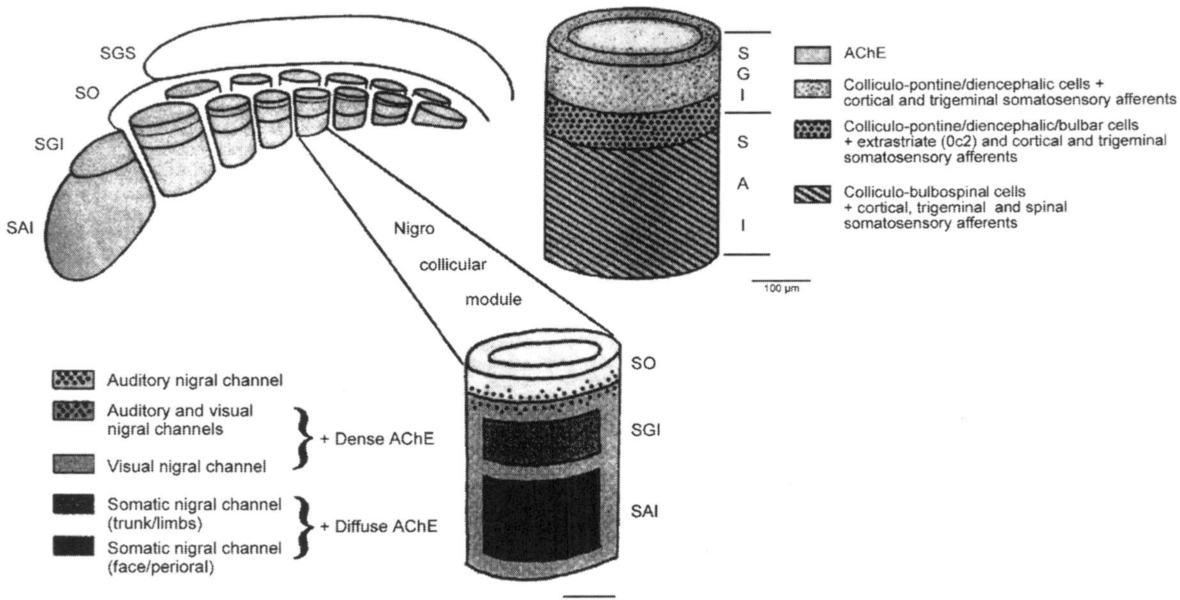

FIGURE 4.17 Drawings on left show cylinders/modules in the rat SC and the distribution of different nigrotectal channels. Drawing on the right illustrates the distribution of inputs from the spinal cord, spinal trigeminal nucleus, somatosensory cortex and the second visual cortical area (Oc2) and how these inputs are spatially related to each other and colliculo-pontine/diencephalic, colliculo-pontine, and colliculo-bulbospinal cells. Abbreviations as in Figure 4.7. (Modified from Mana S., and Chevalier, G., Honeycomb-like structure of the intermediate layers of the rat superior colliculus: afferent and efferent connections, *Neuroscience*, 103, 673, 2001 and Mana S. and Chevalier, G., The fine organization of nigro-collicular channels with additional observations of their relationships with acetylcholinesterase in the rat, *Neuroscience*, 106, 357, 2001. With permission.)

4.12 CONCLUSIONS AND WHERE DO WE GO FROM HERE?

The recent studies in the rat have elevated our understanding of puffs and patches in the SGI/SAI. The vertical/cylindrical architecture of the AChE lattice revealed in tangentially sectioned tissue, and the association of functionally related systems to this vertical architecture, suggest that these cylinders are functional units. However, these individual units must somehow work together to carry out SC functions, as the somas of cells are associated with more than one cylinder and their dendrites cross many cylinders.

It would be interesting to examine the specific cholinergic inputs to the frames and walls of these cylinders, as the PTN and LDTN projections might exhibit specific affiliations. Moreover, studies using complementary markers to AChE (parvalbumin?) might extend and/or modify the cholinesterase findings. Data from already published studies need to somehow be interpreted in light of a cylindrical/modular organization. However, it would be difficult to take these data (most from studies in the cat) and determine which afferents lie in which cylinders. To redo the experiments in the cat seems unwise. Perhaps the best thing to do is to carry out double labeling studies of SC afferents and efferents in the rat using tangentially sectioned colliculi. Only a few afferents have been studied by Mana and Chevalier[26,46] and it might even be possible to actually determine the components of a single cylinder. (This of course depends upon being able to regularly identify a certain cylinder in a specific row).

As discussed earlier, over 20 different cortical areas in the cat distribute in a patchy manner to the SGI and only one area (Oc2) has been examined in the rat.[26] There are three different ladder-like tiers of *visual* nigrotectal input (only the dorsal one was targeted by Oc2) and it would be of interest to determine if visual cortical areas other than Oc2 overlap these nigral tiers.

Then there is the question regarding how to interpret Illing's[34] findings showing that the somas of tectofugal neurons lie in the AChE-poor regions while their dendrites penetrate the AChE- rich zones. Mana and Chevalier[26] failed to see many efferent cells in the hollow AChE-free centers of the cylinders and Harting and colleagues[41] suggest that both somas and dendrites are targeted by trigeminal and cortical inputs in AChE-poor zones. Studies focused upon the role of cortical and brain stem inputs of the same modality to the same cylinder will prove interesting. As mentioned earlier, cortical inputs to each cylinder appear to convey signals regarding a more limited part of the sensory sheet (a single whisker) than that conveyed by the brain stem trigeminal fibers (several whiskers). Finally, the biggest challenge for future studies is to find more physiological and some behavior correlates of the cylindrical organization. While there have been hints of a clustering of functionally related cells,[28,35,50] others have found no such relationship to the AChE domains.[37] There are many interesting questions remaining and lots of fun experiments to carry out. Puffs and patches are here to stay!

ACKNOWLEDGMENTS

This work was supported by grants EY 01277 and EY12869. I thank David Van Lieshout for his many contributions to the studies discussed in this review.

REFERENCES

1. Weyand, T. and Gafka, A.C., Visuomotor properties of corticotectal cells in area 17 and posteromedial lateral suprasylvian (PMLS) cortex of the cat, *Vis. Neurosci.*, 18, 77, 2001.
2. Wise, S.P. and Jones, E.G., Cells of origin and terminal distribution of descending projections of the rat somatic sensory cortex, *J. Comp. Neurol.*, 175, 129, 1977.
3. Graybiel A.M., Anatomical organization of retinotectal afferents in the cat; an autoradiographic study, *Brain Res.,* 96, 1, 1975.
4. Graybiel A.M., Evidence for banding in the cat's retinotectal connection, *Brain Res.,* 114, 318, 1976.
5. Harting, J.K. and Guillery, R.W., Organization of retinocollicular pathways in the cat, *J. Comp. Neurol.,* 166, 133, 1976.
6. Graybiel, A.M., Organization of the nigrotectal connection: an experimental tracer study in the cat, *Brain Res.,* 143, 339, 1978.
7. Kunzle, H., Akert, K., and Wurtz, R.H., Projection of area 8 (frontal eye field) to superior colliculus in the monkey. An autoradiographic study, *Brain Res.*, 117, 487, 1976.
8. Huerta, M.F. and Harting, J.K., Connectional organization of the superior colliculus, *Trends Neurosci.,* 7, 286, 1984.
9. Huerta, M.F., Frankfurter, A., and Harting, J.K., Studies of the principal sensory and spinal trigeminal nuclei of the rat: projections to the superior colliculus, inferior olive and cerebellum. *J. Comp. Neurol.*, 220, 147, 1983.
10. Huerta, M.F. and Harting, J.K., The projection from the nucleus of the posterior commissure to the superior colliculus of the cat: patch-like endings within the intermediate and deep grey layers, *Brain Res.*, 238, 426, 1982.
11. Kawamura, S. et al., The cerebellar projections to the superior colliculus and pretectum in the cat — an autoradiographic and horseradish-peroxidase study, *Neuroscience*, 7, 1673, 1982.
12. Huerta, M.F. and. Harting, J.K., The mammalian superior colliculus: studies of its morphology and connectivities, in *Comparative Neurology of the Optic Tectum*, Vanegas, H., Ed., Plenum, New York, 1984, pp. 687.
13. Victorov, I.V., Neuronal structure of anterior corpora quadrigemina in the cat, *Arkh. Anat. Gistol. Embriol.*, 55, 51, 1968.
14. Stein, B.E., Mafgalhaes-Castro, B., and Kruger, L., Relationship between visual and tactile representations in cat superior colliculus, *J. Neurophysiol.*, 39, 401, 1976.
15. Hashikawa, T., and Kawamura, K., Identification of cells of origin of tectopontine fibers in the cat superior colliculus: an experimental study with the horseradish peroxidase method, *Brain Res.,* 130, 65, 1977.
16. Kawamura, K. and Hashikawa, T., Cell bodies of origin of reticular projections from the superior colliculus of the cat: an experimental study with the use of horseradish peroxidase as a tracer, *J. Comp. Neurol.*, 181, 1, 1978.
17. Huerta, M.F., Frankfurter, A., and Harting, J.K., The trigeminocollicular projection in the cat: patch-like endings within the intermediate grey, *Brain Res.*, 211, 1, 1981.
18. Illing, R.-B. and Graybiel, A.M., Convergence of afferents from frontal cortex and substantia nigra onto acetylcholinesterase-rich patches of the cat's superior colliculus, *Neuroscience*, 14, 455, 1985.
19. Illing, R.-B. and Graybiel, A.M., Complementary and non-matching afferent compartments in the cat's superior colliculus: innervation of acetylcholinesterase-poor domain of the intermediate gray layer, *Neuroscience,* 18, 373, 1986.

20. Guitton, D. and Mandl, G., Frontal oculomotor area in alert cat. I. Eye movements and neck activity evoked by stimulation, *Brain Res.,* 149, 295, 1978.
21. Guitton, D. and Mandl, G., Frontal oculomotor area in alert cat. II. Unit discharges associated with eye movements and neck muscle activity, *Brain Res.,* 149, 313, 1978.
22. Hikosaka, O. and Wurtz, R.H., Visual and oculomotor functions of monkey substantia nigra pars reticulata. I. Relation of visual and auditory responses to saccades, *J. Neurophysiol.*, 49, 1230, 1983.
23. Joseph, J.P. and Boussaoud, D., Role of the cat substantia nigra pars reticulata in eye and head movements. I. Neural activity, *Exp. Brain Res.*, 57, 286, 1985.
24. Clemo, H.R. and Stein, B.E., Organization of a fourth somatosensory area of cortex in the cat, *J. Neurophysiol.,* 50, 10, 1983.
25. Harting, J.K. et al., Neuroanatomical studies of the nigrotectal projection in the cat, *J. Comp. Neurol.,* 278, 615, 1988.
26. Mana S., and Chevalier, G., Honeycomb-like structure of the intermediate layers of the rat superior colliculus: Afferent and efferent connections, *Neuroscience*, 103, 673, 2001.
27. Wallace, M.N. and Fredens, K., Relationship of afferent inputs to the lattice of high NADPH-diaphorase activity in the mouse superior colliculus, *Exp. Brain Res.*, 78, 435, 1989.
28. Dräger, U.C. and Hubel, D.H., Response to visual stimulation and relationship between visual, auditory and somatosensory inputs in mouse superior colliculus, *J. Neurophysiol.,* 38, 690, 1975.
29. Hall, W.C. et al., Cholinergic innervation of the superior colliculus in the cat, *J. Comp. Neurol.,* 287, 495, 1989.
30. Jeon, C.-J. and Mize, R.R., Choline acetyltransferase immunoreactive patches overlap specific efferent cell groups in the cat superior colliculus, *J. Comp. Neurol.*, 337, 127, 1993.
31. Harting, J.K. and Van Lieshout, D.P., Spatial relationships of axons arising from the substantia nigra, spinal trigeminal nucleus and pedunculopontine tegmental nucleus within the intermediate gray of the cat superior colliculus, *J. Comp. Neurol.,* 305, 543, 1991.
32. Morizumi, T. et al., Topographic projections from the basal ganglia to the nucleus tegmenti pedunculopontinus compacta of the cat with special reference to the pallidal projections, *Exp. Brain Res.*, 71, 298, 1988.
33. Kobayashi, Y. et al., Contribution of pedunculopontine tegmental nucleus neurons to performance of visually guided saccade tasks in monkeys, *J. Neurophysiol.*, 88, 715, 2002.
34. Illing, R.- B., Association of efferent neurons to the compartmental architecture of the superior colliculus, *Proc. Natl. Acad. Sci. USA*, 89, 10900, 1992.
35. Peck, C.K., Saccade related neurons in the cat superior colliculus: pandirectional movement cells with postsaccadic responses, *J. Neurophysiol.*, 52, 1154, 1984.
36. Cohen, B. and Buttner-Ennever, J.A., Projections from the superior colliculus to a region of the central mesencephalic reticular formation (cMRF) associated with horizontal saccadic eye movements, *Exp. Brain Res.,* 57, 167, 1984.
37. Ma, T.P., Graybiel, A.M., and Wurtz, R.H., Location of saccade-related neurons in the macaque superior colliculus, *Exp. Brain. Res.*, 85, 21, 1991.
38. Harting, J.K., Huerta, M.F., Updyke, B.V., and Van Lieshout, D.P., Corticotectal projections in the cat: anterograde transport studies of twenty-five cortical areas, *J. Comp. Neurol.*, 324, 379, 1992.

39. Stein, B.E., Spencer, R.F., and Edwards, S.B., Corticotectal and corticothalamic efferent projections of S-IV somatosensory cortex in the cat, *J. Neurophysiol.*, 50, 896, 1983.
40. Yin, T.C.T. and Greenwood, M. Visuomotor interactions in the middle and lateral suprasylvian cortices of the behaving cat, *Exp. Brain Res.*, 88, 15, 1992.
41. Harting, J.K., Feig, S., and Van Lieshout, D.P., Cortical somatosensory and trigeminal inputs to the cat superior colliculus: light and electron microscopic analyses, *J. Comp. Neurol.,* 388, 313, 1997.
42. Harting, J.K. and Van Lieshout, D.P., Projections from the rostral pole of the inferior colliculus to the cat superior colliculus, *Brain Res.,* 881, 244, 2000.
43. Meredith, M.A. and Clemo, H.R., Auditory cortical projections from the anterior ectosylvian sulcus (Field AES) to the superior colliculus in the cat: an anatomical and electrophysiological study, *J. Comp. Neurol.,* 289, 687, 1989.
44. Higo, S. et al., Differential projections to the superior collicular layers from the perihypoglossal nuclei in the cat, *Brain Res.*, 599, 19, 1992.
45. May, P.J., Sun, W., and Hall, W.C., Reciprocal connections between the zona incerta and the pretectrum and superior colliculus of the cat, *Neuroscience*, 77,1091, 1997.
46. Mana S. and Chevalier, G., The fine organization of nigro-collicular channels with additional observations of their relationships with acetylcholinesterase in the rat, *Neuroscience*, 106, 357, 2001.
47. Chevalier, G. and Mana, S., Honeycomb-like structure of the intermediate layers of the rat superior colliculus, with additional observations in several mammals: AChE patterning, *J. Comp. Neurol.*, 419, 137, 2000.
48. Rhoades, R.W. et al., Organization of the projections from the trigeminal brainstem complex to the superior colliculus in the rat and hamster: anterograde tracing with *Phaseolus vulgaris* Leucoagglutinin and intra-axonal injection, *J. Comp. Neurol.*, 289, 641, 1989.
49. Westby, G.W.M. et al., Output pathways from the rat superior colliculus mediating approach and avoidance have different sensory properties, *Exp. Brain Res.*, 81, 626, 1990.
50. Kassell, J., Superior colliculus projections to tactile areas of rat cerebellar hemispheres, *Brain Res.*, 202, 291, 1980.

5 Structure–Function Relationships in the Superior Colliculus of Higher Mammals

Alexej A. Grantyn and Adonis K. Moschovakis

CONTENTS

5.1 INTRODUCTION

One of the central goals of the neurosciences is to describe perception, action, and thought in terms of brain processes. Near the middle of the 19th century it was felt that such a reductionist research program could not hope to succeed unless the behavior of any one animal were completely described in terms of its neurons (this is a demand that bears the name of Thomas Huxley). So far, the most successful affirmative answer to Huxley's demand rests on evidence collected in invertebrates, in particular the marine snail *Aplysia californica.*[1] Experimental work in this species offers several advantages (eloquently summarized in the early chapters of Reference 1) including a fairly simple central nervous system made of a relatively small number of neurons. Mammalian neurons are more numerous (by seven orders of magnitude or more) and display a large number of interconnections and discharge patterns. It is therefore not surprising that progress in satisfying Huxley's demand has been and remains much slower in mammals.

0-8493-0097-5/04/$0.00+$1.50

One of the best understood complex circuits of the mammalian brain subserves saccades. Saccades are extremely fast eye movements, which reorient the line of sight toward interesting features of the world. It is now known that neurons responsible for saccades are distributed within more than ten brain regions, each one of which is heterogeneous in terms of activity, connections, and input-output characteristics (reviewed in Reference 2). Our goal is to provide a summary of the relevant evidence as it pertains to one of these brain areas, namely the superior colliculus (SC). Three issues are considered: (1) What is the data to support its causal role in orienting movements (including saccades)? (2) How are the discharge patterns and connections of the neurons it contains related to orienting movements? (3) How does it generate the signals that it sends to its targets? The last section of this report is devoted to models that organize psychophysical, neurological, anatomical, and physiological findings in a quantitative account of its involvement in the control of saccades.

5.2 THE CAUSAL RELEVANCE OF SC SIGNALS

Lesion and stimulation methods have been long used to test whether necessity and sufficiency criteria are satisfied before a certain brain area is pronounced responsible for a certain behavioral function. The first of these criteria is certainly fulfilled in subprimate species. In cats, ablation[3] or cooling[4] of the SC abolish orienting toward targets in the contralateral visual field. The syndrome consists in a visual deficit in attention-perception and a motor deficit of eye, head and body movements. The latter can be reproduced by selective lesions of the crossed tecto-bulbo-spinal tract, without inducing a perceptual neglect. Selective impairment of visuospatial discrimination and of orienting towards discrete targets has also been amply demonstrated in rodents after SC ablations or transections of its crossed efferent pathway to the brain stem gaze centers.[5,6] In monkeys, electrolytic and surgical lesions of the SC, often produced only subtle and transient behavioral deficits.[7] This mild deficit is presumably due to the redundancy of components in the primate brain; the dysfunctional SC apparently can be replaced by extracollicular structures. More recently, chemical inactivation of the SC produced robust, albeit partial, deficits of rapid eye movements.[8,9] Visually triggered saccades had longer latencies, slower velocities, and reduced accuracy. The primate SC appears therefore necessary for the optimal performance of the saccadic system during rapid and accurate shifts of the line of sight toward salient stimuli. In particular, it is necessary for the execution of saccades with extremely short latencies (*express* saccades) which are irreversibly eliminated following SC lesion or ablation.[10] However, a reasonable behavioral performance is ensured by other oculomotor structures.

The sufficiency criterion is fulfilled in all species investigated, as repeatedly demonstrated in stimulation experiments conducted since the pioneering work of Adamük (see Reference 11) and Ferrier.[12] Electrical stimulation of the SC has been shown to generate saccades[13] and sequences of fast and slow eye movements in head-fixed animals,[14] head movements in head-free animals,[15–18] and complex sequences of orienting eye, head, and body movements.[19] Figure 5.1 illustrates several examples of such orienting movements evoked in response to the electrical stimulation of the SC in cats and monkeys. Besides demonstrating that SC activation is sufficient for the generation of saccades and other orienting movements, electrical stimulation of the SC generates a number of phenomena with important implications for models of the saccadic system. First, the amplitude and direction of elicited saccades is influenced by the location of the stimulating electrode in both the cat[20] and the monkey.[13] Stimulation of the rostral SC causes small saccades and stimulation of the caudal SC large saccades while stimulation of medial sites leads to upward saccades and stimulation of lateral sites to downward saccades. These data support the notion that saccade vectors are represented in an orderly manner over the mediolateral and anteroposterior extent of the SC to form a map of oculomotor space. Second, long trains of pulses often evoke more than one saccade (a staircase of saccades) in both the cat[14] and the monkey.[13] Third, the amplitude and direction of saccades evoked in response to simultaneous stimulation of two SC sites is equal to the average and not to the vectorial sum of the saccades produced when each of the sites is stimulated separately.[13] Finally, the size of saccades evoked in response to stimulation of the feline SC[21–25] and of the caudal primate SC[15,16,26] depends on the position of the eyes at the time of saccade onset. An example of the interplay between the initial horizontal position of the eyes (H_1) and the size of the horizontal components (ΔH) of saccades evoked in response to identical stimuli is illustrated in Figure 5.2A, while Figure 5.2B is a plot of this and other such data obtained from a single site in the intermediate layers of the left SC of a cat. The linear regression line through the data (solid) obeys the expression,

$$\Delta H = S_H - k_H H_1 \tag{5.1}$$

The intercept S_H is equal to the horizontal component of the saccades (–7.2° in Figure 5.2B) that would have been produced had the movement started from the primary position. Following McIlwain,[27] we use the term horizontal component of the *characteristic* vector of the evoked saccades to refer to this constant. The coefficient k_H is a measure of the influence of the initial position of the eyes upon the size of the movement. In principle, it ranges between –1 (a value indicative of the extreme position sensitivity of movements, which should bring the eyes to the same *goal* from any point of departure) and 0 (a value indicative of movements, which are not at all sensitive to the initial position of the eyes and which thus always have the same *fixed vector* size). In the example of Figure 5.2B, k_H is equal to –0.52, i.e., to a value intermediate between these two extremes.

Data such as this has been traditionally thought to support *position control* theories. These assume that the brain controls the final position, the goal, to be reached by the movement. Once this is specified, the mechanical properties of the moving body segment determine its trajectory without any need for the brain to elaborate dynamical control signals associated with it. As formulated for the saccadic system, the *goal hypothesis* assumes that it is the final, postsaccadic position of the eyes in the orbit that is place coded in the SC,[25] instead of eye displacements of particular amplitude and direction (a *fixed-vector*). To explain how the retinal error signal (r_e) reaching the SC is transformed into postsaccadic eye positions corresponding to goals (E_2), McIlwain[27] assumed that the SC works as a saccadic controller whose output is equal to E_2, and whose front stage adds a signal proportional to initial eye position (E_1) onto r_e. McIlwain suggested that an electrical stimulus applied to the SC reproduces the r_e signal but also interferes with the E_1 signal (as if it were multiplied by a gain factor, $g < 1$). Accordingly, the output (E_2) of the electrically stimulated SC is equal to $r_e + gE_1$, and because the burst generator (BG) implements the subtraction $E_2 - E_1$, BG output (ΔE) is equal to $r_e + gE_1 - E_1$. After rearranging, one obtains $\Delta E = r_e - (1 - g) \cdot E_1$. This expression is the same as Equation 5.1 if one substitutes r_e by S_H and $(1 - g)$ by k_H.

Despite its ability to formally account for the position sensitivity of evoked saccades, McIlwain's[27] formulation of the goal hypothesis does not account for a number of experimental observations. First, it does not account for the fact that the position sensitivity of saccades is stronger when evoked from caudal SC sites and weaker or absent when evoked from rostral SC sites.[23] Because the rostral SC encodes small saccades while the caudal SC encodes big saccades the position sensitivity of evoked saccades should increase with the size of their characteristic vector. The existence of such a relationship (an example is illustrated in Figure 5.2C) has been repeatedly documented in the feline[14,24] and primate[28] SC and in the frontal and supplementary eye fields of the monkey.[29] Second, the saccadic controller employed by goal hypotheses is not consistent with experimental data (discussed in References 2 and 30 to 32). Saccadic controllers, which are supported by experimental data (such as the one described in the final section of this review), generate a fixed pulse of collicular activity, and thus fixed saccadic vectors, in response to SC stimulation. How could this be reconciled with data such as that illustrated in Figure 5.2A? To answer this question, consider first that, in addition to saccades, electrical stimulation of the feline SC gives rise to slow drifts[14,17,24,25] (Figure 5.2A, arrowheads), the metrics of which are sensitive to the initial position of the eyes. Due to their time course and the fact that their size depends on stimulation parameters

---

FIGURE 5.1 (See facing page.) Gaze shifts evoked by stimulation of the superior colliculus in cats and monkeys. (A) and (B) Photographs illustrating evoked *synergistically coordinated* orienting movements. (Aa) and (Ab) Symmetrical contraversive head movements evoked in response to stimulation of the left (a) or of the right (b) SC. Stimulation also evoked movements of the eyes in the same direction and at shorter latencies as well as reorientation of the vibrissae and sometimes of the ears. (Ac) Effect of stronger and/or more prolonged stimulation: maximal deviation of the head and extension of the contralateral foreleg initiating reorientation of the body axis. (Ba-c) Similar gradation of contraversive, upward head movements in response to stimulation of the medial half of the SC. (From Hess, W.R., Bürgi, S., and Bucher, V., *Psychiat. Neurol.*, 112, 1, 1946. With permission.) (C) Records of eye, head and gaze movements evoked in response to stimulation of rostral (a, b) and caudal (c, d) sites of the cat SC. (From Paré, M., Crommelinck, M. and Guitton, D., *Exp. Brain Res.*, 101, 123, 1994. With permission.) Coordinated eye and head movements are evoked from both rostral and caudal sites but in the former case gaze displacement is accomplished mainly by a saccadic eye movement and in the latter case by a head movement (E, G, H, vertical (v) and horizontal (h) eye, gaze and head positions, respectively). (D) Coordinated eye-head gaze shifts evoked in response to stimulation of rostral (a) and caudal (b) sites of the monkey SC (upward arrows indicate the onset and offset of the stimulation. (Modified from Figure 1 of Freedman, E.G., Stanford, T.R., and Sparks, D.L., *J. Neurophysiol.*, 76, 927, 1996. With permission.)

(duration and frequency), it has been argued that slow drifts represent the expected step response of the oculomotor plant, possibly due to the activation of oligosynaptic projections of the SC onto extraocular motoneurons either directly or through premotoneuronal structures such as the neural integrators.[14,24] Due to these projections, high frequency, long duration SC stimulation (such as the one usually employed in SC stimulation studies) would force motoneurons to discharge at a high rate for the duration of the stimulation. Since motoneuron firing rate and eye position are well correlated,[33] the eyes would tend to move to the same position in the orbit (E) whatever their point of departure (E_1). It is clear that the magnitude of eye displacement ($E - E_1$) must in this case depend on the initial position of the eyes (E_1).

The same mechanism could also account for the position sensitivity of evoked saccades, a postulate that bears the name *superposition hypothesis.*[24] Since slow

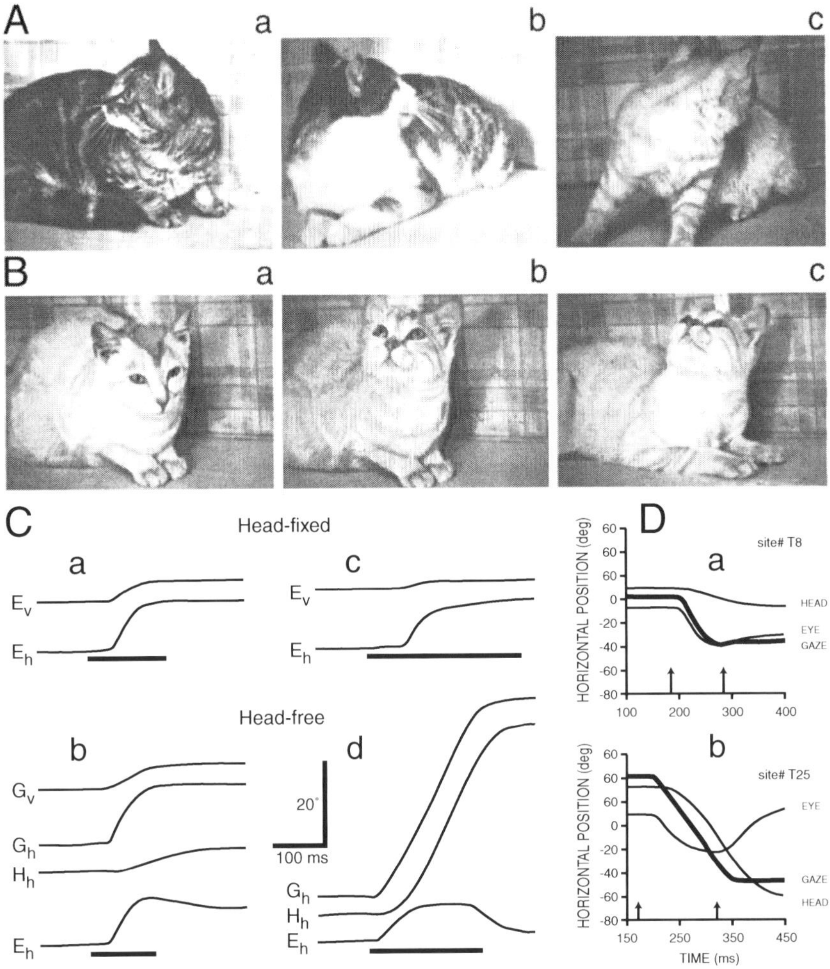

FIGURE 5.1

drifts follow most of the saccades that are electrically evoked from the SC,[14] the latencies of the two kinds of evoked movements might overlap considerably. Accordingly, part of the eye displacement during evoked saccades would be due to superimposed slow drifts as diagrammatically illustrated in Figure 5.2D (stippled). If the eye executed just the slow drift starting from position E_1, it would reach position E within a period of time roughly equal to three times the time constant of the oculomotor plant (τ). For saccades superimposed on slow drifts, the total displacement of the eyes ($E_2 - E_1$) equals the sum of the position invariant saccadic displacement (S_H) plus the slow drift displacement (G) accomplished within a period

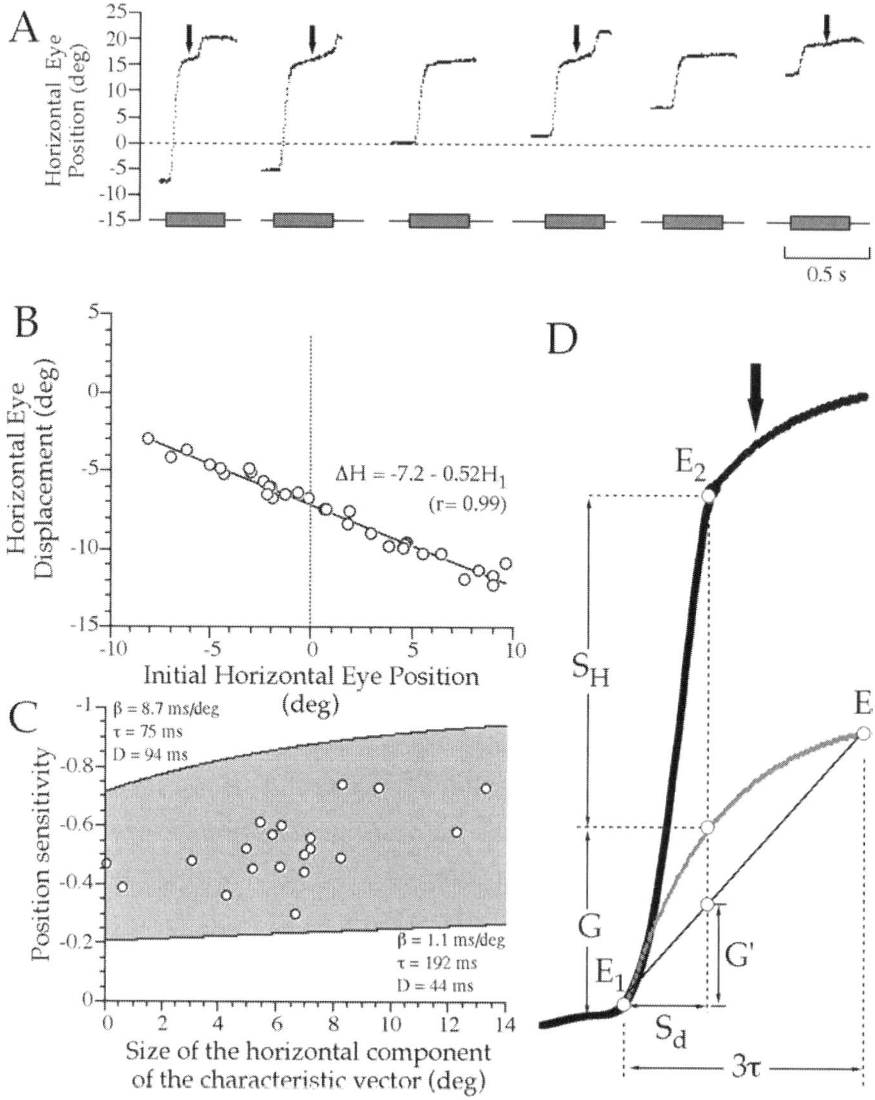

FIGURE 5.2

of time equal to saccade duration (S_d). To simplify the argument, G is replaced by G' from the linear approximation (Figure 5.2D, thin solid line) to the slow drift (a derivation of the exact relationship between k_H and S_H, $k_H = e^{-(D+\beta S_H)/\tau} - 1$, can be found in Reference 24). There follows that,

$$G'/S_d = (E - E_1)/3\tau \tag{5.2}$$

From the slope of the main sequence curve (β), S_d can be expressed as $\beta \cdot S_H$. Substituting and rearranging one obtains,

$$G' = (E - E_1) \cdot \beta \cdot S_H/3\tau = E \cdot \beta \cdot S_H/3\tau - E_1 \cdot \beta \cdot S_H/3\tau \tag{5.3}$$

As argued above, the end point of slow drifts is always the same (E) and therefore the first term on the right hand side of Equation 5.3 is a constant (as is S_H in Equation 5.1). The second term ($-\beta \cdot S_H/3\tau$) is a coefficient of proportionality of E_1 (as is k_H in Equation 5.1). The fact that S_H appears in its numerator explains why the position sensitivity of evoked saccades increases with the size of their characteristic vector. Its remaining terms lead one to expect that the position sensitivity of evoked saccades would increase together with the slope of the main sequence curve (β) and inversely with the time constant (τ) of the oculomotor plant. Indeed, as shown in Figure 5.2C, the values that these constants obtain in the cat are consistent with the experimentally determined values of k_H in this species (they cluster around 0.5). The much smaller

FIGURE 5.2 (See facing page.) The *superposition hypothesis* accounts for the position sensitivity of electrically evoked saccades. (Modified from Moschovakis, A.K., et al., *J. Neurophysiol.*, 80, 3373, 1998. With permission.) (A) Examples of eye movements evoked in response to the electrical stimulation of a single site in the left SC of the cat with identical stimulus parameters (current intensity: $2 \times T$; frequency: 200 Hz; number of pulses: 70; train duration: 350 ms). The dashed line indicates zero horizontal eye position. Boxes on the time axis indicates the onset and duration of the stimulus trains. The presence of slow drifts is indicated by arrows. (B) Plot of initial horizontal eye position (H_1, abscissa) versus horizontal eye displacement (ΔH, ordinate) of saccades evoked in response to the electrical stimulation of the SC. The solid line is the least squares regression line through the data and obeys the equation displayed. (C) Plot of the horizontal position sensitivity (k_H) of saccades evoked in response to stimulation of the feline SC (at $2 \times T$) against the horizontal component of their characteristic vectors (S_H). Each open circle is from a different stimulation site and often from a different animal. Solid lines obey the expression: $1 - e^{-\frac{D+\beta \cdot S_H}{\tau}}$ calculated from the extreme values of D, β and τ encountered in our experiments and/or in the literature (indicated on the plot). The dashed region corresponds to curves that would be obtained with intermediate values of D, β and τ. (D) Typical eye movement evoked in response to the electrical stimulation of the SC (thick solid line) and a simulated slow drift (stippled line) that starts from the same position and is superimposed on the rapid movement. Abbreviations: E_1, initial eye position; E_2, eye position at the end of the saccade; E, final eye position to be reached by slow drifts of long duration in the absence of a saccade; G, contribution of the slow drift to saccade size; G', linear approximation to slow drift displacement; S_d, saccade duration; τ, time constant of the drift.

value of β (by a factor of 5 or more[34]) and the larger value of τ (by a factor of 2[34,35]) in the monkey, accounts for the fact that the position sensitivity of saccades evoked in response to electrical stimulation of the SC is too small to notice in primates, unless areas responsible for large saccades are activated.

5.3 THE DISCHARGE PATTERN OF SC NEURONS

As expected of a structure that causes orienting movements, neurons of the intermediate and deep layers of the primate SC discharge, often intensely, during visuomotor tasks. Early descriptions of saccade related cells emphasized the emission of bursts preceding all saccades whether visually guided, spontaneous, quick phases of vestibular and optokinetic nystagmus,[36, 37] with the head free to move or fixed.[38] Until recently, studies of SC cell discharge focused mostly on saccadic eye movements executed by animals with their heads fixed. This trend was set in early recording[38] and stimulation[39] experiments in head-free monkeys, which led to the conclusion that the SC is not a source of motor commands guiding the head during orienting movements.

The best studied subclass of SC movement cells comprises saccade related burst neurons (in short SRBNs), which represent about 30% of all movement cells of the primate SC.[40] SRBNs are characterized by high frequency bursts (as high as 1200 spikes/s) emitted shortly (i.e., 20 ms on the average) before saccade onset and by low levels of spontaneous activity.[40,41] Their spatial tuning has been repeatedly studied.[36,42,43] SRBNs can be very selective about the direction and amplitude of the saccades their discharges precede (collectively defining the cell's movement field). The literature contains examples of cells discharging intensely prior to saccades of small size, less intensely for saccades of other sizes and not at all for much bigger saccades.[44] SRBN movement fields[43] are consistent with the metrics of saccades evoked in response to electrical stimulation of the SC region that contains them.[36] Cells that prefer small saccades are located rostrally and cells that prefer big saccades are located caudally while cells discharging before upward saccades are located medially and cells discharging before downward saccades are located laterally. The early preamble of SRBN discharge also encodes the desired horizontal and vertical amplitude of ensuing saccades.[45] No other parameter of discharge (e.g., number-of-spikes, burst duration) has been consistently related to the size of optimal saccades. Also, the relationship between cell discharge and the execution of preferred saccades is not as obligate as originally thought. For example, cell discharge does not reflect the actual displacement of the eyes during saccades that have been adapted[46] or that are executed toward remembered targets.[47] The same is true of saccades that participate in gaze (eye + head) shifts.[48] In all these cases, the discharge of SC movement cells is better related to the metrics of desired saccades (i.e., those that would have foveated the target had they been executed) rather than to the metrics of the saccades that are actually executed.

In addition to cells emitting compact bursts for saccades, the SC contains neurons whose activity increases about 80 to 100 ms before saccade onset, reaches a peak value preceding saccades by 10 to 20 ms, and then wanes. Such neurons were originally called type II SRBNs by Sparks and his colleagues[44] and together with type I SRBNs displaying strong early prelude responses, they were later renamed build-up neurons (in short BUNs).[49] Given the looseness of the criteria used to define them, it is not surprising that the relative encounter frequency of BUNs and SRBNs varies from laboratory to laboratory and even from study to study of the same laboratory (from 2:1[44,50] to roughly 1:1,[51] and even 1:2[52]). Other efforts to classify SC movement cells relied on the temporal relationship between burst offset and saccade end.[53] In this manner, a distinction was proposed between (1) clipped cells, which do not discharge beyond the end of the saccade, (2) partially clipped cells, which display some residual discharge after the end of the saccade, and (3) unclipped cells, which display discharge patterns akin to those of the type II SRBNs described above. It is more appropriate, however, to describe such classes as forming a continuum rather than three distinct populations of neurons. Moreover, they do not correspond to morphologically distinct classes of cells (Section 5.4). Other intermediate and deep layer neurons have been seen to emit discrete bursts in response to visual stimuli as well as before saccades to the same targets, but not before spontaneous saccades of the same metrics.[40,54] These neurons have been called visually triggered movement cells (VTMs) and are preferentially located near the border between the superficial and the intermediate tectal layers. Besides neurons that burst for saccades, the primate SC contains *fixation cells* (see Chapter 3 by Munoz and Schall in this volume), which fire tonically during stable gaze direction and pause for saccades[55]; they are preferentially located in the rostral SC.

In the cat, the only nonprimate species systematically studied with the extracellular recording technique, two distinct classes of cells discharging before and during either eye saccades or head movements (each comprising about 20% of all units found in the deeper SC layers) were described.[56] These authors did not encounter cells discharging like the primate SRBNs. In contrast, they found cells emitting discharges with long lead times and durations outlasting, in a fairly unpredictable manner (by 0.2 to 2 s), the duration of saccades. Discharges accompanied contraversive saccades if the eyes crossed the vertical meridian and terminated in the contralateral hemi-field. Considering their large movement field size and the extremely variable temporal relationship between cell discharge and eye movements Straschill and Schick[56] concluded that these cells can not account for the *foveation* of targets. They also found cells discharging in relation to head movement parameters such as velocity (burst cells), position (tonic cells) or both (burst-tonic cells). Since the reported relationships were qualitative rather than quantitative, claims of direct SC involvement in head movement control were premature. Subsequent work confirmed the existence of feline SC cells specifically related to movements of the head, but their encounter frequency was low (4/44) and their long-lead discharge was only loosely related to movement parameters.[17] Harris[17] noted that such cells were activated only prior to vigorous head movements, i.e., those attaining amplitudes of at least 5 to 10° and velocities higher than 15 to 20°/s. However, minimal values such as these are rather modest if one considers the amplitude-velocity relationship of head movements executed by food-deprived cats during either visually triggered or spontaneous, searching gaze shifts.[57] Peck[58] also observed a few head movement units activated in the absence of ocular saccades (5/156 movement-related cells encountered in the SC). But, their discharge was usually initiated after the onset of the head movement and continued well past its end. In contrast to the cells described by Harris, they displayed no selectivity for contraversive movements.

These two studies also are not in agreement when it comes to saccade related neurons. Harris[17] described only one variety of cells located in the deeper layers of the rostral SC (an area representing 2 to 15° of horizontal eccentricity). His neurons were visually responsive and emitted weak bursts for eye movements, which terminated within the receptive field or overshot it. They could therefore encode saccade direction with rather low precision, and saccade amplitude with even lower precision. However, his cells did not discharge when saccades of the right metrics occurred in the dark. Even though burst timing in relation to saccade onset was not reported in this study, the characteristics of the cells described by Harris resemble those of the VTMs rather than those of the SRBNs or the BUNs of the monkey. To our knowledge, Peck's[58] study is the only one to affirm the existence of feline neurons with response properties akin to those of primate SRBNs. In her study, cells classified as SRBNs (13.5% of all movement related cells) were found in the rostral SC (an area representing horizontal eccentricities smaller than 10°) and were characterized by brisk bursts starting 13 to 34 ms before saccade onset and by movement fields with definable or *closed* borders, compatible with vectorial coding of saccades. They discharged equally well for saccades of the right metrics, even in the absence of a target (i.e., in the dark or for saccades to memorized targets). Only minor differences from the primate SRBNs were reported: lower intraburst frequencies, increased incidence of background discharge, absence of correlation between saccade offset and burst end. More numerous (about 30% of all movement related cells) were neurons discharging with still longer leads. These cells bore some resemblance to primate BUNs but required the presence of a visual target to be reliably activated with saccades in their movement fields, as did the cells described by Harris. Also, about 20% of the movement related cells found by Peck displayed long-lead discharges during gaze shifts made either by eye movements alone, by head movements alone or a combination of the two. Although the relationship between neuron discharge and the metrics of gaze shifts was not explored, the discharge of these cells is roughly reminiscent of later descriptions of primate gaze related SC neurons.[48]

In summary, it is not easy to reconcile existing descriptions of the discharge pattern of movement-related SC neurons recorded extracellularly in alert cats. The following schemes can be entertained: (1) The SC generates output signals that control orienting movements of the head.[56] This view is contradicted by the rarity and ill-defined coding properties of *head movement cells* described in the studies of Harris[17] and Peck.[58] (2) Just as in the monkey, the less numerous SRBNs and BUNs of the feline SC determine the metrics of ocular saccades.[58] This view is not supported by the other two studies, which, instead, favor the idea that the SC coarsely controls the direction of only visually triggered saccades, which redirect the eyes towards broad areas of the visual world. (3) The SC controls combined gaze shifts through eye-head cells such as those described by Peck, but missed by Straschill and Schick and by Harris. (4) The SC plays a dual functional role with saccade-related cells located rostrally and head-related cells caudally. A more unified view of the role of the feline SC in the control of orienting movements emerges from studies of the discharge pattern of physiologically and/or morphologically identified projection neurons of the SC.

5.4 THE CONNECTIONS OF FUNCTIONALLY IDENTIFIED SC NEURONS

Before exploring this issue it is instructive to briefly summarize the morphology of cell types encountered in the SC of cats and monkeys. Early efforts to morphologically classify the large variety of neurons contained in the mammalian SC relied on somatodendritic features revealed in Golgi material.[59–67] But because axonal trajectories provide clues to function, our brief summary emphasizes descriptions obtained with the help of the intracellular HRP-staining technique in the cat[68,69] and the monkey.[70] Several major cell classes, designated with the capital letters of the English alphabet (X, T, I, A, and U) have so far been found in the intermediate and deep layers of the feline SC. The first three (X, T, and I) have been found often enough to permit a confident description of their morphology. X neurons have fairly big somata and several stout dendrites, which bifurcate profusely up to 1 mm from the cell body. Their axons are also stout (often more than 5 μm in diameter). They course around the periaqueductal gray and cross the midline with Meynert's decussation to join the predorsal bundle (PDB). T neurons are smaller with fewer and sparser dendrites. Their axons are also thinner (about 2.5 μm on the average); they issue several recurrent collaterals with bushy terminal fields and branch into two fibers of roughly equal thickness one of which invades the contralateral SC while the second fiber follows a varied trajectory (some join the PDB, while others project to the dorsolateral pontine nuclei or elsewhere). Finally, I neurons are small cells with a few slender, long and sparsely ramifying dendrites and a thin axon (about 1 μm in diameter), which emits long and slender recurrent collaterals. The extent and orientation of the dendritic trees of all three cell types exist in fairly stereotyped forms to warrant their subdivision into subtypes (indicated by numeral subscripts, which in the case of X cells range from 1 to 5).

X and T neurons have also been found in the primate SC.[70] By comparison to their counterparts in the cat, primate cells are smaller and their dendritic trees are not as complex and expansive. The pattern of axonal terminations of primate and feline neurons is not identical either. Unlike the feline ones, axons of primate X neurons often issue recurrent collaterals. Also in comparison to their feline counterparts, PDB participation is much more common among fibers arising from primate T cells. To explore the connectivity of identified tectal projection neurons, intraaxonal injections of HRP were combined with studies of their discharge pattern in alert behaving cats[71,72] and monkeys.[73,74]

5.4.1 X Cells of the Cat: Physiology

Figure 5.3 gives a representative example of the results obtained in the cat. The axon illustrated in Figure 5.3A was identified as originating from the contralateral SC (direct response to electrical stimulation with a latency <0.5 ms) and as reaching the C2 to C3 segments of the spinal cord (antidromic response). Intraaxonal HRP injection at a midpontine level (arrow) revealed widespread collateral connections in the pons and medulla, confirming the results obtained in acute experiments[68] and justifying the term *tecto-reticulo-spinal neuron* (in short TRSN). We first summarize the behavioral response properties of such cells and then we provide an overview of their connections.

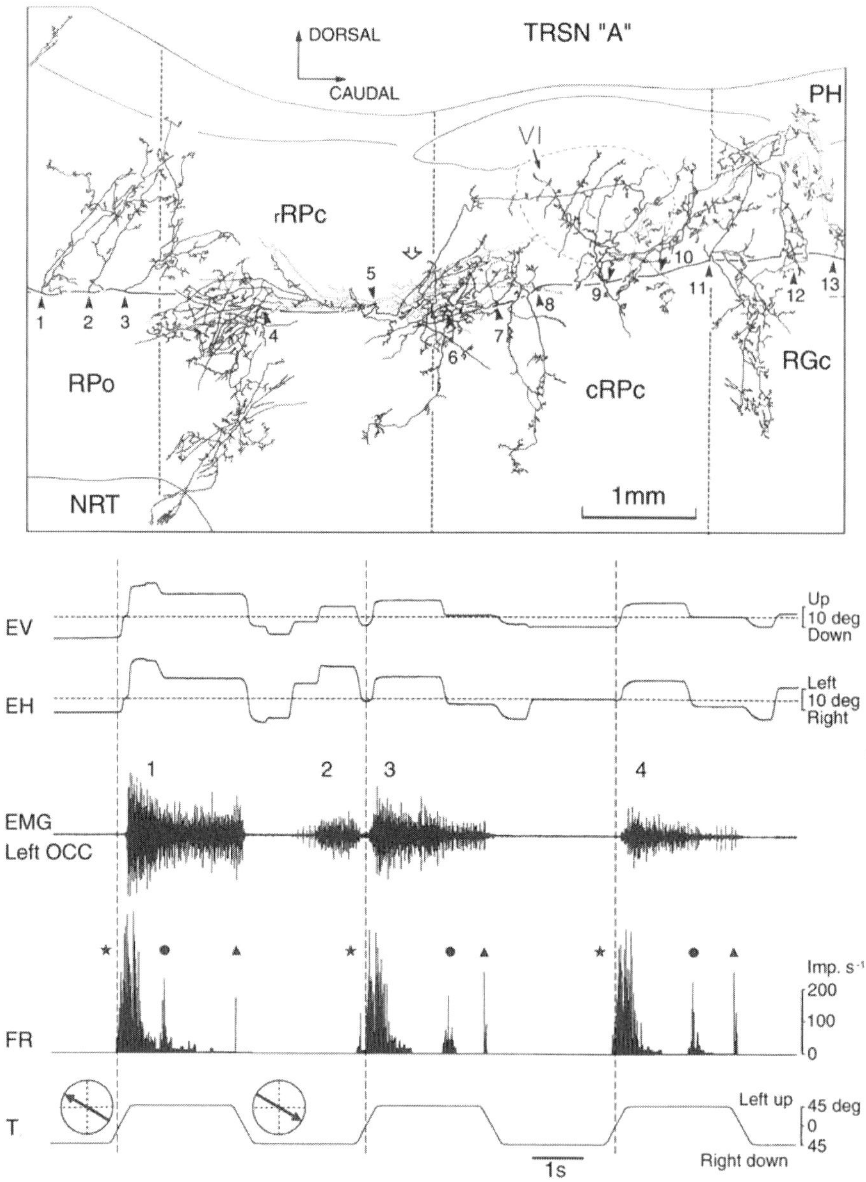

FIGURE 5.3

With the exception of a small fraction of silent neurons, all TRSNs are activated by moving visual stimuli. Responses are directionally selective for contraversive motion and broadly tuned. Preferred directions usually do not differ from horizontal by more than 45°, but a small fraction of cells are selective for nearly vertical target motion. The majority of TRSNs (about 2/3) were characterized as visuomotor[71] because their discharges are strongest when cats shift their gaze while attempting to follow a moving visual target. The lead time of bursts is variable and usually long (30 to 200 ms) with respect to gaze movement onset. Movement related discharges are in register with the visual responses they provide in the absence of tracking movements. The intensity of firing is strongly influenced by the motivation of untrained animals to attend to and to follow targets, while habituation of orienting responses to periodically moving targets is usually accompanied by the disappearance of the motor enhancement of bursts. Spontaneous saccades across a homogeneous visual background are not accompanied by bursts, even when their direction corresponds to the preferred direction of visuomotor bursts. This nonobligatory relationship to saccades is reminiscent of descriptions provided by Harris and by Peck (summarized in Section 5.3).

The temporal characteristics of visuomotor bursts generated by the TRSN population are not homogeneous. For example, the bursts of some neurons contain only 3 to 10 interspike intervals smaller than 10 ms while the bursts of other neurons may contain as many as 80 to 160. The percentage of interspike intervals shorter than 5 ms varies from zero to 25%, as judged from nonsequential interval histograms

---

FIGURE 5.3 (See facing page.) Axonal morphology and discharge pattern of a visuomotor TRSN recorded in the alert cat (Table 5.1, TRSN A). (A) Reconstruction of axon collaterals in the pons and in the rostral medulla in the parasagittal plane. (Modified from Grantyn, A., et al., in *Multisensory Control of Movement,* A. Berthoz (Ed.), Oxford University Press, Oxford, 1993, pp. 185. With permission.). Open arrow indicates the point of intraaxonal HRP injection. Dotted portions of collaterals 4 and 5 represent branching domains beyond 2 mm from the midline and extending in the lateral reticular formation. Dotted collaterals 10 and 13 are oriented more laterally than their neighbors and reach the lateral periphery of the abducens and prepositus nuclei. Vertical dashed lines separate segments of the reticular formation used for superposition of terminals from another neuron (TRSN B, Figure 5.4). Abbreviations, NRTP, nucleus reticularis tegmenti pontis; RGc, nucleus reticularis gigantocellularis; RPc, nucleus reticularis pontis caudalis (r, rostral; c, caudal); RPo, nucleus reticularis pontis oralis; VI, abducens nucleus. (B) Discharge pattern of the same neuron during tracking of a moving target. (Modified from Olivier, E., et al., *J. Physiol.*, 482, 455, 1995. With permission.) Traces, from top to bottom, indicate: vertical (EV) and horizontal (EH) eye position, electromyogram (EMG) of the left obliquus capitis cranialis (OCC) muscle, instantaneous firing rate (FR) of the neuron and position of the target (T). Insets show the direction of target motion. Stars: visuomotor bursts during target motion in the neuron's preferred direction associated with tracking saccades, slow drifts (event 4) and neck muscle contractions. Vertical dashed lines are drawn through the onset of high frequency firing before saccades. Note weaker bursts of presumably visual origin during return saccades (circles) and at the onset of target motion in the null-direction (triangles). Event 2 indicates the absence of discharge during a spontaneous gaze shift in the neuron's preferred direction.

compiled from visuomotor discharges. This variability could be due to the heterogeneous membrane properties of TRSNs documented in acute experiments in the cat.[75] For example, it was found that current thresholds for rhythmic firing vary between 0.5 and 6 nA and that the slopes of frequency-current relationships can differ by a factor of four. Differences in excitability may also partly account for the different ranges of burst durations displayed by individual TRSNs. Some cells generate bursts whose duration rarely exceeds the sum of lead time and saccade duration (up to 300 to 400 ms) while others generate sustained bursts lasting 0.5 to 2 s. Figure 5.3B gives an example of a TRSN displaying particularly prolonged firing during tracking of a target in its preferred direction (stars). Initial portions of the bursts reflect a visual response to target motion. Increments of the spike rate to 150 to 200 s^{-1} precede tracking saccades by about 100 ms, but firing continues beyond saccade end (e.g., Figure 5.3B, event 3) and persists during sequences of saccades (event 1) and postsaccadic slow eye movements (event 4). Late portions of the bursts accompany the dynamic overshoot of the EMG of contralateral neck muscles participating in attempted head movements. For some neurons of this type, a significant positive correlation was found between the duration of postsaccadic firing and the duration of EMG transients and/or postsaccadic eye drifts.[72] Records illustrated in Figure 5.3B also contain examples of bursts dissociated from orienting movements in the preferred direction and accompanying saccades in the opposite direction (circles) or visual off-responses to moving stimuli (triangles).

The responses of feline SC projection neurons have been also recorded extracellularly following antidromic stimulation to ascertain that their axons course with the contralateral predorsal bundle and, for some neurons, that they reach the first spinal segment.[76,77] Munoz and coworkers[76,77] used cats trained for food reward in a paradigm approaching natural orienting behavior in head-free and head-fixed conditions while recording with floating electrodes, which permitted a more complete quantitative evaluation of sensory and movement-related discharges. Despite the use of dissimilar techniques, their extracellular study confirmed most of the response properties of visuomotor TRSNs elucidated with intraaxonal methods.[71] Two kinds of projection neurons were found in the extracellular study: fixation neurons (fTR(S)N) located near the rostral pole of the SC and *orientation* neurons (oTR(S)N) located throughout the remaining SC. Since the discharges of oTR(S)Ns resemble those of the visuomotor TRSNs described above, we briefly describe response properties that were not revealed by intraaxonal recordings in untrained animals. First, virtually all oTR(S)Ns tested with visual, auditory and somesthetic stimuli display a multisensory convergence, consistent with the responses of antidromically identified TRSNs in anesthetized cats.[78,79] Moving visual stimuli produce much stronger responses than stationary ones, but the latter are useful for the measurement of response latencies, which cluster in a narrow range (48 to 58 ms) in the majority of oTR(S)Ns. Second, about 95% of oTR(S)Ns generate phasic discharges before and during gaze shifts terminating in their movement fields. Their high frequency portions precede movement onset by 40 to 90 ms while their low frequency prelude can lead by as much as 200 ms. Maximal intraburst frequencies and burst offset roughly coincide with the start and the end of gaze shifts, respectively. The prolonged movement related discharges reported by Olivier et al.[72] were not observed in either head-free or head-fixed conditions. The relationship between firing rate and the speed of orienting gaze shifts suggests that the bursts of oTR(S)Ns are causally relevant for eye as well as head acceleration.[77] Of particular interest in this regard is the demonstration that increments of intraburst firing rate are tightly coupled to increases of eye (10 to 12 ms later) and head (24 to 32 ms later) velocity. Similar temporal relationships are observed when short stimulus trains are applied to the SC during ongoing gaze movements. Munoz and coworkers[77] were the first to study the spatial properties of movement related oTR(S)N discharges. They found that maximal firing is always associated with a gaze shift of optimal amplitude and direction. However, all neurons emit some spikes for smaller or larger gaze displacements, i.e., they have very large movement fields and may lack clear boundaries in the direction of larger than optimal displacements. This description of their movement fields is reminiscent of the results obtained by Harris, and extends them to cells located throughout the rostrocaudal length of the SC. It also supplied one of the arguments of Munoz et al.,[80] in favor of the moving mountain hypothesis. According to it, a wave of intracollicular activity starts from a location that defines the desired displacement of the eyes and *moves* through the SC to end at its rostral pole (a region that corresponds to the representation of the fovea and zero amplitude saccades). Because the instantaneous location of the population of active cells on the motor map of the SC traces the trajectory of the movement, cells discharging prior to small saccades should also discharge during bigger movements. A second argument relied on the timing of bursts during gaze shifts of varying amplitude. However, there is considerable evidence indicating that the active SC zone does not move during saccade either in cats[81] or primates.[50,82,83]

Besides phasic movement-related discharges, about 75% of oTR(S)Ns emit sustained discharges during the preparation of gaze shifts towards space locations roughly corresponding to the neuron's visual receptive field. This type of activity could be demonstrated with cats participating in the *barrier paradigm*, which mimics some of the features of *memorized saccades* and of the gap paradigm often used in monkeys. Varying the size and the orientation of the barrier, allowed evaluation of the intensity of preparatory discharges of several oTR(S)Ns as a function of gaze position error (GPe, the angular separation between visual axis and predicted target location).[84] When a criterion of 25% maximal firing is used to delineate GPe fields, their size is large — *up to 70° horizontally and 40° vertically* — and translation invariant when recalculated in SC neural space. Optimal GPe vectors are perfectly correlated to optimal gaze shifts determined from phasic discharges.

The above studies of two research groups on identified TRSNs of the cat are in good enough agreement to conclude that the results describe one and the same neuronal population. Compared to primate SRBNs, these cells display more complex response properties (e.g., multisensory convergence, strong dependence of discharge intensity on the behavioral context, nonobligatory coupling to saccades). The identified tectal projection neurons studied to date in the cat are reminiscent of descriptions provided by Harris,[17] except that their bursts are not necessarily weak. Consistent with ideas expressed early on by Hess et al.,[19] movement-related discharges of visuomotor TRSNs could cause a fairly widespread facilitation of premotor extracollicular circuits whose engagement reorients the head and the eyes toward a crudely defined region of space. As shown below, the divergent TRSN projections to ponto-bulbar targets support this notion.

5.4.2 X Cells of the Cat: Axonal Morphology

The connections of feline X cells were first studied in acute experiments.[68,69,85,86] Later, the pattern of branching and termination of fibers arising from functionally identified visuomotor TRSNs was studied in the alert cat by Olivier et al.[72] There is no doubt that the same cell type (TRSN or X neuron) was studied in both acute and chronic experiments. The pattern of axonal terminations of this cell type is well established. In contrast, knowledge of feline T cells is limited to axonal segments confined to the SC and neighboring structures[68,69,86] such as the dorsal mesencephalic reticular formation (MRF). The conduction velocities of the tecto-bulbar portions of intracellularly recorded axons were too high (range: 30 to 65 m/s)[71] to correspond to the calculated velocities of the small diameter of fibers arising from T cells (1.7 to 4 μm).[69] In contrast, about 65% of the conduction velocities of oTR(S)Ns (range: 4 to 40 m/s) studied by Guitton and Munoz[76] were lower than 30 m/s. This disagreement remains unresolved and the question whether the feline X and T neurons differ in terms of their descending projections and their participation in eye and head movements (as expected from the discharge pattern of the primate X- and T-cells[73]) remains open.

The diameters of TRSN axons range between 3 to 8 μm at their exit from the SC but more than 90% of them are thicker than 4 μm. Before decussating in Meynert's decussation to join the contralateral PDB, they all emit a long branch coursing longitudinally to form the ventral ascending bundle (Av). Occasionally, a second similarly oriented collateral is emitted just after the decussation. Rostrally, these branches reach the fields of Forel, including a medial portion homologous to the rostral interstitial nucleus of the medial longitudinal fasciculus (riMLF) of primates. Other collaterals originate either from the main axon on its course to the decussation or from the Av branch. The MRF is supplied by terminals throughout its rostro-caudal extent. Its central portion (the cMRF) is the most common convergence area of X terminal fields. The density of innervation decreases from medial to lateral within this area, and only a small fraction of fibers reaches the laterally placed paralemniscal area. The cuneiform area next to the SC is free of TRSN terminals, except for its most medial stripe, which borders the central gray. Other tegmental regions can be called *facultative* targets of X fibers because they are supplied by some and not by others. The supraoculomotor region of the central gray receives projections from about one third of TRSNs, whereas collaterals terminating in the interstitial nucleus of Cajal (NIC) have been observed rarely, and only in the periphery of the nucleus. Finally, fibers reach the nucleus of posterior commissure (NPC) and other pretectal nuclei, but incomplete labeling precludes any conclusions about the quantity of terminals deployed therein. The identity of midbrain targets of SC efferents has been studied by intraaxonal HRP injections of neurons responding monosynaptically to electrical stimulation of the SC.[85] Only one of 14 neurons displayed the collateral connections to the inferior rectus subdivision of the III nucleus required for disynaptic transmission to extraocular motoneurons. Six candidates for tri-synaptic transmission were identified, and these terminated in the NIC and the riMLF. Interestingly, the main axons of about two thirds of the neurons examined descended to the cervical spinal cord ($N = 7$) or at least to the rostral pons ($N = 2$).

To summarize, the functional connectivity of the SC with the upper brain stem has been the object of few studies and the anatomical strength of connections with different populations of target neurons has not been evaluated quantitatively. Nevertheless, targets of the feline SC have been shown to include midbrain regions housing the vertical saccade generator (riMLF, NIC, NPC[87]), other oculomotor related structures (the cMRF and the supraoculomotor central gray) and neurons projecting to the spinal cord and participating in the control of the vertical components of head movements (fields of Forel[88]).

Importantly, the X cells that participate in these projections through their Av fibers also participate in the SC projections to horizontal movers through their PDB fibers. Intraaxonal labeling of such fibers in anesthetized cats[68] demonstrated their extensive branching and identified the medial pontobulbar reticular formation, in particular its dorsal tier, as their main target. Weaker connections were traced to the abducens nucleus (AbdN), the tegmental reticular nucleus (NRTP) and even fewer to the nucleus prepositus hypoglossi (PH). In this study, successful labeling was limited to particularly thick axons (7 to 10 μm). On the other hand, the axons of orienting-related visuomotor TRSNs identified in alert cats are thinner (4.6 to 7 mm) and their trajectory differs somewhat.[71,72] Reconstructions of four such TRSNs confirmed that collaterals arise at all rhombencephalic levels. A complete labeling, suitable for bouton counts, has been obtained in the caudal pons and rostral medulla for three cells. We chose to illustrate TRSN *A* (Figure 5.3A) because its most profuse branching supplied regions targeted by all four visuomotor TRSNs. The first of these regions (caudal extension of collateral 3 and dorsal ramification domain of collateral 4) is the rostral pole of the caudal pontine reticular nucleus (RPc), midway between its dorsal and ventral limits. The second (collaterals 5 to 8) is at approximately the same dorsoventral position, just rostral to the AbdN. The third region (collaterals 9 and 10) corresponds to the caudal extremity of the dorsal RPc, where the projections of all visuomotor TRSNs are particularly concentrated ventromedially to the AbdN before entering it. The same pattern of terminations is displayed in the frontal reconstruction of fibers arising from a different TRSN (Figure 5.4).

Other tegmental nuclei are contacted by visuomotor TRSNs in a facultative manner (Table 5.1). Individual differences can be appreciated by comparing TRSNs A (Figure 5.3) and B (Figure 5.4). In the rostral RPc, the branching domain of neuron A strongly favors the ventral tier of the PPRF, with the longest branches reaching the NRTP. The collaterals of neuron B display the opposite bias at this level (Figure 5.4A). In the caudal RPc, innervation of the periabducens area is identical in the two neurons, but TRSN A is the only one issuing ventral collaterals. Finally, the central and ventral tiers of the rostral gigantocellular reticular nucleus (RGc) are supplied from both neurons, but its dorsal tier receives input only from neuron A whose terminals extend into the PH. In agreement with bulk tracer studies,[89–91] superposition of all four identified cells reveals that, as a group, visuomotor TRSN fibers supply the entire medial reticular formation of the caudal pons and rostral medulla. However, connections of individual neurons are more restricted and selective for some regions. The *signature* of individual fibers is evident in projections to facultative targets such as the NRTP, the PH and the abducens and facial motor nuclei. The strength of their projections to areas containing excitatory (Figure 5.4A) and inhibitory (Figure 5.4C) medium lead burst neurons also differs among different visuomotor TRSNs (Table 5.1).

Intraaxonal labeling clearly demonstrates the presence of TRSN terminals in the AbdN (Figure 5.3A and 5.4B). Recordings of postsynaptic responses to collicular

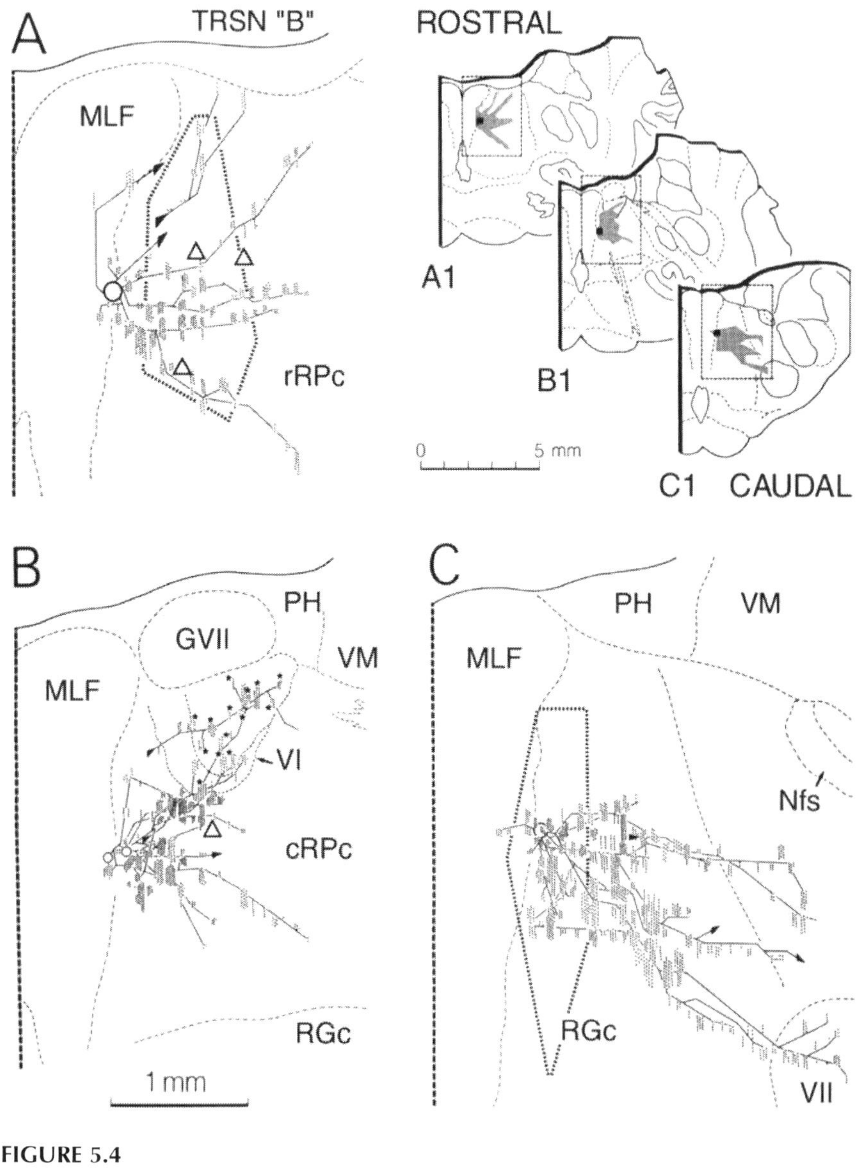

FIGURE 5.4

TABLE 5.1
Differential Ponto-Bulbar Projections of Visuomotor TRSNs

TRSN	Rostral RPc	EBN Area	Caudal RPc	OPN Area	IBN Area	Rostral RGc	VI Nucleus	PH Nucleus	VII Nucleus
"A"	++++	++	+++	++	++	++	++	+	–
"B"	++	++	++	+	++	+++	+	–	+
"C"	+	+	++	+	+	+	+	–	–

Note: Number of boutons deployed within different target areas by three intraaxonally labeled TRSNs ("A", "B", "C") with complete HRP labeling of terminal ramifications at brainstem levels indicated in Figure 5.3A. Branching patterns of neurons "A" and "B" are illustrated in Figures 5.3A and 5.4, respectively. The number of crosses indicates the relative strength of connections: (–) no connection, (+) < 200 boutons, (++) 201 to 1500 boutons, (+++) 1501–2000 boutons, (++++) > 2000 boutons. The number of boutons in the OPN area was counted within 0.8 mm from the midline. The length of the *rostral RGc* segment is two times smaller than of the two RPc segments. Counts in this column are useful for comparisons between neurons but not between different segments of the reticular formation. See legends of Figures 5.3 and 5.4 for abbreviations.

stimulation indicate, however, that the efficacy of a direct excitatory link is low[92] or negligible.[90] The same studies agree that main connections to abducens motoneurons and internuclear neurons are disynaptic. Contralateral excitatory effects are mediated by neurons in the PPRF, including identified medium lead burst neurons[93,94] and eye-neck reticulospinal neurons[95] (Figure 5.4A and 5.4B, triangles). A direct projection of TRSNs to the paramedian reticular formation just caudal to the AbdN (Figure 5.4C) is at the origin of the disynaptic reciprocal inhibition of abducens motoneurons through inhibitory burst neurons.[90,96] A similar innervation pattern has been demonstrated for internuclear neurons of the AbdN, except that the proposal of a monosynaptic SC projection to these cells[97] was recently rejected.[90] This dis-

FIGURE 5.4 (See facing page.) Branching and terminations of visuomotor TRSN B (Table 5.1) reconstructed in the frontal plane (From Olivier, E., et al., *Exp. Brain Res.*, 93, 435, 1993. With permission.) Comparison with TRSN A (Figure 5.3A) reveals a similarly profuse branching pattern and certain differences in the topography of terminal domains. (A-C) Superposition of collaterals and boutons present in segments rRPc, cRPc and RGc of Figure 5.3A. Open circles correspond to the mean (A, C) or individual (B) positions of the origins of collaterals. Arrows at proximal ends of collaterals denote continuation in a more rostral segment. Arrows at distal ends denote entry in a more caudal segment. Stars: bouton clusters in the abducens nucleus. Triangles: locations of eye-neck reticulospinal neurons.[108] Thick stripped lines delimit the EBN (A) and IBN (C) areas.[94,140] (A1 to C1) Low magnification drawings to show positions of the main axon and terminal domains in sections through the brainstem segments illustrated in (A to C). Abbreviations: VII, facial nucleus; GVII, genu of the facial nerve; MLF, medial longitudinal fasciculus; Nfs, nucleus of the solitary tract; PH, nucleus prepositus hypoglossi; VM, medial vestibular nucleus. Other abbreviations as in Figure 5.3.

agreement does not invalidate the general conclusion that TRSN connections are appropriate for producing a coordinated pattern of excitation and suppression of oculomotor nuclei participating in conjugate contraversive movements in the horizontal plane.

Similarly, the existence of monosynaptic connections between TRSNs and neck motoneurons has been clearly demonstrated in only one intracellular study.[98] Results of other studies were ambiguous[99,100] or negative.[101] Huerta and Harting[89] and Rose et al.,[102] described terminations of the tectospinal tract in the motor nuclei of neck muscles. However, when intraaxonal HRP labeling of tectospinal axons was combined with retrograde labeling of motoneurons, no tecto-motoneuronal contacts were found by microscopic evaluation of the material.[103] As shown in the same study, morphologically identified interneurons located in laminae V to VIII can mediate a disynaptic connection. Intracellular studies in anesthetized cats have repeatedly demonstrated that reticulospinal neurons of the medial pontine and bulbar reticular formation mediate disynaptic responses of neck motoneurons following stimulation of the SC.[99,101,104–106] These studies described in detail the topography of reticulospinal neurons (RSNs) receiving monosynaptic EPSPs from the contralateral SC, the trajectory and branching of their axons in the upper cervical segments, and the divergence of connections to different motor pools. A contribution of these RSNs to orienting head movements of tectal origin can not be inferred from this work because, as documented in alert cats,[107] 75.6% of the RSNs that receive monosynaptic excitatory input from the contralateral SC do not generate any movement-related discharges. The occurrence of monosynaptic EPSPs is higher in functionally identified orienting RSNs (92.3%), but this difference does not permit a reliable identification of such neurons in acute experiments. Functional identification, together with the identification of synaptic inputs from the SC and identification of their projections to the spinal cord, is available only for caudal pontine RSNs[108,109] (Figure 5.4A and 5.4B, triangles). These neurons generate phasic and phasic-decremental discharges correlated with several components of orienting synergies (saccades, neck muscle EMG, head movement). Some of them discharge tonically during attempted eccentric gaze in head-fixed but not in head-free animals. Intraaxonal labeling has shown that phasic-decremental discharges are typical for neurons that send collaterals with conspicuous terminations in the AbdN, the PH and the vestibular nuclear complex. Such connections are particularly suited for functional coupling of eye and head movements. Phasic and tonic RSNs have predominantly spinal projections and thus may represent cell groups involved in the extraction of head-movement-related signals from the composite *gaze signals* transmitted from the SC.

There are two possible views about the functional role of feline X neurons. The first one underscores their common functional and morphological features. Their firing patterns are compatible with triggering and facilitation of contraversive orienting synergies, but not with the control of their separate motor components. In terms of models, their output signals may correspond to desired gaze displacement, leaving to postcollicular structures the task of extracting and refining effector-specific signals. The combination of midbrain and rhombencephalic target areas collectively innervated by orienting-related TRSNs supports such a generalized function of the

SC. The second interpretation underscores the unique functional properties and combinations of target areas of individual TRSNs (Table 5.1). This morpho-functional heterogeneity is a clearly established fact and conceivably contributes to the decomposition of gaze-related signals into saccadic (clipped TRSNs) and cephalomotor (unclipped TRSNs) components.

5.4.3 Morphophysiology of Primate X and T Neurons

The discharge pattern of primate X neurons has also been studied intraaxonally in alert monkeys,[73] but no relationship with the position of the eyes or the occurrence of spontaneous saccades could be discerned. In contrast, the discharge of cells belonging to the T class of tectal efferent neurons is related to saccade execution in the monkey. Figure 5.5 illustrates the salient morphological and physiological features of one such neuron, which emitted high frequency bursts (Figure 5.5A) for spontaneous saccades of appropriate amplitude and direction. Because the latency of their bursts is in the long lead range (21 to 46 ms on the average), such cells are called tectal long lead burst neurons (in short TLLBs). The spatial tuning of TLLBs (Figure 5.5B) is as selective as that of SRBNs. As expected of cells causally relevant for saccades, TLLBs are generally silent both between saccades and for saccades of other amplitudes or directions. The movement fields of TLLBs are consistent with the metrics of saccades evoked by electrical stimuli delivered in the same region of the SC.[36] Cells that discharge prior to small saccades are located in the rostral SC while cells that discharge prior to big saccades are located more caudally. Also, cells that discharge before upward saccades are located medially in the SC, whereas cells that discharge before downward saccades are located laterally. Because of their discrete bursts, TLLBs do not correspond to type II SRBNs or BUNs, whose morphology and projections remain unknown. Further, because TLLBs discharge before spontaneous saccades as well as before visually driven ones they do not correspond to VTMs, whose morphology and projections also remain unknown. With the exception of one cell (which displayed a discharge pattern similar to that of clipped cells) the TLLBs that have been recovered display a discharge pattern similar to that of *partially clipped* cells.[74] Given the discharge pattern and axonal projections of TLLBs (considered below), it is a largely partially clipped signal that the SC sends to the burst generators.

The shape and small or medium size of TLLB somata (cross sectional area ranging from 265 to 650 μm^2) allows their assignment to the T class of tectal efferent neurons.[73] This identification is also consistent with their fairly superficial location in the SC, namely in the ventral stratum opticum and the dorsal stratum griseum intermediale.[73] The relatively dorsal location of TLLBs was later verified with the help of extracellular recording techniques.[110] Further, it is consistent with the dendritic trees of TLLBs, which are of moderate complexity and have their long axis oriented normally to the surface of the SC. Moreover, it is consistent with the relatively thin diameter (2.5 to 3.5 μm) of their axons (which render attempts to impale and record from them a rather painstaking endeavor). Finally, it is consistent with the fact that unlike TRSN axons of the cat,[68,69] TLLB axons give rise to recurrent (Figure 5.5C, arrowhead) and commissural (Figure 5.5C, arrow) fibers.

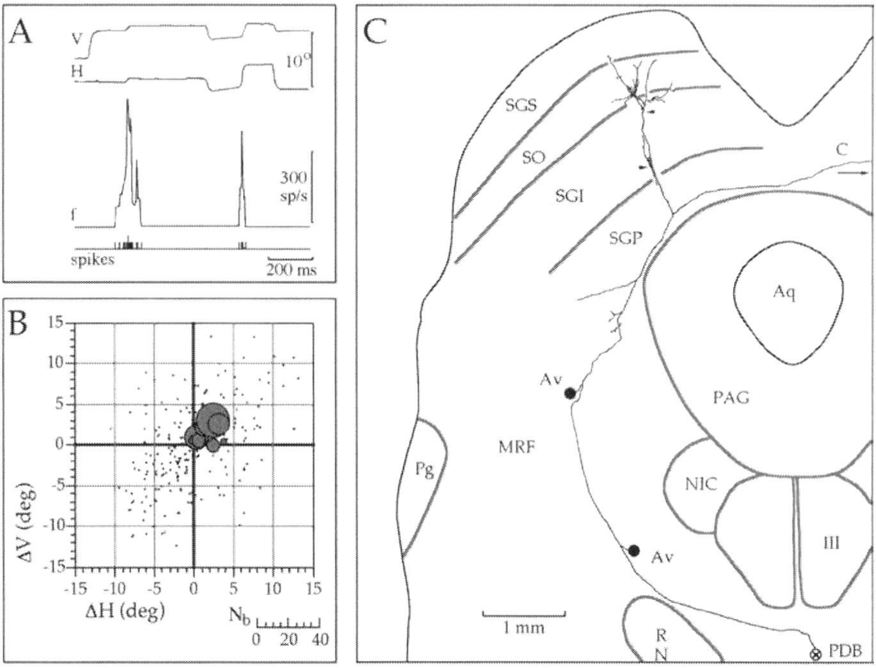

FIGURE 5.5 Salient morphological and physiological features of TLLBs. (Modified from Moschovakis, A.K., Karabelas, A.B., and Highstein, S.M., *J. Neurophysiol.*, 60, 263, 1988; and Scudder, C.A., et al., *J. Neurophysiol.*, 76, 332, 1996. With permission.) (A) Saccade related discharge pattern. Abbreviations: V, vertical eye position; H, horizontal eye position; f, instantaneous firing rate. (B) Bubble diagram of the neuron's movement field. Circles are centered over the end points of saccades (in retinotopic coordinates) and their diameter is proportional to the number of spikes in accompanying bursts (N_b). Dots indicate saccades not accompanied by spikes. (C) Camera lucida reconstruction of the neuron's somatodendritic and ipsilateral axonal system. Arrowheads point to the origin of recurrent collaterals. Solid circles indicate the point where major axonal branches assume rostral trajectory. (D) **(See facing page)** Distal pattern of axonal trajectory and projections of three TLLBs in the sagittal plane. Abbreviations: III, oculomotor nucleus; IV, trochlear nucleus; Aq, aqueduct; Av, Ventral ascending fiber; C, commissural fiber; IO, inferior olive; MRF, mesencephalic reticular formation; NIC, interstitial nucleus of Cajal; PAG, periaqueductal grey; PDB, predorsal bundle; Pg, parabigeminal nucleus; PT, pyramidal tract; RN, red nucleus; SGI, stratum griseum intermediale; SGP, stratum griseum profundum; SGS, stratum griseum superficiale; SO, stratum opticum. Other abbreviations as in Figures 5.3 and 5.4.

As shown in Figure 5.5C, the main axons of TLLBs travel along the contour of the periaqueductal grey and cross to the opposite side in the dorsal tegmental decussation of Meynert. Before crossing, they emit one or two major branches that follow the Av tectofugal fiber bundle on its way toward the riMLF. They also emit several thin collaterals that deploy terminal fields in the region of the central mesencephalic reticular formation, where the somata of reticulo-tectal long-lead burst neurons are known to reside.[73] Figure 5.5D illustrates the distal trajectory

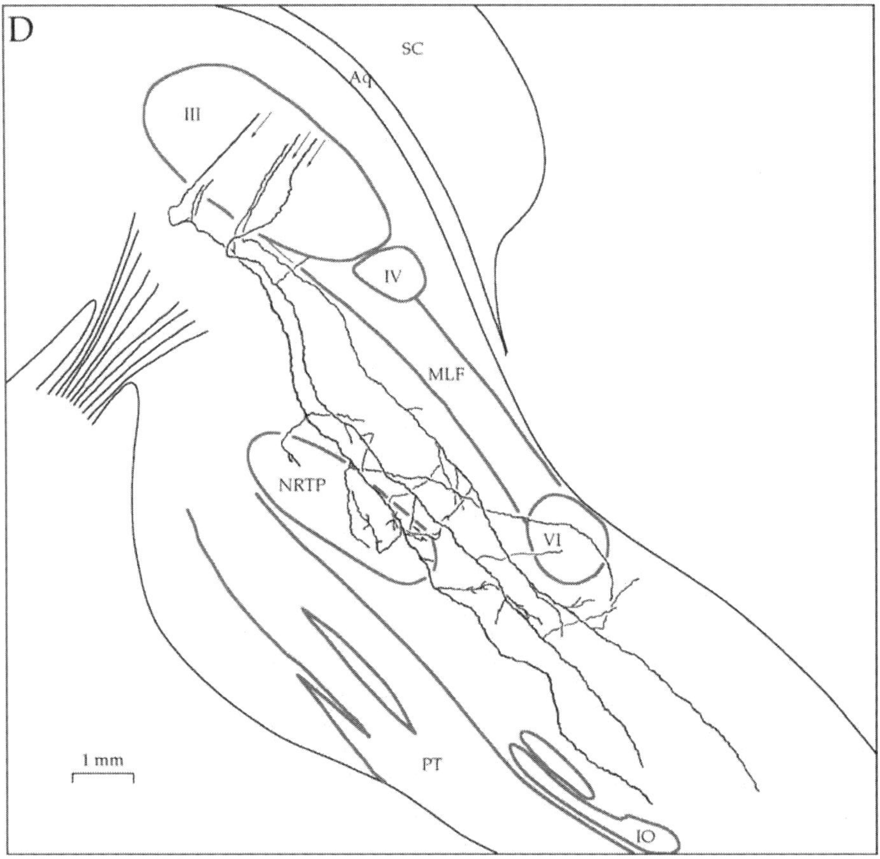

FIGURE 5.5 Continued.

and patterns of termination of PDB branches of single identified TLLBs elucidated with intraaxonal recording and HRP injection in alert monkeys.[74] Their distal targets include the nucleus reticularis tegmenti pontis, the nuclei reticularis pontis oralis and caudalis, the nucleus paragigantocellularis dorsalis, and the nucleus raphe interpositus. A description of the discharge pattern and connections of cells residing in nuclei targeted by Av and PDB fibers is beyond the scope of this chapter. Suffice it to say that they contain most of the premotoneuronal interneurons that comprise the burst generators.[2] To enable comparisons with the projections of feline X neurons, Table 5.2 summarizes the strength of TLLB projections to pontomedullary nuclei.

5.5 BURST GENERATION

The presaccadic bursts of tectal efferent neurons are in part due to the roughly similar bursts they receive from the cortical frontal eye fields (FEF[111]). However, since saccades can still be executed after lesions of the FEF,[112] the SC must be able to

TABLE 5.2
Differential Strength of Descending Projections of Intraaxonally Labeled TLLBs

TLLB	RPO_d	RPO_v	RTP	RPc	rip	PGd	NRG
18-1	++	–	++	++	+++	+++	++
21-3	+	++	++	+++	+	++	++
21-4	–	++	+++	+	+	+	+
25-1	+++	+	++	++	+	–	+
26-2	–	+	+	++	–	–	+
26-3	+	+	++	++	++	+	+
26-4	+	+++	++++	+	++	+	+++
26-5	+	+++	+++	+	+	+++	+++
26-6	+++	+	++	+	+	–	+
26-7	++	++	++	++	+	–	++
26-8	+++	+	+	++	+	–	+
26-10	+	++	++	+	++	++	+++

Note: Columns show neuron identification and relative innervation of seven regions. RPO_d, pars dorsalis of the nucleus reticularis pontis oralis, RPo_v, pars ventralis of the nucleus reticularis pontis oralis, NRTP, nucleus reticularis tegmenti pontis, RPc, nucleus reticularis pontis caudalis (including the EBN area), rip, nucleus raphe interpositus, PGd, nucleus paragigantocellularis dorsalis (including the IBN area), NRG, nucleus reticularis gigantocellularis.

Modified from Scudder, C.A., et al., Anatomy and physiology of saccadic long-lead burst neurons recorded in the alert squirrel monkey. I. Descending projections from the mesencephalon, *J. Neurophysiol.*, 76, 332, 1996. With permission.

generate presaccadic bursts from signals it receives from other sources and need not be either phasic or saccade related. Three, possibly synergistic mechanisms are considered in this section: (1) the projection of the superficial tectal layers to the deeper, (2) a combination of membrane conductances and transmitter receptors promoting bursting discharges, and (3) the recurrent excitatory network deployed by TLLBs.

The existence of a projection from the superficial to the deeper tectal layers remained controversial till the late 1980s. It was unequivocally demonstrated for the first time in adult monkeys with the intrasomatic injection of HRP in neurons of their superficial SC layers.[70, 113] It originates from a particular class of superficial SC cells, the L neurons. Figure 5.6 provides two typical examples of such cells and illustrates the range of their morphological characteristics. Typical of such cells is the fact that one of them was found in the superficial gray layer (SGS) and the other in the stratum opticum (SO). Also typical of such cells is their somatodendritic morphology, which can be characterized as narrow field vertical in the case of cell I and wide field vertical in the case of cell II. As with all L cells, the axons of the neurons illustrated in Figure

5.6 issue several intratectal collaterals. Some of these collaterals deploy terminal fields in the superficial tectal layers (SGS and SO), while others deploy terminal fields in the intermediate gray layer (SGI). These terminal fields contain a fairly large number of boutons (up to 3000 and sometimes more) and are largely confined to a cylinder somewhat less than 1 mm in diameter that is oriented normally to the surface of the SC. The projection of the superficial layers to the deeper is organized in a columnar fashion, as indicated by the fact that the dendritic territories of L neurons overlie their axonal terminations.[70] Finally, typical of L cells is the fact that one of their axonal branches projects to the parabigeminal nucleus.[70,113] Since primate tectoparabigeminal cells respond to visual stimuli[114] it is reasonable to conclude that L neurons carry visual information from the superficial to the deeper tectal layers. As argued before, such direct conveyance of visual signals from the superficial to the deeper SC could underlie the short latency of *express* saccades.[2,70]

The existence of a projection from the superficial to the deeper tectal layers was later confirmed in the cat[115] and several other mammals (i.e., hamster,[116] tree shrew,[117] and rat[118]). Although the activity and connections of SC neurons in these species may depart from those in cats and monkeys, we will summarize results obtained in slices of their SC *in vitro*, in a series of experiments that warrant replication in higher mammals. Brief electrical stimulation of the superficial tectal layers of frontally cut slices of the SC of young tree shrews evokes monosynaptic excitatory postsynaptic potentials (EPSPs) in SGI neurons.[119] Synaptic responses can have multiple peaks and their duration, which can outlast the duration of the stimuli by 100 fold or more, is not related to the duration or the intensity of the stimulus. The response has thus been attributed to a regenerative all-or-nothing process. In young rats, EPSPs evoked in a similar manner are abolished by AMPA- and NMDA-type antagonists,[118,120] thus indicating that this pathway uses glutamate as a neurotransmitter. The negative slope conductance of the NMDA receptors could account for the nonlinear all-or-nothing responses of SGI neurons.

Depolarization is a potent means to switch on regenerative processes due to NMDA receptors. In slices, this can be accomplished with bath application of the GABAergic antagonist bicuculline, which removes background synaptic inhibition. In experiments such as this, SGI cells respond to superficial layer stimulation with prolonged bursts of action potentials riding on top of increased amplitude EPSPs.[118,120] The source of GABA in such preparations is probably local interneurons.[121] The SC of intact animals has several additional sources of GABAergic input (summarized in Reference 2). The best studied of these originates in the pars reticulata of the substantia nigra (SNR). The primate and feline SNR contains cells that pause before and during saccades towards visible or remembered targets.[122,123] There is substantial evidence to indicate that these neurons are inhibitory and that their targets include TLLBs. First, nigrotectal terminal fields are particularly dense at the border between the SO and the SGI,[124] where TLLBs are often encountered. Also, electrical stimulation of the SNR has been shown to produce monosynaptic IPSPs in all T neurons studied to date.[70,86] Given the average firing rate of SNR neurons (between 35 and 120 spikes/s, in the cat[123]) and the duration of the IPSPs they generate in the same species (about 15 ms on the average[86]), it is reasonable to conclude that TLLBs are tonically inhibited. Their release from inhibition could

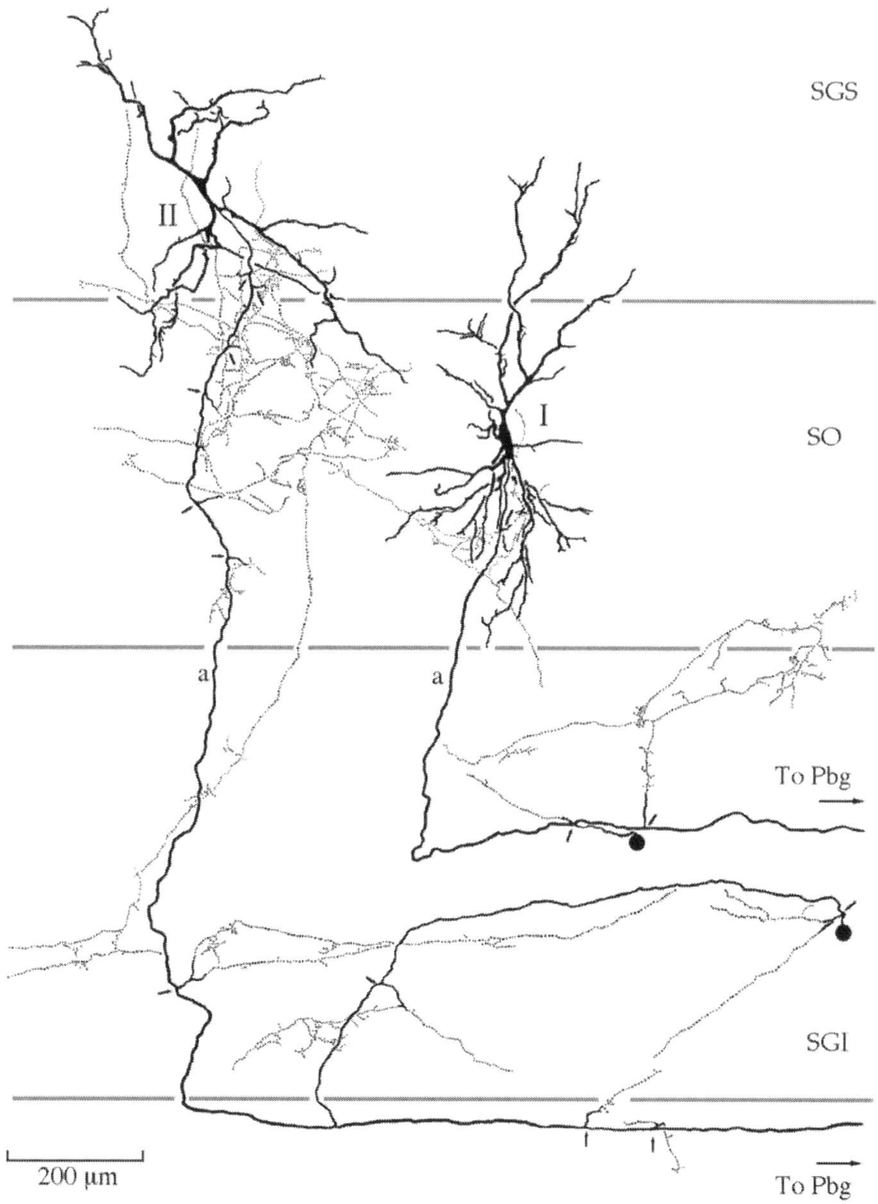

FIGURE 5.6 Complete camera lucida reconstruction of the dendritic (solid) and intratectal axonal ramification patterns (dotted) of two primate L neurons, in the frontal plane (Modified from Moschovakis, A.K., Karabelas, A.B., and Highstein, S.M., *J. Neurophysiol.*, 60, 232, 1988. With permission.) The solid horizontal line indicates the surface of the superior colliculus and the stippled lines indicate borders between layers. Arrowheads point to the origin of intrinsic collaterals. Solid circles indicate the point where major axonal branches assume rostral trajectory. Abbreviations as in Figure 5.5.

engage NMDA dependent regenerative processes and lead to bursts that command saccades. Consistent with this, microinjection of bicuculline has been shown to increase the frequency of express saccades.[8] Besides release from nigral inhibition, entry in a regenerative firing mode can be potentiated by activation of cholinergic receptors. This is indicated by the fact that injection of nicotine in the SC also increases the frequency of express saccades.[125]

The intense bursts of TLLBs could also be due to reciprocal excitatory connections between neighboring TLLBs. Such a plexus of excitatory connections is deployed by T neurons in general and TLLBs in particular.[69,70,73] An example of the distribution of *boutons* deployed by the recurrent collaterals of primate TLLBs is illustrated in Figure 5.7. The notion that TLLBs use glutamate as a neurotransmitter is supported by the fact that their main axon joins the predorsal bundle[73] and the fact that predorsal bundle efferents are glutamatergic.[126] The existence of excitatory glutamatergic interconnections between SGI neurons is further supported by the demonstration of brief inward currents superimposed on long-lasting inward currents in response to photorelease of caged glutamate.[127] Since feline X neurons lack recurrent collaterals, they can not participate in this manner in the generation of TRSN bursts. Alternatively, TRSN bursts could be partly due to lateral interactions between feline SRBNs,[58] with the caveat that such cells have been encountered in only one study[58] and that their morphology (including their axonal system) remains unknown.

Although discussed in the context of express saccades, several of these mechanisms could underlie other kinds of saccades as well. This is particularly true of the SNR mediated disinhibition and the recurrent excitatory plexus deployed by TLLBs. Additional mechanisms influencing the excitability of SC neurons could also play an important role in the genesis of saccade related SC bursts. For example, the intensity of the early prelude response of SRBNs depends on the predictability of impending saccades,[128] their metrics[45] and whether they are directed towards or away from a target.[51] The response magnitude is also well related to the latency of saccades.[129] It has been argued that the latency of saccades is determined by the rate of rise of a decision signal and that saccades are elicited as soon as a certain threshold is surpassed.[130] Such a process, possibly an intracortical one,[131,132] could influence the discharge of SC neurons. The mechanisms that transform it into the saccade related SC bursts that ultimately determine saccade initiation remain unknown.

5.6 MODELS

The wealth of information presently available for the SC severely constrains any model proposed for its role in saccade generation. Clearly, all SC models should replicate the electrically evoked movements summarized in the second section of this chapter. For example, they should replicate the staircases of saccades evoked in response to prolonged stimulation of the SC, as well as the weighted average phenomenon produced when two sites of the SC are stimulated simultaneously. Further, model units should display activation functions reminiscent of the discharge patterns of SC neurons (summarized in Section 5.3) and they should be connected in a manner that respects their patterns of connections (summarized in Section 5.4).

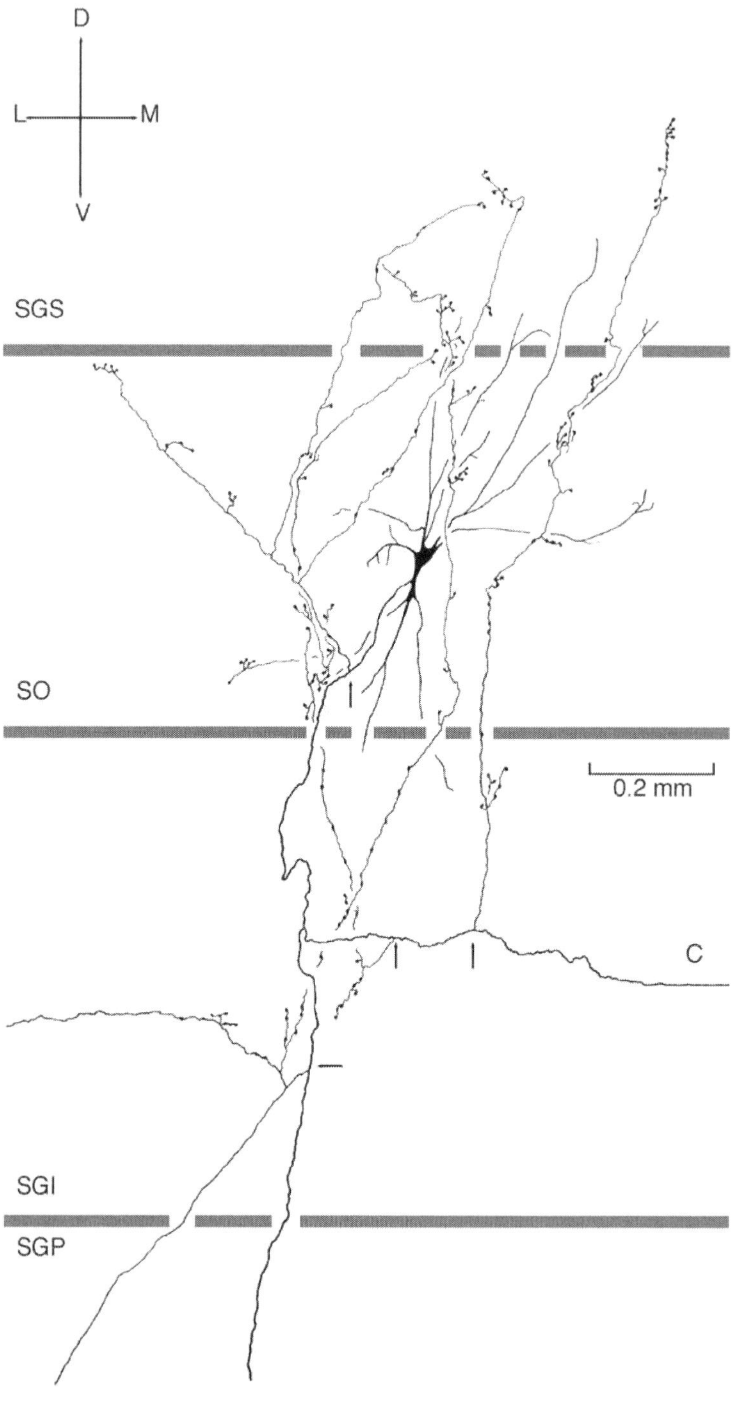

FIGURE 5.7

Finally, the effects of lesions of model units should respect known neurology. For example, saccades somewhat smaller than those corresponding to the lesioned area should be shortened and saccades somewhat larger should be lengthened.[133]

Evaluation of the relative merits of the SC models proposed to date is beyond the scope of this chapter. A simplified version of a previously proposed[134] nonlinear, distributed model of the SC is illustrated in Figure 5.8 to anchor pictorially our discussion of the interplay between theory and experiment. It relies on the following assumptions: (1) Model input is information about target location. It is carried by L neurons of the superficial layers (discussed in the previous section), indexed by their location in an array of similar cells so that neurons with small and large index numbers are activated by small and large retinal errors (distance of target image from the fovea), respectively. (2) Consistent with their pattern of axonal terminations,[70] the outflow of L neurons is routed to TLLBs that are located directly underneath them (and are thus similarly indexed). Dashed lines are used to indicate that L neuron output may influence TLLBs indirectly, possibly through several synaptic steps. (3) Based on evidence summarized in the previous section, TLLBs receive inhibitory input from the SNR. (4) Neither excitation originating from L cells nor the removal of nigral inhibition can fully explain the high frequency bursts of TLLBs. To account for these bursts and in particular for their all-or-nothing nonlinear character, the model of Figure 5.8 relies on the lateral excitatory connections between TLLBs, which were also described in the previous section of this chapter. Spatially constrained connection strengths between TLLBs, experimentally determined from computer assisted reconstructions of the spatial distribution of the recurrent terminal fields of intraaxonally stained TLLBs,[134] are shown in Figure 5.8 as plots of $W_{(TLLBi \rightarrow TLLBj)}$ versus $i - j$. (5) To also account for the eventual end of their burst, TLLBs are reciprocally connected to inhibitory neurons (tllbs) that display a discharge pattern similar to that of TLLBs. This pattern of connections endows the SC with a biological oscillator, leading us to predict that TLLB output is a periodic function of time even when the SC receives a constant stimulus. (6) Consistent with known anatomy and physiology,[73,74] TLLBs carry the output of the SC to the saccadic burst generators, in particular, to its long lead burst (LLB) and *latch* units.[135] The strength of the TLLB→LLB connection ($w_{(TLLB \rightarrow LLB)}$; plotted next to the pathway it concerns in Figure 5.8) depends on the index (i) of TLLBs, so that cells preferring small saccades (small i) project less heavily to the burst generators than cells that prefer bigger saccades (bigger i). This simple device implements the *spatiotemporal* transformation and is consistent with known anatomy.[136] Its existence is of particular importance as it allows a motor circuit (the motor SC) to use the *place* code generally preferred by sensory systems while keeping its motor map in register with its sensory map of visual space. (7) The input-output characteristics of model units display a

FIGURE 5.7 (See facing page.) Somatodendritic and partial axonal camera lucida reconstruction of a primate TLLB in the frontal plane. (Modified from Moschovakis, A.K., Karabelas, A.B., and Highstein, S.M., *J. Neurophysiol.*, 60, 263, 1988. With permission.) Arrows point to the origin of recurrent collaterals. Orientation symbols: D, dorsal; V, ventral; L, lateral; M, medial. Other abbreviations as in Figure 5.5.

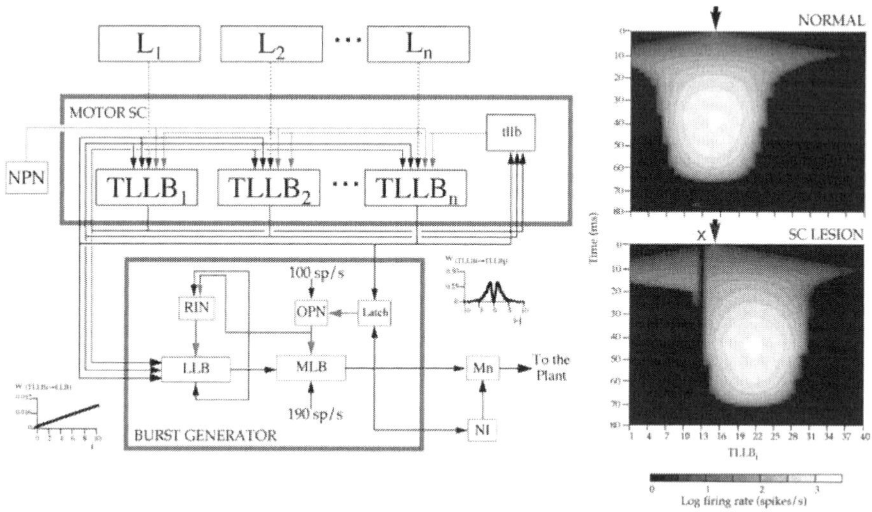

FIGURE 5.8 Model of distributed population coding of saccade metrics in the SC. Stippled lines indicate inhibitory connections. All other connections are excitatory. Insets indicate spatially varying connection strengths (w), plotted as a function of the location of the source (i), or of the target (j) neurons within their respective arrays, or in terms of the distance between them (i to j). Gray scale contour plots illustrate the spatiotemporal profile of the logarithm of the activation function of all TLLB units employed in the model as a function of their index (abscissa) and time (ordinate) for saccades evoked from a normal (top) or a *focally lesioned* (bottom) SC. Arrows point to the spatial location of the activated L cells while (x) points to the spatial location of the *lesion*. Abbreviations: L, L-type neuron; LLB, long-lead burst neuron; MLB, medium-lead burst neuron; Mn, motoneuron; NI, neural integrator; OPN, omnipause neuron, RIN, resettable integrator neuron; NPN, neuron of the substantia nigra pars reticulata that pauses for saccades; TLLB, tllb, excitatory and inhibitory tectal long-lead burst neurons.

linear region (for inputs greater than zero), a threshold equal to zero and a saturating nonlinearity beyond 1000 Hz. (8) At any point in time n, the activation $A_{j(n)}$ of unit j receiving input from m other units with activation A_i, is equal to,

$$A_{j(n)} = \sum_{i=1}^{m} A_{i(n-1)} W_{ij} \tag{5.4}$$

where W_{ij} is the strength of the connection between units i and j.

When constrained in this manner, the model of Figure 5.8 accomplishes the following: (1) It generates a wide range of saccade sizes (1° to about 30°) obeying known psychophysics (e.g., they are consistent with the main sequence curve). (2) The waveforms of its units resemble published curves describing the instantaneous frequency of discharge of saccade related units encountered in the SC of primates.[2] (3) It generates accurate saccades in double step stimulation experiments without making use of eye position or eye velocity information (if endowed with some additional machinery described elsewhere[134]). (4) It generates staircases of saccades in response to prolonged stimuli. (5) The size of saccades evoked in response to stimulation of two SC sites is the average of the saccades that are evoked when the two sites are separately stimulated. In this case, the model predicts that the region of the SC that is maximally activated occupies a position intermediate and largely overlapping with the SC regions that are maximally activated when the two sites are separately stimulated.[134] (6) It replicates the effects of focal SC lesions on saccade metrics[133]; namely, saccades are hypometric if the inactivated region corresponds to saccades bigger than the executed one and hypermetric if the inactivated region corresponds to saccades smaller than the executed one. Further, the model predicts that this counterintuitive result is due to shifts of the focus of excitation within the TLLB array so that it is pushed more caudally (rostrally) when the site of stimulation is caudal (rostral) to the site of the lesion. To illustrate how such lesions affect the evolution of the spatiotemporal profile of TLLB activation functions, Figure 5.8 includes gray scale contour plots of the intensities of discharge of all TLLB units of the model versus time (ordinate) and unit index number (abscissa). The top one illustrates the case for the normal SC. Initially, activated cells include TLLBs located at a considerable distance from the active L cells. Following their release from SNR inhibition, and due to the recurrent plexus that connects them to one another, TLLBs start discharging more and more intensely. The profile of TLLB discharge progressively evolves so that the most intensely activated TLLBs are the ones that are fairly close to the stimulated L cells. Further into the saccade, TLLB activation progressively dies out around these units, and it is around them that activity perseveres the longest. The bottom plot explains why evoked saccades are lengthened when lesions are placed in SC regions (*x*) corresponding to saccades smaller than those coded by the stimulated (arrowhead) area. Following such lesions, the aggregate activity in the TLLB layer is pushed to the right (more caudally), and thus commands a saccade of bigger size than would have been the case in the absence of the lesion.

5.7 FUTURE RESEARCH

Research in the past thirty years has culminated in a fairly detailed quantitative theory of collicular function that provides a coherent framework for hundreds of psychophysical, anatomical, neurophysiological and neurological observations. But a lot remains unknown. For example, the morphological identity and connections of cells other than the primate TLLBs and the feline visuomotor TR(S)Ns (e.g., BUNs, VTMs, fTR(S)Ns, etc.) are unknown. The same is true of the neurons carrying gaze and head displacement commands issuing from the SC, the mechanisms responsible for partitioning putative gaze displacement commands into appropriate eye and head components, their modification into signals driving the movements that are actually executed, and the congruence of the putative separate maps of desired eye and head displacements. Also, information about processes intrinsic to the SC, including the mechanisms that transform the rise of decision related signals into the saccade related SC bursts, is rather sketchy at present. To some extent this lack of knowledge is due to the relative merits and limitations of presently available techniques. Extracellular recordings in primates employ more and more sophisticated behavioral paradigms to disclose important phenomena, but offer few if any explanations about how these phenomena are produced by neurons. The much needed morphological identification of functionally identified cell classes is barely feasible and thus seldom undertaken. *In vitro* studies are needed to study the biophysics of neurons, the integrative properties of dendritic trees and the peculiarities of synaptic transmission, but functional identification is lost together with much of the dendritic tree of the neurons they contain, possibly obscuring the normal operation of the synaptic signals they receive. Finally, present day models are minimalist, relying on simplistic and unverified assumptions. For example, neuron geometry is generally ignored. Addressing these issues, possibly with the help of a combination of novel and classical techniques, will undoubtedly sustain the excitement that comes from participating in efforts to understand how a piece of biological material gives rise to the behavior of higher mammals.

REFERENCES

1. Kandel, E.R., *Cellular Basis of Behavior. An Introduction to Behavioral Neurobiology*, W.H. Freeman & Co., San Francisco, CA, 1976.
2. Moschovakis, A.K., Scudder, C.A., and Highstein, S.M., The microscopic anatomy and physiology of the mammalian saccadic system, *Progr. Neurobiol.*, 50, 133, 1996.
3. Sprague, J.M. and Meikle, T.H., The role of the superior colliculus in visually guided behavior, *Exp. Neurol.*, 11, 115, 1965.
4. Lomber, S.G., Payne, B.R., and Cornwell, P., Role of the superior colliculus in analyses of space: superficial and intermediate layer contributions to visual orienting, and visuospatial discriminations during unilateral and bilateral deactivations, *J. Comp. Neurol.*, 441, 44, 2001.
5. Ellard, C.G. and Goodale, M.A., A functional analysis of the collicular efferent pathways: a dissociation of deficits following lesions of the dorsal tegmental decussation and the ipsilateral collicular efferent bundle in the Mongolian gerbil, *Exp. Brain Res.*, 71, 307, 1988.
6. Milner, A.D. and Taylor, M.J., Auditory orienting and detection in rats following lesions of the superior colliculus, *Behav. Brain Res.*, 37, 293, 1990.
7. Pasik, P., Pasik, T., and Bender, M.B., The pretectal syndrome in monkeys. II. Spontaneous and induced nystagmus and "lightning" eye movements, *Brain*, 92, 871, 1969.
8. Hikosaka, O. and Wurtz, R.H., Modification of saccadic eye movements by GABA-related substances. I. Effect of muscimol and bicuculline in monkey superior colliculus, *J. Neurophysiol.*, 53, 266, 1985.
9. Hikosaka, O. and Wurtz, R.H., Saccadic eye movements following injection of lidocaine into the superior colliculus, *Exp. Brain Res.*, 61, 531, 1986.
10. Schiller, P.H., Sandell, J.H., and Maunsell, J.H.R., The effect of frontal eye field and superior colliculus lesions on saccadic latencies in the rhesus monkey, *J. Neurophysiol.*, 57, 1033, 1987.
11. Donders, F.C., Ueber angeborene und erworbene Association, *Albert von Graefe Arch. Ophthal.*, 18, 152, 1872.
12. Ferrier, D., *The Functions of the Brain*, G.P. Putnam's Sons, New York, 1876.

13. Robinson, D.A., Eye movements evoked by collicular stimulation in the alert monkey, *Vision Res.*, 12, 1795, 1972.
14. Grantyn, A.A., et al., Neuronal mechanisms of two-dimensional orienting movements in the cat. I. A quantitative study of saccades and slow drifts produced in response to the electrical stimulation of the superior colliculus, *Brain Res. Bull.*, 41, 65, 1996.
15. Cowie, R.J. and Robinson, D.L., Subcortical contributions to head movements in macaques: I. Contrasting effects of electrical stimulation of a medial pontomedullary region and the superior colliculus, *J. Neurophysiol.*, 72, 2648, 1994.
16. Freedman, E.G., Stanford, T.R., and Sparks, D.L., Combined eye-head gaze shifts produced by electrical stimulation of the superior colliculus in rhesus monkeys, *J. Neurophysiol.*, 76, 927, 1996.
17. Harris, L.R., The superior colliculus and movements of the head and eyes in cats, *J. Physiol.*, 300, 367, 1980.
18. Roucoux, A., Guitton, D., and Crommelinck, M., Stimulation of the superior colliculus in the alert cat. II. Eye and head movements evoked when the head is unrestrained, *Exp. Brain Res.*, 39, 75, 1980.
19. Hess, W.R., Bürgi, S., and Bucher, V., Motorische Funktion des Tectal- und Tegmentalgebietes, *Mschr. Psychiat. Neurol.*, 112, 1, 1946.
20. McIlwain, J.T., Distributed spatial coding in the superior colliculus: a review, *Vis. Neurosci*, 6, 3, 1991.
21. Guitton, D., Crommelinck, M., and Roucoux, A., Stimulation of the superior colliculus in the alert cat. I. Eye movements and neck EMG activity evoked when the head is restrained, *Exp. Brain Res.*, 39, 63, 1980.
22. McIlwain, J.T., Effects of eye position on saccades evoked electrically from the superior colliculus of alert cats, *J. Neurophysiol.*, 55, 97, 1986.
23. McIlwain, J.T., Topography of eye-position sensitivity of saccades evoked electrically from the cat's superior colliculus, *Vis. Neurosci.*, 4, 289, 1990.
24. Moschovakis, A.K., et al., New mechanism that accounts for position sensitivity of saccades evoked in response to electrical stimulation of superior colliculus, *J. Neurophysiol.*, 80, 3373, 1998.
25. Straschill, M. and Rieger, P., Eye movements evoked by focal stimulation of the cat's superior colliculus, *Brain Res.*, 59, 211, 1973.
26. Segraves, M.A. and Goldberg, M.F., Properties of eye and head movements evoked by electrical stimulation of the monkey superior colliculus, in *The Head-Neck Sensori-Motor System,* Berthoz, A., Vidal, P.P., and Graf, W., Eds., Oxford University Press, Oxford, U.K., 1992, pp. 292.
27. McIlwain, J.T., Effects of eye position on electrically evoked saccades: a theoretical note, *Vis. Neurosci.*, 1, 239, 1988.
28. Azuma, M., et al., Effects of eye position on saccades evoked by stimulation of the monkey superior colliculus, *Neuroreport*, 7, 2287, 1996.
29. Russo, G.S. and Bruce, C.J., Effect of eye position within the orbit on electrically elicited saccadic eye movements: A comparison of the macaque monkey's frontal and supplementary eye fields, *J. Neurophysiol.*, 69, 800, 1993.
30. Moschovakis, A.K. and Highstein, S.M., The anatomy and physiology of primate neurons that control rapid eye movements, *Ann. Rev. Neurosci.*, 17, 465, 1994.
31. Moschovakis, A.K., Neural network simulations of the primate oculomotor system. II. Frames of reference, *Brain Res. Bull.*, 40, 337, 1996.
32. Moschovakis, A.K., The superior colliculus and eye movement control, *Curr. Opin. Neurobiol.*, 6, 811, 1996.

33. Robinson, D.A., Oculomotor unit behavior in the monkey, *J. Neurophysiol.*, 33, 393, 1970.
34. Robinson, D.A., Control of eye movements., in *The Nervous System,* Brooks, V.B., Ed., Williams and Wilkins, Baltimore, MD, 1981, pp. 1275.
35. Zuber, B.L., Eye movement dynamics in the cat: The final motor pathway, *Exp. Neurol.*, 20, 255, 1968.
36. Schiller, P.H. and Stryker, M., Single-unit recording and stimulation in superior colliculus of the alert rhesus monkey, *J. Neurophysiol.*, 35, 915, 1972.
37. Wurtz, R.E. and Goldberg, M.E., Activity of superior colliculus in behaving monkey. III. Cells discharging before eye movements, *J. Neurophysiol.*, 35, 575, 1972.
38. Robinson, D.L. and Jarvis, C.D., Superior colliculus neurons studied during head and eye movements of the behaving monkey, *J. Neurophysiol.*, 37, 533, 1974.
39. Stryker, M.P. and Schiller, P.H., Eye and head movements evoked by electrical stimulation of monkey superior colliculus, *Exp. Brain Res.*, 23, 103, 1975.
40. Sparks, D.L., Functional properties of neurons in the monkey superior colliculus: coupling of neuronal activity and saccade onset, *Brain Res.*, 156, 1, 1978.
41. Kase, M., Nagata, R., and Kato, M., Saccade-related activity of periaqueductal gray matter of the monkey, *Invest. Ophthal. Vis. Sci.*, 27, 1165, 1986.
42. Sparks, D.L., Response properties of eye movement-related neurons in the monkey superior colliculus, *Brain Res.*, 90, 147, 1975.
43. Sparks, D.L. and Mays, L.E., Movement fields of saccade-related burst neurons in the monkey superior colliculus, *Brain Res.*, 190, 39, 1980.
44. Sparks, D.L., Holland, R., and Guthrie, B.L., Size and distribution of movement fields in the monkey superior colliculus, *Brain Res.*, 113, 21, 1976.
45. Glimcher, P.W. and Sparks, D.L., Movement selection in advance of action in the superior colliculus, *Nature (Lond)*, 355, 542, 1992.
46. Frens, M.A. and van Opstal, A.J., Monkey superior colliculus activity during short-term saccadic adaptation, *Brain Res. Bull.*, 43, 473, 1997.
47. Stanford, T.J. and Sparks, D.L., Systematic errors for saccades to remembered targets: Evidence for a dissociation between saccade metrics and activity in the superior colliculus, *Vision Res.*, 34, 93, 1994.
48. Freedman, E.G., and Sparks, D.L., Activity of cells in the deeper layers of the superior colliculus of the rhesus monkey: Evidence for a gaze displacement command, *J. Neurophysiol.*, 78, 1669, 1997.
49. Optican, L.M., Control of the trajectory of saccadic eye movements implied by a computational field theory of the superior colliculus, *Proc. Soc. Neurosci.,* 20, 141, 1994.
50. Anderson, R.W., et al., Two-dimensional saccade-related population activity in superior colliculus in monkey, *J. Neurophysiol.*, 80, 798, 1998.
51. Everling, S., et al., Role of primate superior colliculus in preparation and execution of anti-saccades and pro-saccades, *J. Neurosci.*, 19, 2740, 1999.
52. Munoz, D.P. and Wurtz, R.H., Saccade-related activity in monkey superior colliculus. I. Characteristics of burst and buildup cells, *J. Neurophysiol.*, 73, 2313, 1995.
53. Waitzman, D.M., et al., Superior colliculus neurons provide the saccadic motor error signal, *Exp. Brain Res.*, 72, 649, 1988.
54. Mohler, C.W. and Wurtz, R.H., Organization of monkey superior colliculus: Intermediate layer cells discharging before eye movements, *J. Neurophysiol.*, 39, 722, 1976.
55. Munoz, D.P. and Wurtz, R.H., Fixation cells in monkey superior colliculus. I. Characteristics of cell discharge, *J. Neurophysiol.*, 70, 559, 1993.

56. Straschill, M. and Schick, F., Discharges of superior colliculus neurons during head and eye movements of the alert cat, *Exp. Brain Res.*, 27, 131, 1977.
57. Guitton, D., Douglas, R.M., and Volle, M., Eye-head coordination in cats, *J. Neurophysiol.*, 52, 1030, 1984.
58. Peck, C.K., Neuronal activity related to head and eye movements in cat superior colliculus, *J. Physiol.*, 421, 79, 1990.
59. Cajal, S.R.Y., *Histologie du Système Nerveux de l'Homme et des Vertébrès*, Maloine, Paris, 1911.
60. Laemle, L.K., A Golgi study and cellular morphology in the superficial layers of superior colliculus in man, saimiri, and macaca, *J. Hirnforsch.*, 22, 253, 1981.
61. Laemle, L.K., A Golgi study of cell morphology in the deep layers of the human superior colliculus, *J. Hirnforsch*, 24, 297, 1983.
62. Langer, T.P., Cellular and fiber patterns in the superior colliculus of the cat. A Golgi and degeneration study., University of Washington, Seattle, 1976.
63. Ma, T.P., et al., Intermediate and deep layers of the macaque superior colliculus: a Golgi study, *J. Comp. Neurol.*, 295, 92, 1990.
64. Norita, M., Neurons and synaptic patterns in the deep layers of the superior colliculus of the cat. A Golgi and electron microscopic study, *J. Comp. Neurol.*, 190, 29, 1980.
65. Tokunaga, A. and Otani, K., Dendritic patterns of neurons in the rat superior colliculus, *Exp. Neurol.*, 52, 189, 1976.
66. Valverde, F., The neuropil in superficial layers of the superior colliculus of the mouse. A correlated Golgi and electron microscopic study, *Z. Anat. Entwicklungsgesch.*, 142, 117, 1973.
67. Viktorov, I.V., Neuronal structure of superior colliculi of corpora quadrigemina, *Arkh. Anat. Gistol. Embriol.*, 54, 45, 1968.
68. Grantyn, A. and Grantyn, R., Axonal patterns and sites of termination of cat superior colliculus neurons projecting to the tecto-bulbo-spinal tract, *Exp. Brain Res.*, 46, 243, 1982.
69. Moschovakis, A.K. and Karabelas, A.B., Observations on the somatodendritic morphology and axonal trajectory of intracellularly HRP-labeled efferent neurons located in the deeper layers of the superior colliculus of the cat, *J. Comp. Neurol.*, 239, 276, 1985.
70. Moschovakis, A.K., Karabelas, A.B., and Highstein, S.M., Structure-function relationships in the primate superior colliculus. I. Morphological classification of efferent neurons, *J. Neurophysiol.*, 60, 232, 1988.
71. Grantyn, A. and Berthoz, A., Burst activity of identified tectoreticulo-spinal neurons in the alert cat, *Exp. Brain Res.*, 57, 417, 1985.
72. Olivier, E., et al., The control of slow orienting movements by tecto-reticulo-spinal neurons in the cat: behaviour, discharge patterns and underlying connections, *Exp. Brain Res.*, 93, 435, 1993.
73. Moschovakis, A.K., Karabelas, A.B., and Highstein, S.M., Structure-function relationships in the primate superior colliculus. II. Morphological identity of presaccadic neurons, *J. Neurophysiol.*, 60, 263, 1988.
74. Scudder, C.A., et al., Anatomy and physiology of saccadic long-lead burst neurons recorded in the alert squirrel monkey. I. Descending projections from the mesencephalon, *J. Neurophysiol.*, 76, 332, 1996.
75. Grantyn, R., Grantyn, A., and Schierwagen, A., Passive membrane properties, afterpotentials and repetitive firing of superior colliculus neurons studied in the anesthetized cat, *Exp. Brain Res.*, 50, 377, 1983.

76. Guitton, D. and Munoz, D.P., Control of orienting gaze shifts by the tectoreticulospinal system in the head-free cat. I. Identification, localization, and effects of behavior on sensory responses, *J. Neurophysiol.*, 66, 1605, 1991.
77. Munoz, D.P., Guitton, D., and Pelisson, D., Control of orienting gaze shifts by the tectoreticulospinal system in the head-free cat. III. Spatiotemporal characteristics of phasic motor discharges, *J. Neurophysiol.*, 66, 1642, 1991.
78. Meredith, M.A. and Stein, B.E., Descending efferents from the superior colliculus relay integrated multisensory information, *Science (Wash. DC)*, 227, 657, 1985.
79. Meredith, M.A., Wallace, M.T., and Stein, B.E., Visual, auditory and somatosensory convergence in output neurons of the cat superior colliculus: multisensory properties of the tecto-reticulo-spinal projection, *Exp. Brain Res.*, 88, 181, 1992.
80. Munoz, D.P., Pelisson, D., and Guitton, D., Movement of neural activity on the superior colliculus motor map during gaze shifts, *Science (Wash DC)*, 251, 1358, 1991.
81. Kang, I. and Lee, C., Properties of saccade-related neurons in the cat superior colliculus: patterns of movement fields and discharge timing, *Exp. Brain Res.*, 131, 149, 2000.
82. Keller, E.L. and Edelman, J.A., Effects of electrical stimulation in the omnipause region on cells in the superior colliculus of the monkey, in *Proc. Soc. Neurosci. Abstr.,* 19, 1993.
83. Moschovakis, A.K., Gregoriou, G., and Savaki, H.E., Functional imaging of the primate superior colliculus during saccades to visual targets, *Nature Neurosci.*, 4, 1026, 2001.
84. Munoz, D.P. and Guitton, D., Control of orienting gaze shifts by the tectoreticulospinal system in the head-free cat. II. Sustained discharges during motor preparation and fixation, *J. Neurophysiol.*, 66, 1624, 1991.
85. Grantyn, A., et al., Tectal control of vertical eye movements: a search for underlying neuronal circuits in the mesencephalon, in *Physiological and Pathological Aspects of Eye Movements.,* Roucoux, A. and Crommelinck, M., Eds., Dr. W. Junk, The Hague, 1982, pp. 337.
86. Karabelas, A.B. and Moschovakis, A.K., Nigral inhibitory termination on efferent neurons of the superior colliculus. An intracellular horseradish peroxidase study in the cat, *J. Comp. Neurol.*, 239, 309, 1985.
87. Nakao, S., Shiraishi, Y., and Oikawa, T., Vertical eye movement-related neurons in the medial mesodiencephalic junction: Their firing pattern, location and projection to oculomotor and trochlear nuclei, *Neurosci. Res.*, 3(S), 66, 1986.
88. Isa, T., Itouji, T., and Sasaki, S., Excitatory pathways from Forel's field H to head elevator motoneurones in the cat, *Neurosci. Let.*, 90, 89, 1988.
89. Huerta, M.F. and Harting, J.K., Tectal control of spinal cord activity: neuroanatomical demonstration of pathways connecting the superior colliculus with the cervical spinal cord grey, in *Descending Pathways to the Spinal Cord,* Kuypers, H.G.J.M. and Martin, G.F. Eds., Elsevier, Amsterdam, 1982, pp. 293.
90. Izawa, Y., Sugiuchi, Y., and Shinoda, Y., Neural organization from the superior colliculus to motoneurons in the horizontal oculomotor system of the cat, *J. Neurophysiol.*, 81, 2597, 1999.
91. Kawamura, K., Brodal, A., and Hoddevik, G.H., The projection of the superior colliculus onto the reticular formation of the brain stem. An experimental anatomical study in the cat, *Exp. Brain Res.*, 19, 1, 1974.
92. Grantyn, A.A. and Grantyn, R., Synaptic actions of tectofugal pathways on abducens motoneurons in the cat, *Brain Res.*, 105, 269, 1976.

93. Chimoto, S., et al., Monosynaptic activation of medium-lead burst neurons from the superior colliculus in the alert cat, *J. Neurophysiol.*, 75, 2658, 1996.
94. Sasaki, S. and Shimazu, H., Reticulovestibular organization participating in generation of horizontal fast eye movement, *Ann. NY Acad. Sci.*, 374, 130, 1981.
95. Grantyn, A., Ong-Meang, J., and Berthoz, A., Reticulospinal neurons participating in the control of synergic eye and head movements during orienting in the cat. II. Morphological properties as revealed by intra-axonal injections of horseradish peroxidase, *Exp. Brain Res.*, 66, 355, 1987.
96. Hikosaka, O. and Kawakami, T., Inhibitory reticular neurons related to the quick phase of vestibular nystagmus — their location and projection, *Exp. Brain Res.*, 27, 377, 1977.
97. Grantyn, A. and Berthoz, A., Synaptic actions of the superior colliculus on medial rectus motoneurons in the cat, *Neuroscience*, 2, 945, 1977.
98. Gura, E.V. and Limanski, Y.P., Influence of superior colliculus stimulation on neck muscle motoneurons in the cat, *Neirofiziologiya*, 18, 197, 1986.
99. Anderson, M.E., Yoshida, M., and Wilson, V.J., Influence of superior colliculus on cat neck motoneurons, *J. Neurophysiol.*, 34, 898, 1971.
100. Olivier, E., et al., Post-spike facilitation of neck EMG by cat tectoreticulospinal neurons during orienting movements, *J. Physiol.*, 482, 455, 1995.
101. Alstermark, B., Pinter, M.J., and Sasaki, S., Tectal and tegmental excitation in dorsal neck motoneurons of the cat, *J. Physiol. (Lond)*, 454, 517, 1992.
102. Rose, P.K., MacDonald, J., and Abrahams, V.C., Projections of the tectospinal tract to the upper cervical spinal cord in the cat: A study with the anterograde tracer PHA-L, *J. Comp. Neurol.*, 314, 91, 1991.
103. Muto, N., Kakei, S., and Shinoda, Y., Morphology of single axons of tectospinal neurons in the upper cervical spinal cord, *J. Comp. Neurol.*, 372, 9, 1996.
104. Iwamoto, Y. and Sasaki, S., Monosynaptic excitatory connections of reticulospinal neurons in the nucleus reticularis pontis caudalis with dorsal neck motoneurones in the cat, *Exp. Brain Res.*, 80, 277, 1990.
105. Iwamoto, Y., Sasaki, S., and Suzuki, I., Input-output organization of reticulospinal neurons, with special reference to connections with dorsal neck motoneurones in the cat, *Exp. Brain Res.*, 80, 260, 1990.
106. Kakei, S., Muto, N., and Shinoda, Y., Innervation of multiple neck motor nuclei by single reticulospinal tract axons receiving tectal input in the upper cervical spinal cord, *Neurosci. Lett.*, 172, 85, 1994.
107. Kitama, T., Grantyn, A., and Berthoz, A., Orienting related eye-neck neurons of the medial ponto-bulbar reticular formation do not participate in horizontal canal-dependent vestibular reflexes of alert cats, *Brain Res. Bull.*, 38, 337, 1995.
108. Grantyn, A., et al., Relationship between task-related discharge patterns and axonal morphology of brainstem projection neurons involved in orienting eye and head movements, in *Vestibular and Brain Stem Control of Eye, Head and Body Movements,* Shimazu, H. and Shinoda, Y. Eds., Japan Scientific Societies Press, Tokyo, 1992, pp. 255.
109. Isa, T. and Naito, K., Activity of neurons in the medial pontomedullary reticular formation during orienting movements in alert head-free cats, *J. Neurophysiol.*, 74, 73, 1995.
110. Ma, T.P., Graybiel, A.M., and Wurtz, R.H., Location of saccade-related neurons in the macaque superior colliculus, *Exp. Brain Res.*, 85, 21, 1991.
111. Segraves, M.A. and Goldberg, M.E., Functional properties of corticotectal neurons in the monkey's frontal eye field, *J. Neurophysiol.*, 58, 1387, 1987.

112. Tusa, R.J., Zee, D.S., and Herdman, S.J., Effect of unilateral cerebral cortical lesions on ocular motor behavior in monkeys: Saccades and quick phases, *J. Neurophysiol.*, 56, 1590, 1986.
113. Karabelas, A.B., Moschovakis, A.K., and Highstein, S.M., Morphological classification of intracellularly HRP injected efferent neurons of the deeper layers of the primate superior colliculus, *Proc. Soc. Neurosci.,* 12, 1185, 1986.
114. Marrocco, R.T., McClurkin, J.W., and Young, R.A., Spatial properties of superior colliculus cells projecting to the inferior pulvinar and parabigeminal nucleus of the monkey, *Brain Res.*, 222, 150, 1981.
115. Behan, M. and Appell, P.P., Intrinsic circuitry in the cat superior colliculus: Projections from the superficial layers, *J. Comp. Neurol.*, 315, 230, 1992.
116. Mooney, R.D., et al., The projection from the superficial to the deep layers of the superior colliculus: An intracellular horseradish peroxidase injection study in the hamster, *J. Neurosci.*, 8, 1384, 1988.
117. Hall, W.C. and Lee, P., Interlaminar connections of the superior colliculus in the tree shrew. I. The superficial gray layer, *J. Comp. Neurol.*, 332, 213, 1993.
118. Isa, T., Endo, T., and Saito, Y., The visuo-motor pathway in the local circuit of the rat superior colliculus, *J. Neurosci.*, 15, 8496, 1998.
119. Lee, P.H., et al., Role of intrinsic synaptic circuitry in collicular sensorimotor integration, *Proc. Natl. Acad. Sci.*, 94, 13299, 1997.
120. Özen, G., Augustine, G.J., and Hall, W.C., Contribution of superficial layer neurons to premotor bursts in the superior colliculus, *J. Neurophysiol.*, 84, 460, 2000.
121. Meredith, M.A. and Ramoa, A.S., Intrinsic circuitry of the superior colliculus: Pharmacophysiological identification of horizontally oriented inhibitory interneurons, *J. Neurophysiol.*, 79, 1597, 1998.
122. Hikosaka, O. and Wurtz, R.H., Visual and oculomotor functions of monkey substantia nigra pars reticulata. I. Relation of visual and auditory responses to saccades, *J. Neurophysiol.*, 49, 1230, 1983.
123. Joseph, J.P. and Boussaoud, D., Role of the cat substantia nigra pars reticulata in eye and head movements. I. Neural activity, *Exp. Brain Res.*, 57, 286, 1985.
124. Harting, J.K. and van Lieshout, D.P., Spatial relationships of axons arising from the substantia nigra, spinal trigeminal nucleus, and pedunculopontine tegmental nucleus within the intermediate gray of the cat superior colliculus, *J. Comp. Neurol.*, 305, 543, 1991.
125. Aizawa, H., et al., Injection of nicotine into the superior colliculus facilitates occurrence of express saccades in monkeys, *J. Neurophysiol.*, 82, 1642, 1999.
126. Büttner-Ennever, J.A. and Horn, A.K.E., Neuroanatomy of saccadic omnipause neurons in nucleus raphe interpositus, in *Contemporary Ocular Motor and Vestibular Research: A Tribute to David A. Robinson,* Fuchs, A.F., Brandt, T., Büttner, U., and Zee, D.S., Eds., Thieme Medical Publishers, New York, 1994, pp. 489.
127. Pettit, D.L., et al., Local excitatory circuits in the intermediate gray layer of the superior colliculus, *J. Neurophysiol.*, 81, 1424, 1999.
128. Basso, M.A. and Wurtz, R.H., Modulation of neuronal activity in superior colliculus by changes in target probability, *J. Neurosci.*, 18, 7519, 1998.
129. Dorris, M.C., Paré, M., and Munoz, D.P., Neuronal activity in monkey superior colliculus related to the initiation of saccadic eye movements, *J. Neurosci.*, 17, 8566, 1997.
130. Carpenter, R.H.S. and Williams, M.L.L., Neural computation of log likelihood in control of saccadic eye movements, *Nature*, 377, 59, 1995.

131. Dias, E.C. and Bruce, C.J., Physiological correlate of fixation disengagement in the primate's frontal eye field, *J. Neurophysiol.*, 72, 2532, 1994.
132. Hanes, D.P. and Schall, J.D., Neural control of voluntary movement initiation, *Science (Wash. DC)*, 274, 427, 1996.
133. Lee, C., Rohrer, W.H., and Sparks, D.L., Population coding of saccadic eye movements by neurons in the superior colliculus, *Nature*, 332, 357, 1988.
134. Bozis, A. and Moschovakis, A.K., Neural network simulations of the primate oculomotor system. III. A one-dimensional one-directional model of the superior colliculus, *Biol. Cybern.*, 79, 215, 1998.
135. Moschovakis, A.K., Neural network simulations of the primate oculomotor system. I. The vertical saccadic burst generator, *Biol. Cybern.*, 70, 291, 1994.
136. Moschovakis, A.K., et al., An anatomical substrate for the spatiotemporal transformation, *J. Neurosci.*, 18, 10219, 1998.
137. Hess, W.R., Bürgi, S., and Bucher, V., Motor functions of tectal and tegmental areas, in *Biological Order and Brain Organization,* Akert, K., Ed., Springer-Verlag, Berlin, 1981, pp. 269.
138. Paré, M., Crommelinck, M., and Guitton, D., Gaze shifts evoked by stimulation of the superior colliculus in the head-free cat conform to the motor map but also depend on stimulus strength and fixation activity, *Exp. Brain Res.*, 101, 123, 1994.
139. Grantyn, A., et al., Control of gaze by tectal and reticular projection neurons, in *Multisensory Control of Movement,* Berthoz, A., Ed., Oxford University Press, Oxford, 1993, pp. 185.
140. Hikosaka, O., et al., Direct inhibitory synaptic linkage of pontomedullary reticular burst neurons with abducens motoneurons in the cat, *Exp. Brain Res.*, 33, 337, 1978.

6 The Intracollicular Neuronal Network

Gülden Özen, Matthew C. Helms, and William C. Hall

CONTENTS

6.1 INTRODUCTION

While many types of experiments have confirmed that the superior colliculus contributes to the initiation of orienting movements,[1–7] very little is known at the cellular level about **how** its intrinsic circuitry performs this function. We currently are addressing this question by using *in vitro* whole-cell patch-clamp methods to carry out anatomical and physiological analyses of collicular circuitry. The objective of this work is to understand how premotor cells of the intermediate gray layer integrate inputs from sources external to the superior colliculus with the activity of intrinsic collicular circuitry to generate command signals for orienting movements.[8] During the past few years, we used this *in vitro* approach to demonstrate an excitatory columnar pathway from visuosensory cells of the superficial gray layer (SGS) to the premotor cells of the intermediate gray layer (SGI). We then analyzed the synaptic actions of this pathway using similar methods.

6.2 METHODS

As described previously,[9–11] we use coronal or parasagittal slices prepared from the superior colliculus of 14 to 28 day old rats. To analyze synaptic interactions within the slice, whole-cell patch-clamp recording techniques are used in combination with stimulation of presynaptic cells. We stimulate these cells either by the photoactivation of *caged* neurotransmitters or by electrical stimulation. For anatomical analysis, the

0-8493-0097-5/04/$0.00+$1.50

recorded cells are filled with biocytin dissolved in the internal solution that diffuses into the cell from the patch-clamp pipette. Also, we prelabel SGI cells *in vivo* using retrograde axonal transport, by injecting green or red rhodamine beads into the paramedian pontine reticular formation 2 to 4 days before the patch-clamp experiments.

6.3 THE INTERLAMINAR NEURONAL NETWORK

The juxtaposition of cells with sensory and motor functions in the superior colliculus has made this structure an especially useful model for studies of the mechanisms that underlie sensorimotor integration. Many experiments have demonstrated that cells within SGS respond to visual stimuli, while premotor cells within the subjacent SGI generate *bursts* of action potentials that have high command value for initiating orienting movements of the head and eyes.[4,8,12,13] The premotor cells transmit these command signals to the midbrain and pontine centers of the reticular formation that organize shifts in the direction of gaze.[14] Our first experiments were designed to examine the synaptic interactions between the cells in these two layers.

Our initial *in vitro* results indicated that a brief (0.5 ms duration; 1 to 100μA amplitude) electrical stimulus delivered to the SGS evokes prolonged bursts of excitatory synaptic currents (EPSCs) and action potentials in the premotor cells of SGI (Figure 6.1). In contrast, when the same brief electrical stimulus is applied to SGS while the responses of SGS rather than SGI cells are recorded, most of the SGS cells that project to SGI respond with only a transient current and a single action potential (Figure 6.2A). These results suggest that the signals transmitted from SGS to SGI are too transient to account for the prolonged bursts of EPSCs or action

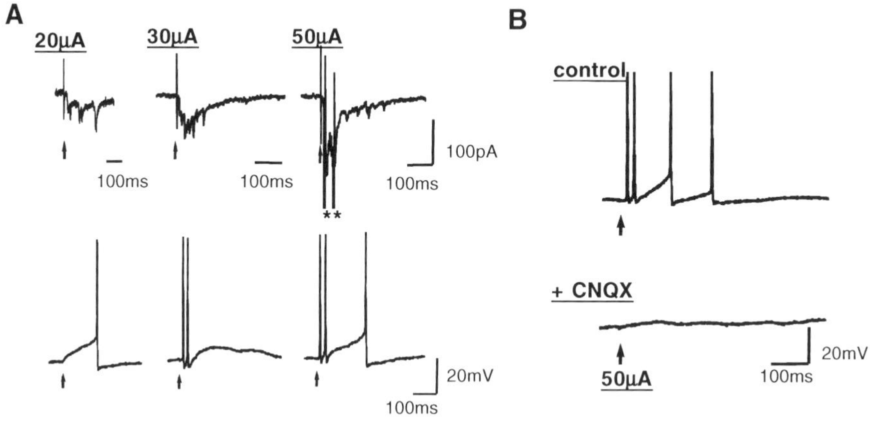

FIGURE 6.1 (A) Brief (0.5 ms) electrical stimuli at increasing intensities in the superficial layer evoke increasing levels of synaptic excitation in the intermediate layer. The upper traces are voltage clamp recordings and the lower traces are current clamp recordings. In the experiment illustrated by the recording at the upper right, a 50 μA SGS stimulus is strong enough to evoke unclamped action currents (*). (B) The evoked responses can be blocked by the addition of the glutamatergic receptor blocker CNQX to the bath, indicating that the responses are synaptically evoked via glutamatergic synapses.

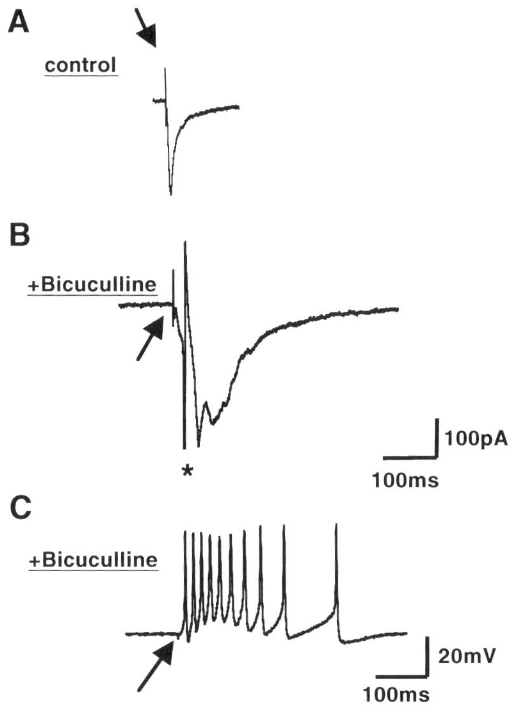

FIGURE 6.2 (A) Brief (0.5 ms) electrical stimulation of the superficial layer evokes a transient inward synaptic current in most SGS wide- and narrow-field neurons (holding potential = -65 mV). (B) and (C) The addition of bicuculline (10 μM) to the bath solution changes the response of this wide-field neuron to a prolonged burst of EPSCs and action potentials, indicating that intracollicular GABAergic inhibition plays a role in shortening the duration of the SGS responses (*).

potentials evoked in SGI. However, when bicuculline (10 μM) is applied to the slice, the responses evoked in the SGS cells change to bursts of ESPCs and multiple action potentials (Figures 6.2B and 6.2C). These results indicate that the transient responses recorded from SGS cells *in vitro* are constructed at least in part by intracollicular GABAergic circuitry. The same inhibitory circuitry may also contribute to the phasic responses of the *novelty detector* cells that are common to SGS *in vivo*.[4]

Early in our *in vitro* experimentation, it became apparent that, due to the persistent problem of activation of axons of passage from unknown sources, traditional electrical stimulation methods were incapable of providing the precision we needed to analyze the functions of local circuits. To overcome this problem, we adopted the method of photostimulation. The goal of photostimulation is to use a high power source of UV light to cleave a molecular *cage* surrounding the active portion of a neurotransmitter molecule. For excitation of neurons, the neurotransmitter of choice is glutamate. Glutamate is the pervasive excitatory neurotransmitter in the central nervous system, and glutamate receptors are found on the dendrites and somas of nearly every CNS neuron. Equally important, they are not present on axons of passage.[15,39,40] Thus, photostimulation allows for precise spatial measurements of the effects of stimulating specific sites in a brain slice without the confounding effects of activating axons of passage that arise elsewhere.

To stimulate by photostimulation, ultraviolet light from an argon ion laser is delivered through a microscope objective and focused on the slice to obtain the photorelease of the caged neurotransmitter. With this arrangement, caged neurotransmitter is photolyzed consistently over an area 50 μm in diameter at the focal plane.[16] An electronic shutter is used to vary the duration of the light pulse (2 to 10 ms) and thus the total amount of photostimulation.

Uncaging glutamate (laser spot diameter ~50μm) in the vicinity of a patch clamped neuron in the premotor layer evokes (1) long-lasting inward currents resulting from direct activation of glutamate receptors expressed by the patch clamped neuron itself (Figure 6.3A) and (2) brief, synaptically mediated currents caused by photo activation of neurons presynaptic to the patch-clamped neuron (Figure 6.3B and 6.3C). Direct responses are only seen when the photostimulus is near the soma or dendrites of the recorded neuron, that is, near processes of the recorded cell that express glutamate receptors. In contrast, the brief synaptic currents can be evoked by stimulation at appropriate sites far removed from the recorded cell and provide a means for mapping the synaptic relationships among cells in the slice. Figure 6.3 also illustrates the precision of the method. At stimulation site A, which is directly over the recorded cell, a large direct response is evoked, whereas a stimulus at site B, which is only 50 μm away, evokes only synaptically mediated currents. At site C, which is 50 μm from site B, no responses are evoked.

Our ability to distinguish experimentally between the direct and synaptic responses is critical for the interpretation of the experiments and therefore is illustrated in Figure 6.4. In this experiment, both inhibitory and excitatory synaptic responses are eliminated by including 1 μ*M* TTX in the bath. TTX blocks action potentials in the presynaptic cells and thus eliminates the rapid, synaptically mediated currents. Then, the addition of the glutamate receptor blockers APV (50 μ*M*) and CNQX (10 to 20 μ*M*) eliminates the large direct response by blocking the receptors on the recorded cell.

By using photostimulation methods to study the circuitry that links the superficial and intermediate layers, we are able to provide convincing confirmation of the earlier electrical stimulation studies.[9,10,17] In the photostimulation experiments, large and prolonged clusters of EPSCs that lasted up to several hundred milliseconds are evoked by photostimulation at sites in SGS as far as 1300 μm from the recorded SGI cell (site # 19, Figures 6.5A and 6.5B). These evoked responses are eliminated by the addition of TTX to the slice bath, confirming that they are synaptically mediated (site # 14, Figure 6.5B). These results demonstrate that the intracollicular pathway from the visual to the premotor layer is robust, columnar, and excitatory.

The prolonged clusters of evoked EPSCs observed in SGI neurons following stimulation in SGS (Figure 6.1A) are also evidence that the membrane properties of the SGI cells are not responsible for their prolonged responses to transient signals from SGS. Instead, these clusters indicate that circuitry intercalated between SGS and SGI enhances and prolongs the responses of premotor cells to transient signals. The preservation of this circuitry in the slice gives us an opportunity to analyze

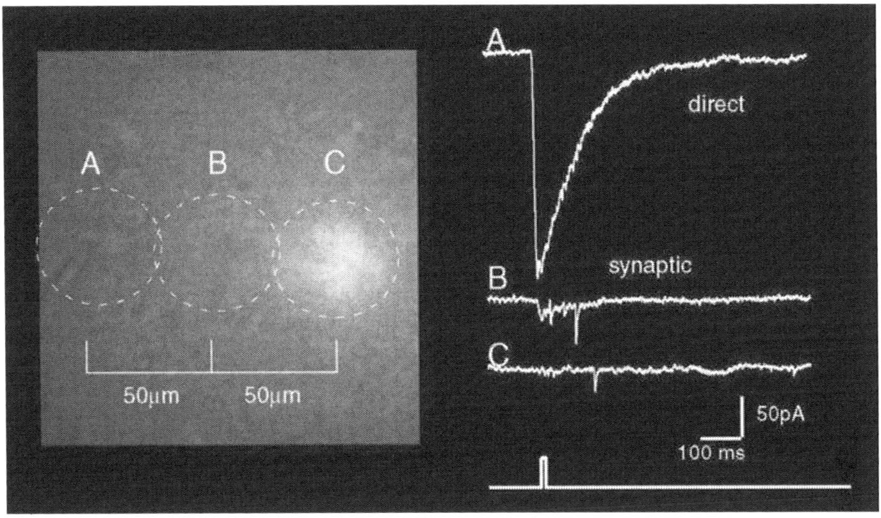

FIGURE 6.3 (See color insert following page 176.) (A) Photostimulation with uncaged glutamate directly activates glutamate receptors on the soma and dendrites of the recorded cell and evokes a large, slow response. In this photograph of the living slice, the micropipette can be seen in contact with the cell body. The dashed circle indicates the size of the laser spot, which was photographed using a long exposure time after the experiment in C. (B) In contrast to stimulation at site A, stimulation at site B, which is only 50 µm from A, evoked brief synaptic events, but little or no direct response. (C) Stimulation at this site failed to evoke a response.

mechanisms responsible for generating the buildup of the SGI cell responses with a precision that was never before possible.

If transient signals are transmitted from SGS to SGI, the responses of SGI cells could be amplified by positive feedback mediated by the recurrent collaterals arising from SGI cells. Local collaterals have been demonstrated for SGI cells and may synapse on both the recorded and neighboring cells[11] to strengthen and prolong the evoked responses. Our results from *in vitro* patch-clamp experiments are consistent with this alternative. As Figure 6.6 illustrates, experiments using photostimulation indicate that excitatory responses are mostly evoked by stimulation within 300 µm of the soma horizontally, and only occasionally by stimulation at sites 500 µm or more lateral to the postsynaptic cell, even though stimulation superficial to the same cell can evoke responses from distances greater than 1300 µm (Figures 6.5 and 6.6). Thus, these results support the view that positive feedback contributes to the buildup of command signals within a spatially restricted region of the motor map in SGI.

6.4 HORIZONTAL INTERACTIONS IN THE INTERMEDIATE GRAY LAYER

We have used the same approach to compare excitatory and inhibitory connections within the intermediate gray layer (Figure 6.6). These experiments were designed

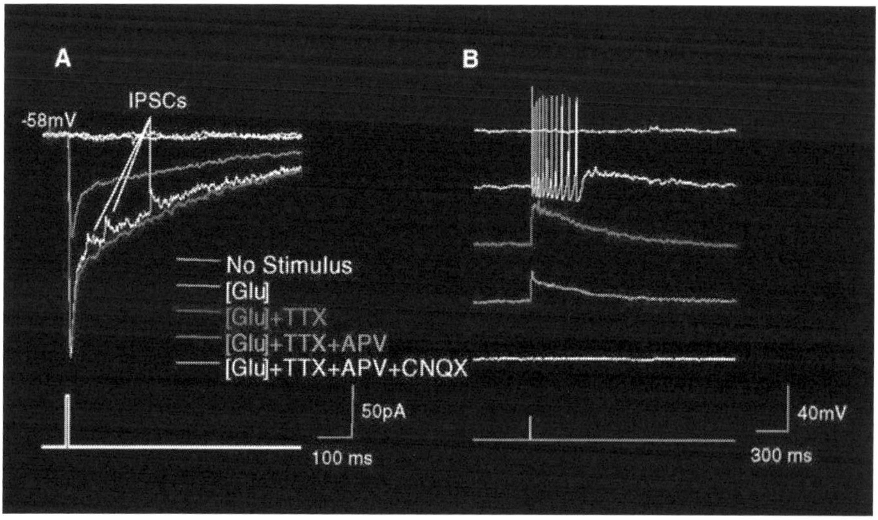

FIGURE 6.4 (See color insert following page 176.) (A) Photostimulation over the cell body elicits a direct response with outward synaptic currents superimposed (IPSCs, yellow). These IPSCs are evoked by inputs from local inhibitory interneurons that are presynaptic to the patch-clamped cell and are excited by the photostimulus to fire action potentials. The synaptic mediation of the IPSCs is indicated by the observation that they are blocked by bath application of TTX, which eliminates action potentials generated by presynaptic cells (red). The response that remains after application of TTX is evoked by direct activation of receptors on the recorded cell, and therefore can be blocked by application of NMDA and AMPA receptor blockers (gray). (B) Current clamp responses evoked by uncaging glutamate directly over the soma illustrates the efficacy of photostimulation in producing action potentials in slice neurons. Photostimulation at the soma is sufficient to elicit a high frequency train of action potentials in this neuron (yellow). TTX blocks the action potentials and the glutamatergic blockers APV and CNQX eliminate the remaining direct response.

to test the idea that short range recurrent excitation and long range inhibition mediate a *winner-take-all* competition among different populations of premotor cells that generate the spatially encoded command signals for orienting movements to competing sensory targets. Models of collicular function hypothesize that long-range inhibitory projections attenuate the activity of cells that otherwise would initiate unwanted saccades to the less salient targets in the visual field.[18–20]

In fact, several types of evidence have been presented in support of the idea that such a widespread inhibitory mechanism operates within the premotor layer of the superior colliculus. The first experimental evidence for intercollicular inhibition was Sprague's demonstration that transection of the tectal commissure restores visually guided responses toward the opposite hemifield after they had been eliminated by unilateral ablations of the visual cortex.[21] In another type of study, glutamate was applied to one site in the colliculus of rats and cats to increase background-firing rate. Then, other sites in the ipsilateral and contralateral colliculi were electrically stimulated.[22] Inhibition, measured at the site where glutamate was applied as a decrease in background-firing rate, was obtained after stimulation throughout both colliculi. A

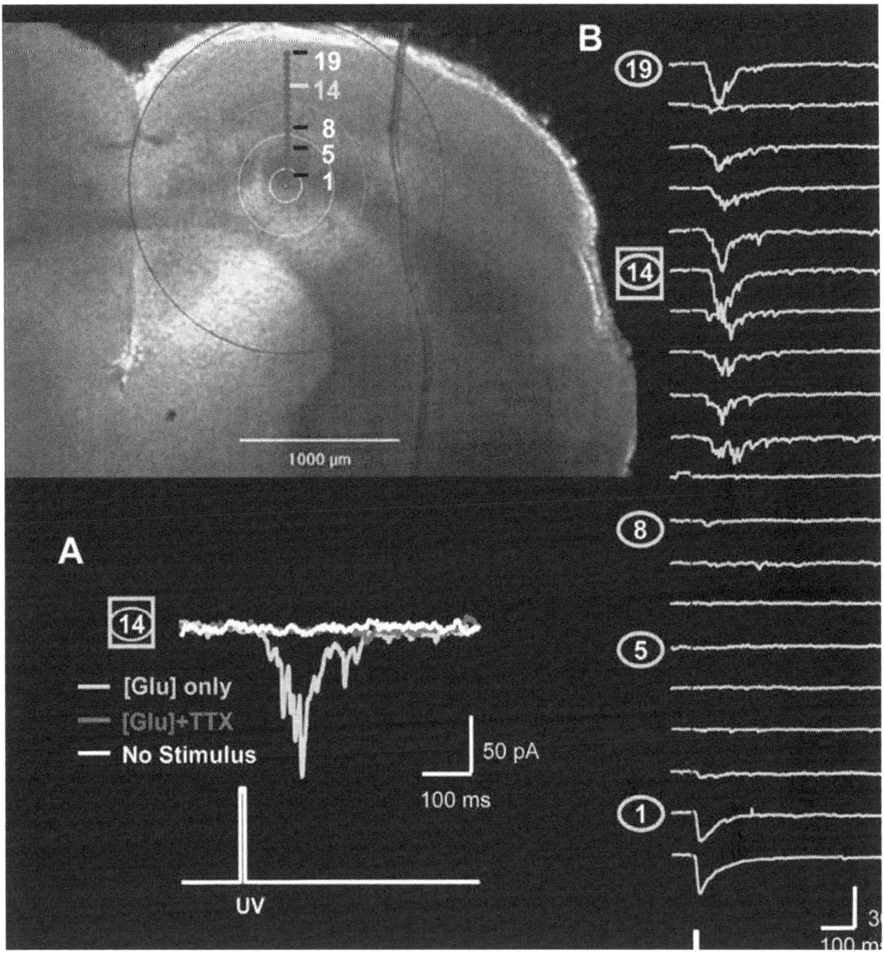

FIGURE 6.5 (See color insert following page 176.) Synaptic circuitry between superficial and intermediate layers is demonstrated by short and long distance (6 to 1000 µm) photostimulation. The patch-clamped cell is located in SGI, at the yellow dot just below stimulation site #1. (A) The responses evoked in the SGI cell can be blocked by TTX, indicating that they are mediated through the activity of presynaptic cells that are activated by the photostimulus in SGS. (B) Robust EPSCs can be evoked from stimulation sites as far away from the patch clamped cell as the upper superficial gray layer (site #19).

similar result was obtained when simultaneous recordings were made from the two colliculi while cats produced horizontal and oblique saccades — the background rate of premotor layer activity increased preceding contralateral saccades and decreased preceding ipsilateral saccades.[23] Also in the cat, stimulation of one colliculus evoked monosynaptic IPSPs in premotor cells on the opposite side.[24,25] In the monkey, the discharges of premotor cells also could be inhibited by electrically stimulating the opposite superior colliculus.[26] Anatomical tracing experiments combined with immunohistochemical methods have provided further support for intercollicular inhibition

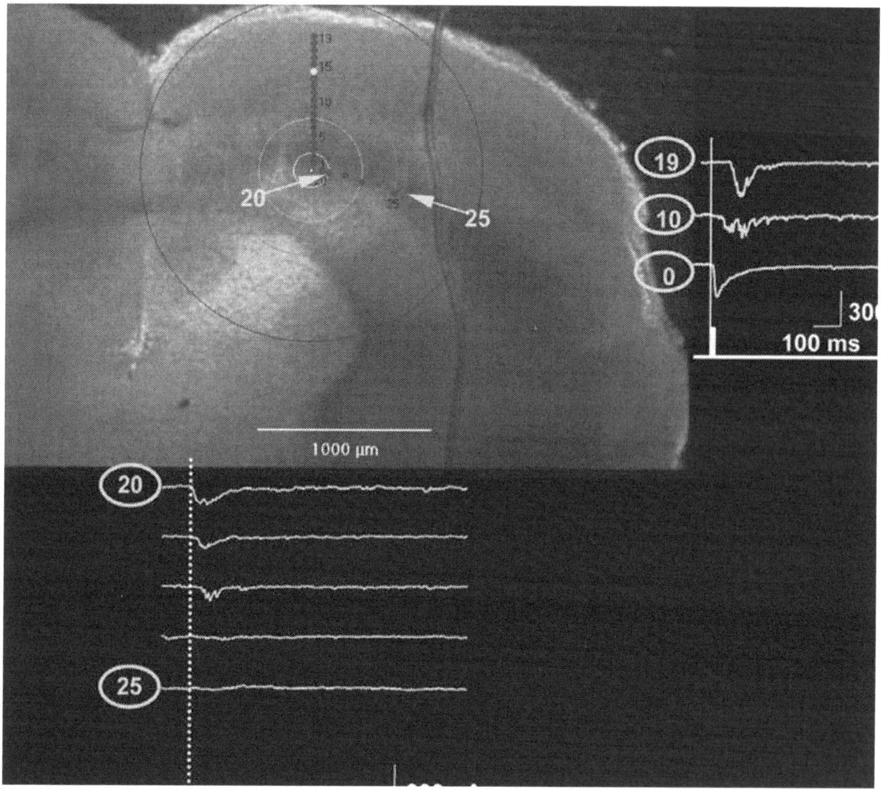

FIGURE 6.6 (See color insert following page 176.) In this experiment, responses were evoked in an SGI cell from vertical (sites # 0 to 19) and horizontal (sites # 20 to 25) sequences of photostimulation sites. The cell, whose soma location is indicated by the small yellow circle at the origin, was excited by overlying vertical sites in the optic and superficial layers that were as far away as 800 μm, but was excited by horizontal sites only as distant as 200 μm (yellow arrows).

by showing that the cells of origin for many commissural fibers are GABAergic.[27,28] As the two colliculi serve to orient the animal toward opposite hemifields, this putative intertectal inhibition might serve to reduce competing commands for orienting movements toward targets located in opposite halves of the visual field.

A similar mechanism may account for the inverse relationship between activity in the rostral pole of the colliculus, which has been variously proposed by different authors to be a *fixation*[29,30] or *microsaccade*[31] zone, and more caudal, saccade-generating regions. For example, stimulation of the rostral pole evokes short latency inhibition in more caudal premotor cells, and stimulation of the rostral premotor layer raises the threshold for evoking saccades electrically.[32,33] Similarly, as the build-up of responses of premotor cells in the caudal colliculus increases, the activity of cells in the rostral zone decreases, and vice versa.[34] Finally, if rostral pole cells are inhibited with muscimol, monkeys are less able to suppress saccades and make more very short latency *express* saccades.[29]

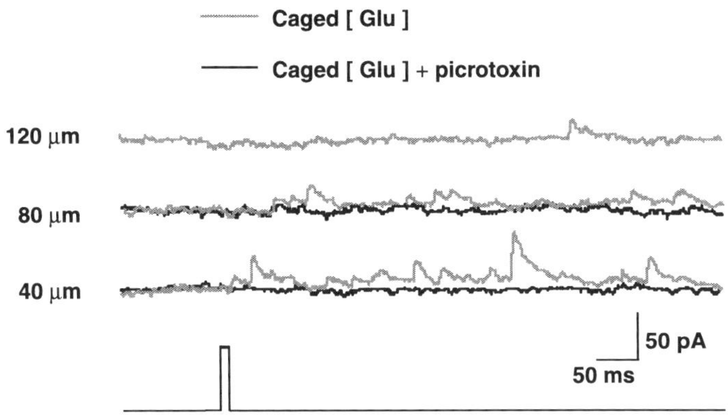

FIGURE 6.7 In this experiment, stimulation at sites 40 and 80 μm from the recorded cell evoked clusters of inhibitory synaptic currents (IPSCs). The IPSCs could be blocked by bath application of the selective chloride channel blocker picrotoxin, indicating that they are mediated by $GABA_A$ receptors.

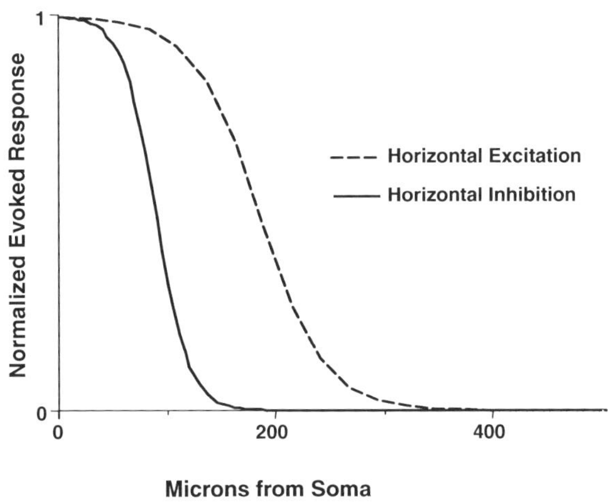

FIGURE 6.8 Schematic summary of the responses evoked in all intermediate layer neurons (N = 14) by stimulation within the intermediate layer. Horizontal synaptic excitation and inhibition are only evoked within a few hundred microns of the soma.

Although experiments of the type just summarized comprise an argument for widespread inhibitory influences in the colliculus, technical difficulties and other data raise questions about the interpretation of the results. The most serious difficulty is the potential contribution of axons of passage to the results of both the physiological and the anatomical experiments. For example, in the experiments in which sites are activated by electrical stimulation, the inhibitory responses could have been evoked by stimulating fibers of passage that arise from structures outside of the superior colliculus. One potential source of such fibers is the GABAergic subdivision of substantia nigra pars reticulata, which projects throughout the ipsilateral premotor layer and, through the collicular commissure, to the contralateral colliculus as well.[35–37] The demonstration that lesions in the substantia nigra contralateral to the unilateral visual cortical lesion reproduce the Sprague effect[37] is consistent with this alternative, and serves to emphasize the possibility that an extrinsic structure like substantia nigra might be responsible for most or all of the global inhibitory effects that have been observed after electrical stimulation in the colliculus. The anatomical experiments also are subject to the same potential source of error, as retrograde tracers can be picked up and transported by fibers of passage. It is therefore possible that the GABAergic cells labeled by retrograde transport following collicular injections give rise to axons that cross the commissure, but terminate in structures outside of the colliculus. One such potential destination is the contralateral vertical gaze center of the rostral midbrain.[38]

Our results using photostimulation in slices indicate that inhibitory synaptic circuitry within SGI is even more restricted spatially than the excitatory. In these experiments, prominent IPSCs, which could be blocked by the addition of the chloride channel blocker picrotoxin, could be evoked by stimulation within 100 μm of the recorded cell (Figure 6.7). However, even when the reversal potential of chloride (E_{Cl^-}) was lowered to –82 mV by changing the composition of the patch pipette solution, IPSCs could not be detected at holding potentials changing from –98 mV to –53 mV after uncaging glutamate at sites over 150 μm from the cell (Figures 6.4, 6.7, and 6.8). These results suggest that local inhibition may play a role in shortening, or *clipping*, the duration of the premotor cell responses, but are not consistent with a role for intratectal inhibition in a wide ranging competitive mechanism.

6.5 SUMMARY AND FUTURE DIRECTIONS

Multiple pathways, which arise from sources both intrinsic and extrinsic to the superior colliculus, utilize multiple neurotransmitters and converge within the motor layers. Their integrative activity provides the mechanism responsible for the selection and generation of the premotor commands for orienting responses that are transmitted from the colliculus to the brainstem gaze centers. *In vitro* whole cell patch-clamp recording in combination with photostimulation provides a powerful approach for examining how the circuits in the superior colliculus contribute to this example of sensorimotor integration. While the experiments summarized in this chapter utilized the photolysis of caged glutamate to activate cells in the colliculus, other caged neurotransmitters also are available. These tools provide a means to assess how pathways utilizing various transmitters interact to govern the activity of an identified population of premotor cells with the well-established function of initiating orienting movements. These approaches thus offer new opportunities to study the neural mechanisms underlying sensorimotor integration with a precision never possible before.

ACKNOWLEDGMENTS

Supported by NIH grant EI 08233. We thank Dr. Adonis Moschovakis for his helpful comments.

REFERENCES

1. Mohler, C.W. and Wurtz, R.H., Organization of monkey superior colliculus: Intermediate layer cells discharging before eye movements, *J. Neurophysiol.,* 39, 722, 1976.
2. Robinson, D.A., Eye movements evoked by collicular stimulation in the alert monkey, *Vision Res.,* 12, 1795, 1972.
3. Schiller, P.H., Sandell, J.H., and Maunsell, J.H.R., The effect of frontal eye field and superior colliculus lesions on saccadic latencies in the rhesus monkey, *J. Neurophysiol.,* 57, 1033, 1987.
4. Schiller, P.H. and Stryker, M., Single unit recording and stimulation in superior colliculus of the alert monkey, *J. Neurophysiol.,* 35, 915, 1972.
5. Schmidt, M. et al., Disinhibition in rat superior colliculus mediated by $GABA_c$ receptors, *J. Neurosci.,* 21, 691, 2001.
6. Schneider, G.E., Two visual systems: brain mechanisms for localization and discrimination are dissociated by tectal and cortical lesions, *Science,* 163, 895, 1969.
7. Wurtz, R.H. and Goldberg, M.F., Activity of superior colliculus in behaving monkey. III. Cells discharging before eye movements, *J. Neurophysiol.,* 35, 575, 1972.
8. Sparks, D.L., Functional properties of neurons in the monkey superior colliculus: coupling of neuronal activity and saccade onset, *Brain Res.,* 156, 1, 1978.
9. Lee, P.H. et al., Role of intrinsic synaptic circuitry in collicular sensorimotor integration, *Proc. Natl. Acad. Sci. U.S.A.,* 94, 13299, 1997.
10. Özen, G., Augustine, G.J., and Hall, W.C., Contribution of superficial layer neurons to premotor bursts in the superior colliculus, *J. Neurophysiol.,* 84, 460, 2000.
11. Pettit, D.L. et al., Local excitatory circuits in the intermediate gray layer of the superior colliculus, *J. Neurophysiol.,* 81, 1424, 1999.
12. Munoz, D.P., Guitton, D., and Pélisson, D., Control of orienting gaze shifts by the tectoreticulospinal system in the head-free cat. III. Spatiotemporal characteristics of phasic motor discharges, *J. Neurophysiol.,* 66, 1642, 1991.
13. Peck, C.K., Saccade-related burst neurons in cat superior colliculus, *Brain Res.,* 408, 329, 1987.
14. Keller, E.L., Colliculoreticular organization in the oculomotor system, *Prog. Brain Res.,* 50, 725, 1979.
15. Callaway, E.M. and Katz, L.C., Photostimulation using caged glutamate reveals functional circuitry in living brain slices, *Proc. Natl. Acad. Sci. U.S.A.,* 90, 7661, 1993.
16. Pettit, D.L. et al., Chemical two-photon uncaging: a novel approach to mapping glutamate receptors, *Neuron*, 19, 465, 1997.
17. Isa, T., Endo, T., and Saito, Y., The visuo-motor pathway in the local circuit of the rat superior colliculus, *J. Neurosci.,* 18, 8496, 1998.
18. Arai, K., Keller, E.L., and Edelman, J.A., Two-dimensional neural network model of the primate saccadic system, *Neural Networks,* 7, 1115, 1994.
19. Das, S., Keller, E.L., and Arai, K., A distributed model of the saccadic system: the effects of internal noise, *Neurocomp.,* 11, 245, 1996.
20. Van Gisbergen, J.A.M., Van Opstal, A.J., and Tax, A.A.M., Collicular ensemble coding of saccades based on vector summation, *Neuroscience,* 21, 541, 1987.
21. Sprague, J.M., Interaction of cortex and superior colliculus in mediation of visually guided behavior in the cat, *Science,* 153, 1544, 1966.
22. Douglas, R.M. and Vetter, M., Widespread inhibition and target selection in the superior colliculus, *Soc. Neurosci. Abstr.*, 12, 458, 1986.

23. Infante, C. and Leiva, J., Simultaneous unitary neuronal activity in both superior colliculi and its relation to eye movements in the cat, *Brain Res.,* 381, 390, 1986.
24. Maeda, M., Shibazaki, T., and Yoshida, K., Monosynaptic inhibition evoked in superior colliculus neurons following contralateral collicular stimulation, in *Integrative Control Functions of the Brain*, Ito, M., Ed., Elsevier, Amsterdam, 1979, pp. 68–71.
25. Moschovakis, A.K. and Karabelas, A.B., Tectotectal interactions in the cat, *Proc. Soc. Neurosci.,* 8, 293, 1982.
26. Munoz, D.P. and Istvan, P.J., Lateral inhibitory interactions in the intermediate layers of the monkey superior colliculus, *J. Neurophysiol.,* 79, 1193, 1998.
27. Appell, P.P. and Behan, M., Sources of subcortical GABAergic projections to the superior colliculus in the cat, *J. Comp. Neurol.,* 302, 143, 1990.
28. Olivier, E. et al., Evidence for glutamatergic tectotectal neurons in the cat superior colliculus: a comparison with GABAergic tectotectal neurons, *Eur. J. Neurosci.*, 12, 2354, 2000.
29. Munoz, D.P. and Wurtz, R.H., Role of the rostral superior colliculus in active visual fixation and execution of express saccades, *J. Neurophysiol.,* 67, 1000, 1992.
30. Munoz, D.P. and Wurtz, R.H., Fixation cells in monkey superior colliculus. I. Characteristics of cell discharge, *J. Neurophysiol.,* 70, 559, 1993.
31. Krauzlis, R.J., Basso, M.A., and Wurtz, R.H., Shared motor error for multiple eye movements, *Science,* 276, 1693, 1997.
32. Meredith, M.A. and Ramoa, A.S., Intrinsic circuitry of the superior colliculus: pharmacophysiological identification of horizontally oriented inhibitory interneurons, *J. Neurophysiol.,* 79, 1597, 1998.
33. Munoz, D.P. and Wurtz, R.H., Fixation cells in monkey superior colliculus. II. Reversible activation and deactivation, *J. Neurophysiol.,* 70, 576, 1993.
34. Munoz, D.P. and Wurtz, R.H., Saccade-related activity in monkey superior colliculus. I. Characteristics of burst and buildup cells, *J. Neurophysiol.,* 73, 2313, 1995.
35. Bickford, M.E. and Hall, W.C., The nigral projection to predorsal bundle cells in the superior colliculus of the rat, *J. Comp. Neurol.,* 319, 11, 1992.
36. May, P.J. and Hall, W.C., Relationships between the nigrotectal pathway and the cells of origin of the predorsal bundle, *J. Comp. Neurol.*, 226, 357, 1984.
37. Wallace, S.F., Rosenquist, A.C., and Sprague, J.M., Ibotenic acid lesions of the lateral substantia nigra restore visual orientation behavior in the hemianopic cat, *J. Comp. Neurol.,* 296, 222, 1990.
38. Edwards, S.B., The commissural projection of the superior colliculus in the cat, *J. Comp. Neurol.,* 173, 23,1977.
39. Dalva, M.B. and Katz, L.C., Rearrangements of synaptic connections in visual cortex revealed by laser photostimulation, *Science,* 265, 255, 1994.
40. Katz, L.C. and Dalva, M.B., Scanning laser photostimulation: a new approach for analyzing brain circuits, *J. Neurosci. Meth.,* 54, 205, 1994.

7 Dynamic Modulation of Signal Transmission through Local Circuits

Tadashi Isa, Yasushi Kobayashi, and Yasuhiko Saito

CONTENTS

7.1 INTRODUCTION

Visually guided saccades reorient our gaze to objects in the visual field. Only one saccade is executed at a time, even when there are several potential targets present simultaneously in the visual field. Upon target fixation, saccades can be released or not released at will. Moreover, the timing of visually guided saccades depends on the behavioral context in which they are released. As first described by Fischer and colleagues,[1,2] the probability of occurrence of saccades called *express* due to their extremely short (80 to 120 ms) reaction time (SRT), dramatically increases when a time gap lasting for a few hundred milliseconds (gap period) separates the time when the stimulus being fixated is turned off and a new stimulus, the saccade target, is turned on. Therefore, it is reasonable to think that the neural processes that determine *when* and *where* to make saccades engage components of brain circuits responsible for decision making (for review see Reference 3).

Where to make saccades is presumably determined by the location of the activated cell population in the spatial map of saccade vectors within the superior colliculus (SC).[4–8] On the other hand, *when* to make saccades is determined by the

0-8493-0097-5/04/$0.00+$1.50

abrupt onset of the movement-related bursts of deeper layer neurons. One typical example is shown in Figure 7.1. Figure 7.1A illustrates the discharge of a superficial layer (SGS) neuron and an intermediate layer (SGI) neuron for *regular saccades* (SRTs > 150 ms) in the no-gap condition. Figure 7.1B illustrates the discharge of the same neurons for express saccades (in a gap task with period equal to 170 ms). As shown here, the discharge of the SGS neuron was virtually the same for regular and express saccades. In both cases, it exhibited phasic visual responses, with latencies ranging between 40 and 50 ms. Thus, the execution of express saccades cannot be ascribed to the modification of the discharge of SGS neurons. On the other hand, the SGI neuron did not exhibit the same pattern of activation for regular and express saccades. For regular saccades, the SGI neuron exhibited a weak discharge with a latency of 40 to 50 ms, i.e., similar to that of the visual responses of SGS neurons. This weak, phasic, *visual response* did not appear to be strong enough, by itself, to trigger the saccade burst generator circuits downstream of the SC (for descriptions of the saccade burst generators and the input they receive from the SC, see References 9, 10, and 11). Then, 50 to 80 ms later, the SGI neuron exhibited a strong burst of spikes (*premotor burst*), which appeared to trigger saccades. In contrast, for express saccades, a gradual increase of discharge (*buildup activity*) was observed during the gap period (arrow in Figure 7.1B). The buildup was followed by a strong unimodal burst of spikes with a latency equal to that of visual responses that occur prior to regular saccades. In the gap paradigm, this unimodal burst appears to trigger saccades. This observation is consistent with the idea that the visual and motor bursts fuse in SGI neurons to trigger express saccades (Dorris et al.,[12] Edelman and Keller[13] and Sparks et al.[14]). Since SRTs are negatively correlated with the firing frequency of the buildup activity just preceding the visual response,[12] the data also suggest that when SGI neurons exhibit higher buildup activity, their weak visual responses (Figure 7.1A) can be transformed into bursts strong enough to trigger saccades. Saccades could also be triggered if buildup activity surpasses a threshold for the generation of bursting discharges.

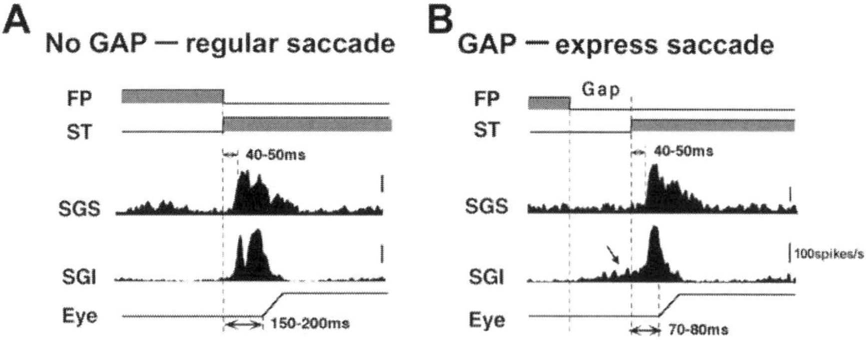

FIGURE 7.1 Activity of SC neurons during saccades in the monkey. Spike density functions of exemplary neurons are shown. One was located in the superficial layer (SGS) of the SC, and one in the intermediate layer (SGI). (A) Responses during regular saccades in the NO GAP paradigm. (B) Responses during express saccades in the GAP paradigm. FP, presentation of fixation point; ST, presentation of saccade target; Eye, eye position.

This chapter focuses on neuronal mechanisms intrinsic to SC local circuits, which could be responsible for the generation of the saccade-related bursts by SGI neurons. It also addresses the origin of the visual signals received by SGI neurons. The presence or absence of an interlaminar SC projection is crucial for addressing the second one of these issues. Robinson measured the saccades evoked by electrical stimulation to first describe the topographical map of saccade vectors in the deeper layers of the SC[5] and in the same year Schiller and Stryker[6] demonstrated that the motor map was in register with the map of visual space represented in the superficial layers. Following these demonstrations, it was reasonable to suppose that the superficial and deeper layers of the SC are organized in a columnar fashion by interlaminar connections that link them. However, the existence of such a connection and its functional implications have been matters of long-lasting debate in the field. Evidence against the existence of interlaminar connections was presented by Edwards[15] on the basis of anatomical studies. In addition, Mays and Sparks[16] showed that during certain saccades, such as, a saccade to the locus of a target that disappeared during the course of a preceding saccade, the bursting discharges of SGI neurons need not be preceded by activation of overlying SGS neurons. On the other hand, Maeda et al.[17] showed that short latency EPSPs can be induced in SGI neurons following electrical stimulation of the optic nerve in anesthetized cats. The existence of an interlaminar connection was unequivocally demonstrated anatomically by Moschovakis and his colleagues in the monkey[18,19] and by Rhoades and his colleagues in the hamster[20–22] and later confirmed in additional species.[23,24] The anatomy of this pathway is briefly summarized in Chapter 5.

This chapter focuses on the functional implications of this interlaminar pathway and on the identity of the neurotransmitters it employs. Our discussion primarily relies on recent experimental observations from both *in vitro* and *in vivo* preparations in our laboratory. We employed whole-cell patch-clamp recording and intracellular staining techniques, which offer powerful approaches for studying the dynamic properties of local brain circuits. The SC is an excellent choice for studies of slice preparations, since the input-output relationships of its primary afferent and efferent pathways, and the information encoded by individual neurons has been well defined in behaving animals.[8,25,26] Slices obtained from rodents were used to study in detail the fundamental structure of the SC local circuits, that is, the synaptic relationships between the neurons it contains and the mechanisms responsible for their modulation. The information obtained from the *in vitro* studies was then used to design *in vivo* experiments in monkeys. Combining these two experimental approaches has opened new vistas for understanding how the cognitive behavior of the awake, behaving animal is influenced by the dynamic regulation of signal transmission in local SC circuits.

7.2 THE INTERLAMINAR CONNECTION OF THE MAMMALIAN SUPERIOR COLLICULUS

7.2.1 EXPERIMENTS IN THE *IN VITRO* SLICE PREPARATION

Figure 7.2A is a pictorial summary of the design of our experiments in slices obtained from rats 17 to 22 days of age. Stimulating electrodes were placed in the SGS and

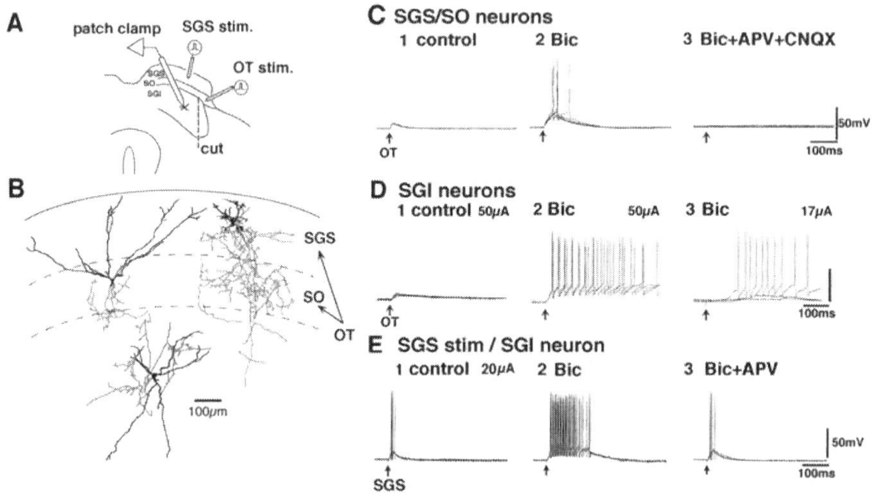

FIGURE 7.2 Effect of the OT and SGS stimulation on the SGI neurons studied in the slice preparation of the rat SC. (A) Experimental arrangement. Bipolar stimulating electrode was placed either in the most dorsal portion of the SO to stimulate the bundle of the optic fibers or in the SGS. In the former case, a cut was added to the slice to eliminate the possibility of current spread outside the OT to activate the SGI neurons. (B) Examples of morphology of recorded cells in the SGS, SO, and SGI, stained with intracellular injection of biocytin. (C) Effects of the OT stimulation on a SGS neuron in the control solution (1) and effect of application of 10 μ*M* bicuculline (2) and further addition of 10 μ*M* CNQX and 50 μ*M* APV. Basically similar results were obtained from SO neurons. (D) The effects of the OT stimulation on a SGI neuron in the control solution (1) and after application of 10 μ*M* bicuculline (2). At the threshold current intensity (17 μA), the bursting response occurred in an all-or-none manner (3). (E) Effects of stimulation of the SGS onto a SGI neuron in the control solution (1), after application of bicuculline (10 μ*M*) (2) and application of 10 μ*M* APV largely abolished the long-lasting depolarization (3).

in the optic tract (OT) near the lateral border of the optic layer (SO) where the OT comprises a bundle of fibers. The responses of SGS, SO, and SGI neurons evoked by stimulation at these sites was recorded through patch pipettes.[27] Single electrical pulses delivered through the OT electrode induced short latency monosynaptic EPSPs (Figure 7.2C1) in both narrow field vertical cells of the SGS, which project extensively to the SO and in wide field vertical cells of the SO, which projects to the SGI (Figure 7.2B). The EPSPs were enhanced by application of the $GABA_A$ receptor antagonist bicuculline (Figure 7.2C2) and were completely suppressed by application of the glutamate receptor antagonists CNQX and APV (Figure 7.2C3). These findings demonstrate that both SGS and SO neurons receive glutamatergic monosynaptic excitation from the OT. To judge from their longer latency and fluctuating onset latencies, the EPSPs evoked in SGI neurons following stimulation of the OT (Figure 7.2D1) were di- or oligosynaptic. These responses were markedly enhanced by application of $GABA_A$ receptor antagonists such as bicuculline (Figure 2D2) or SR95531; the bursting spike discharges evoked in the SGI cells were superimposed on large clusters of EPSPs, which could last longer than 1 s, even when single brief electrical pulses were delivered through the OT electrode. The long lasting depolarization and bursting spike discharges were evoked in an all-or-nothing fashion at threshold stimulus intensities (Figure 7.2D3). Stimulation of the SGS induced monosynaptic EPSPs in SGI neurons, which were again amplified into bursting spike discharges superimposed on long lasting depolarizations following application of bicuculline (Figure 7.2E). These results confirm the existence of an excitatory pathway from the OT to SGI neurons, presumably mediated by SGS or SO neurons, as previously demonstrated by Lee et al.,[28] using whole-cell recording in SC slices from the tree shrew. The latter authors also argued that release from $GABA_A$ receptor-mediated inhibition is needed for SGI neurons to exhibit vigorous bursting responses to SGS stimulation. The bursting responses evoked in SGI neurons by OT stimulation observed *in vitro* following the blockade of $GABA_A$ receptor-mediated synaptic transmission could correspond to the presaccadic bursts of SGI neurons recorded in behaving cats and monkeys. In these species, substantia nigra pars reticulata (SNr) efferents have been shown to contact[29] and inhibit[19,30] SGI neurons via $GABA_A$-receptor synapses. A similar pattern of inhibition may be mediated by neurons in the rostral pole of the SC (fixation neurons).[31,32] Both classes of cells are tonically active and pause firing just before saccade onset in monkeys,[32–35] thus, releasing SGI neurons from tonic $GABA_A$ receptor-mediated inhibition before the onset of saccades. In support of a role for disinhibition in the generation of the bursts, Pettit et al.[36] reported that while bicuculline is not necessary for SGI neurons to emit bursts, their depolarization is far more vigorous and long lasting when the inhibitory synapses are blocked.

7.2.2 Experiments in Primates

In our *in vitro* experiments, we could also explore whether neurotransmitters might modulate the threshold above which SGI neurons emit bursts in response to synaptic drive. Acetylcholine (ACh) is a good candidate for modulating the threshold of SGI cells, as it is employed by pathways to the SGI from the parabrachial region of the midbrain (the pedunculopontine and laterodorsal tegmental nuclei).[37—40] In rat slice experiments, we found that application of ACh primarily activates nicotinic ACh receptors (nAChRs) of SGI neurons and causes their depolarization. We further observed that depolarization caused by nAChR activation lowered the threshold for eliciting bursting response to excitatory synaptic inputs, such as those from the SGS. Based on the observations of the effect of nAChR activation on SGI neurons, we microinjected nicotine locally into the SGI of awake monkeys while they were engaged in a visually guided saccade task.[41] As shown in Figure 7.3, nicotine induced a stepwise reduction in SRTs from that of regular saccades (150 to 160 ms) to that of express saccades (about 100 ms). Interestingly, the SRTs exhibited a clear bimodal distribution after the onset of nicotine injection and even during recovery from it. As shown in Figure 7.3, few saccades of intermediate SRTs were elicited. Increase of the dose of nicotine caused no further reduction in SRTs. Our results suggest that express saccades may be produced by the local increases in the excitability of SGI neurons. They also suggest that the nicotine-induced movements rely on the exist-

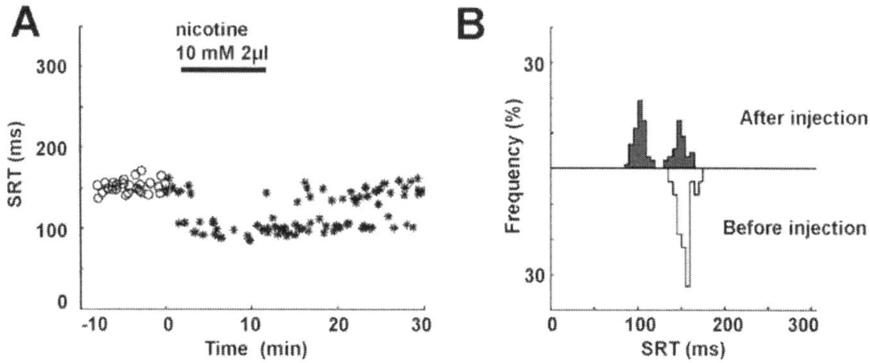

FIGURE 7.3 Effects of nicotine injection (10 m*M*, 2 μl) into the monkey superior colliculus on saccade reaction time (SRT) during NO GAP (A, B) and GAP (170 ms; C, D) visually guided saccade paradigms. Amplitude of saccades was 7 in horizontal and 7 in vertical directions. Horizontal bars in A and C indicate period of nicotine injection (10 min). (A and C) SRTs are plotted against time. Open circles, trials before injection; asterisks, trials during and after nicotine injection. (Modified from Aizawa, H., Kobayashi, Y., Yamamoto, M., and Isa, T., Injection of nicotine into the superior colliculus facilitates occurrence of express saccades in monkeys, *J. Neurophysiol.*, 82, 1642, 1999. With permission.)

ence of a short transmission time pathway, probably within the SC, for their execution. The interlaminar projection of the superficial to the deeper tectal layers provides such a pathway when the signal flow through it is enabled, so that visual signals that reach the SGS are transformed into the bursts of action potentials that comprise motor commands.

7.3 NEURAL MECHANISMS RESPONSIBLE FOR BURST GENERATION

In principle, the presaccadic bursts of SC neurons could rely on two alternative mechanisms for their generation. First, they could be due to the properties of circuits intrinsic to the SC. Alternatively, they could be generated by reverberating positive feedback loops through structures outside the SC and transmitted to the SGI. The results just summarized support the former alternative.

7.3.1 NMDA Receptor-Mediated Excitatory Synaptic Transmission

The bursting responses of neurons could be due to several alternative mechanisms; (1) low threshold spikes causing transient bursting activity in burst spiking neurons,[42] (2) Ca^{2+}-dependent plateau potentials,[43] (3) a hyperpolarization-activated current (Ih) such as the one shown to cause rhythmic bursting activity in other regions of the central nervous system[44] and observed in wide field vertical cells of the SGI, (4) intrinsic membrane properties consistent with a nonlinear frequency-input (f-I) relationship such as that demonstrated in feline SC neurons by Grantyn and her colleagues,[45] (5) NMDA-type glutamate receptor-mediated synaptic transmission. To explore this issue, it is essential to know the intrinsic firing properties of individual identified tectofugal cells, which would be responsible for the first four alternatives listed above. We examined these properties in a previous study.[46] Most of the crossed tecto-reticular SGI cells identified by retrograde labeling with a fluorescent tracer exhibited regular spiking properties and a quasi-linear f-I relationship (Saito, Y., and Isa, T., unpublished observation). Also, only a small population of tectofugal neurons exhibited burst spiking or plateau potentials. Furthermore, Ih could be found only in wide field vertical cells, which project primarily to the lateral pulvinar and to SO plus SGI.[47] On the other hand, as shown in Figure 7.2D, the bursts emitted by SGI neurons in response to stimulation of the SGS in the presence of bicuculline, were suppressed by application of 50 μM APV and thus the bursts depended on activation of NMDA-type glutamate receptors. It is well known that NMDA receptors have a J-shaped current-voltage relationship. Due to Mg^{2+}-block,[48] the NMDA-type glutamate receptors admit inward currents only when the cell is sufficiently depolarized. Once the membrane potential exceeds the value necessary for activation of the NMDA receptor, a regenerative process ensues that further enhances their depolarization. Such nonlinear activation of NMDA receptors can account for the all-or-nothing character of the bursts emitted by SGI neurons.

7.3.2 Lateral Excitatory Interaction

Presaccadic neurons of the primate SC are known to have recurrent collaterals, which ramify in the neighborhood of the parent somata.[49] Their appearance and use in models of the SC[50] is discussed in Chapter 5. To further investigate whether a local circuit including such neurons could support their long lasting depolarization and bursting activity, we obtained simultaneous, dual, whole cell records from pairs of adjacent SGI neurons. Figure 7.4 illustrates an example obtained from a pair of neurons horizontally separated by less than 100 μm. The existence of direct connections between the two cells was evaluated by measuring the responses of one of the cells to current injected in the other in order to induce its single or repetitive firing. In this manner, we established that the two neurons of Figure 7.4B were not directly connected with each other. This result applied to most of the cases we examined (over 95%). When we applied 10 μ*M* bicuculline (or SR95531) and reduced the extracellular concentration of Mg^{2+} from 1.0 m*M* to 0.1 m*M*, the SGI neurons exhibited bursting spike activity superimposed on repetitive, spontaneous, depolarizing potentials. Interestingly, the spontaneous depolarization and the bursting spike activity occurred almost simultaneously in both neurons. Since the spiking discharges of two adjacent presaccadic burst neurons are synchronous,[51] synchronization of SC neuron discharges could underlie the generation of their presaccadic bursts *in vivo*. We further found that activation of NMDA-type glutamate receptors is essential for such synchronous depolarization to occur since it was completely abolished by APV. A small piece of the intermediate layer (0.3 mm thick and measuring about 0.5 mm by 1 mm in the dorsoventral and mediolateral directions, respectively) was adequate to support the synchronous depolarization and bursting of pairs of neurons. Figure 7.5 illustrates the responses of a pair of neurons that

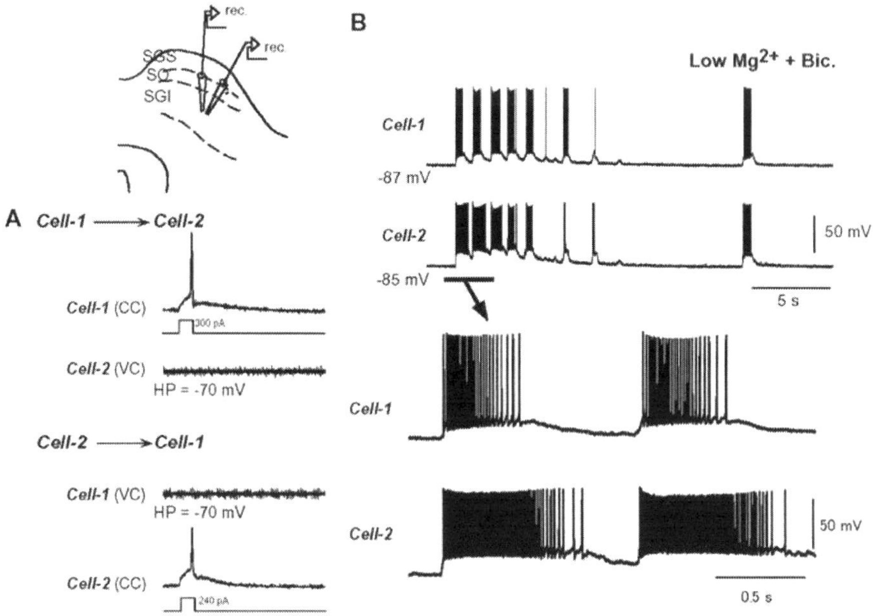

FIGURE 7.4 Simultaneous recordings from a pair of SGI neurons. (A) Recordings of current responses from one cell in voltage clamp mode (VC) after inducing a single action potential in the other cell by current injection in current clamp mode (CC). (B) Spontaneous membrane potentials in the presence of 10 μ*M* Bic and low (0.1 m*M*) $Mg.^{2+}$ Lower records show the faster sweep records of the membrane potentials underlined in the upper panel.

were separated by 36 μm and were connected directly. As shown in Figure 7.5A, an action potential in cell-1 was followed within 0.8 ms by an 13.5 pA excitatory postsynaptic current (EPSC) in cell-2 (held at –70 mV). Repetitive firing of cell-1 induced repetitive EPSCs in cell-2. In contrast, spikes emitted by cell-2 induced no response in cell-1 (Figure 7.5B). Figure 7.5C1 shows the spontaneous membrane potential fluctuations of these two cells in the presence of bicuculline and reduced concentration of $Mg.^{2+}$ In this case, the onset of depolarization of cell-1 sometimes preceded (Figure 7.5C2 right and C3 right) and sometimes followed that of cell-2 (Figure 5C2 left and C3 left). Taken together, our results suggest that the SGI contains a population of neurons that share loose but extensive local excitatory connections, to form a reverberating excitatory network displaying synchronous depolarization and bursting discharges. The existence of such a local excitatory network is consistent with the observation of abundant local collaterals arising from SGI neurons (Figure 7.2A).

7.4 CONCLUSIONS

The following answers to the questions posed in Section 7.1 can be formulated in the light of our experimental results.

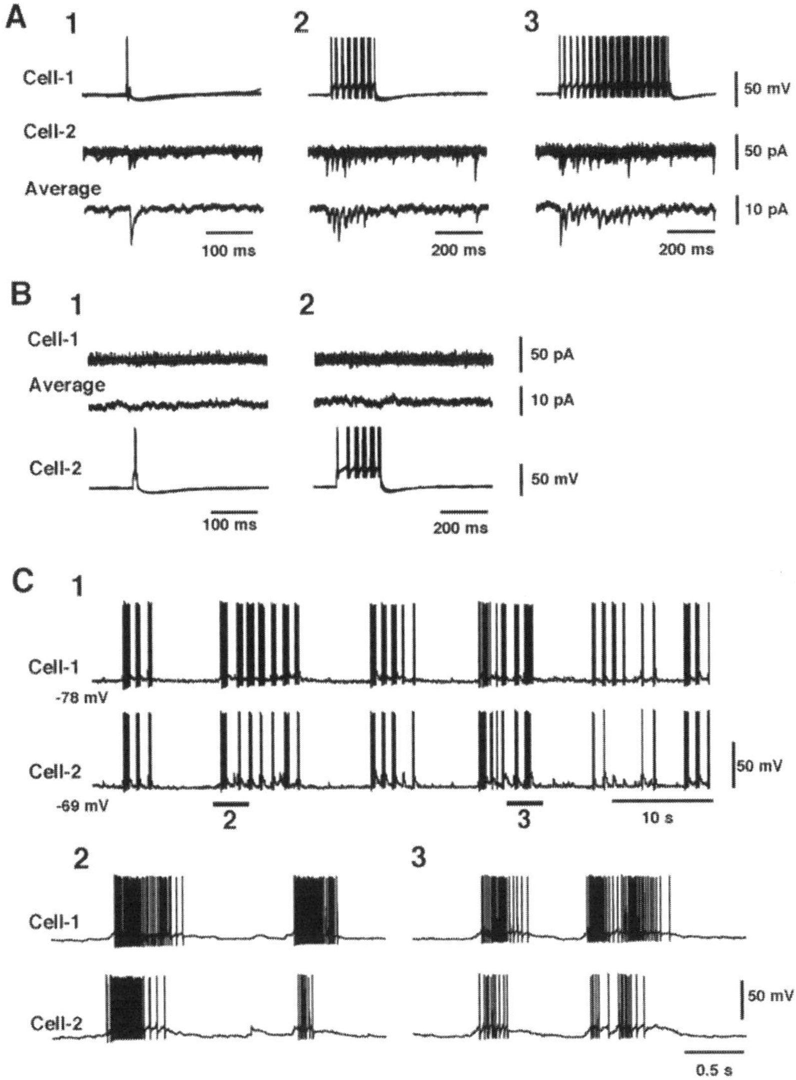

FIGURE 7.5 Relation between a directly connected neuron pair in the SGI. The two neurons (cell-1 and cell-2) shown in this figure are separated by 36 μm in the SGI of the rat SC slice. (A) Action potentials of the cell-1 generated by injection of current pulse induced timely locked EPSCs in cell-2. The top panel indicates the firing response of cell-1, the middle panel indicates the simultaneously recorded current response (under voltage clamp) of cell-2 and the lowest panel indicates average of 10 records. Panels 1, 2 and 3 indicate effects of single spike, 7 to 8 spike trains and 16 to 17 spike trains of the cell-1, respectively. (B) Effects of single (panel 1) and repetitive firing (panel 2) of cell-2 on the current records in cell-1. No effects were observed in these cases. (C) Relationship of spike firing during spontaneous depolarization of the cell-1 and cell-2 under the presence of 10 μ*M* bicuculline and reduced extracellular Mg^{2+} concentration (0.1 m*M*) as shown in Figure 7.4. Panels 2 and 3 are expanded records of parts in the panel 1, indicated by bars with corresponding numerals. Note that in panels 2 and 3, in some cases, cell-1 started firing earlier than cell-2, while vice versa in other cases.

1. Visual information is conveyed to SGI neurons, at least in part, via a relay in the SGS. The existence of the interlaminar connection has been proven in several species, including primates. Our present findings suggest that the signal transmission through it is not robust, but can be modulated by GABAergic inhibition and cholinergic facilitation.
2. The bursting discharges of SGI neurons can be generated within a local SGI circuit. Excitatory SGI neurons comprise a loosely interconnected cell population linked by local collaterals. Our results demonstrate that, when released from $GABA_A$ receptor-mediated inhibition, this excitatory network emits synchronous bursting discharges in response to excitatory inputs. The negative slope conductance of the NMDA receptor channels endows these bursts with an all-or-nothing character. Properties intrinsic to local SC circuits may thus contribute to the initiation of presaccadic bursts in the deeper layers of the SC. However, this does not exclude the possibility that neural structures outside the SC may also contribute to the generation of such bursting discharges. Moreover, our results demonstrate that when the circuit remains disinhibited the depolarization and bursting discharge of SGI neurons can continue for periods of time longer than 1 s. Accordingly, the local excitatory SC network needs additional inhibitory inputs to terminate its bursting responses. These inputs could originate either from structures located outside the SC such as the SNr or from GABAergic interneurons intrinsic to the SC. The organization of GABAergic influences upon SC output cells is obviously an important issue and should continue to attract experimental efforts to address it.

ACKNOWLEDGMENTS

We thank Michi Seo for technical assistance in preparing the manuscript. Our studies were supported by grants from the Ministry of Education, Culture, Sports, Science and Technology, CREST, JST and Mitsubishi Foundation to T.I.

REFERENCES

1. Fischer, B. and Boch, R., Saccadic eye movements after extremely short reaction times in the monkey, *Brain Res.*, 260, 21, 1983.
2. Fischer, B., Boch, R., and Ramsperger, E., Express-saccades of the monkey: effect of daily training on probability of occurrence and reaction time, *Exp. Brain Res.*, 55, 232, 1984.
3. Glimcher, P.W., Making choices: the neurophysiology of visual-saccadic decision making, *TINS*, 24, 654, 2001.
4. Lee, C., Rohrer, W.H., and Sparks, D.L., Population coding of saccadic eye movements by neurons in the superior colliculus, *Nature*, 332, 357, 1988.
5. Robinson, D.A., Eye movements evoked by collicular stimulation in the alert monkey, *Vision Res.*, 12, 1795, 1972.
6. Schiller, P.H. and Stryker, M., Single-unit recording and stimulation in superior colliculus of the alert rhesus monkey, *J. Neurophysiol.*, 35, 915, 1972.

7. Sparks, D.L., Holland, R., and Guthrie, B.L., Size and distribution of movement fields in the monkey superior colliculus, *Brain Res.*, 113, 21, 1976.
8. Wurtz, R.H. and Albano, J.E., Visual-motor function of the primate superior colliculus, *Ann. Rev. Neurosci.*, 3, 189, 1980.
9. Fuchs, A.F., Kaneko, C.R.S., and Scudder, C.A., Brainstem control of saccadic eye movements, *Ann. Rev. Neurosci.*, 8, 307, 1985.
10. Moschovakis, A.K., Scudder, C.A., and Highstein, S.M., The microscopic anatomy and physiology of the mammalian saccadic system, *Progr. Neurobiol.*, 50, 133, 1996.
11. Scudder, C.A., Kaneko, C.R.S., and Fuchs, A.F., The brainstem burst generator for saccadic eye movements. A modern synthesis, *Exp. Brain Res.*, 142, 439, 2002.
12. Dorris, M.C., Paré, M., and Munoz, D.P., Neuronal activity in monkey superior colliculus related to the initiation of saccadic eye movements, *J. Neurosci.*, 17, 8566, 1997.
13. Edelman, J.A. and Keller, E.L., Activity in visuomotor burst neurons in the superior colliculus accompanying express saccades, *J. Neurophysiol.*, 76, 908, 1996.
14. Sparks, D., Rohrer, W.H., and Zhang, Y., The role of the superior colliculus in saccade initiation: a study of express saccades and the gap effect, *Vision Res.*, 40, 2763, 2000.
15. Edwards, S.B., The deep cell layers of the superior colliculus: their reticular characteristics and structural organization, in *The Reticular Formation Revisited*, Hobson, A. and Brazier, M. Eds., Raven, New York, 1980, pp. 193.
16. Mays, L.E. and Sparks, D.L., Dissociation of visual and saccade-related responses in superior colliculus neurons, *J. Neurophysiol.*, 43, 207, 1980.
17. Maeda, M., Shibazaki, T., and Yoshida, K., Labyrinthine and visual inputs to the superior colliculus neurons, *Prog. Brain Res.*, 50, 735, 1979.
18. Karabelas, A.B., Moschovakis, A.K., and Highstein, S.M., Morphological classification of intracellularly HRP injected efferent neurons of the deeper layers of the primate superior colliculus., *Proc. Soc. Neurosci.*, 12, 1185, 1986.
19. Moschovakis, A.K., Karabelas, A.B., and Highstein, S.M., Structure-function relationships in the primate superior colliculus. I. Morphological classification of efferent neurons, *J. Neurophysiol.*, 60, 232, 1988.
20. Mooney, R.D., Huang, X., and Rhoades, R.W., Functional influence of interlaminar connections in the hamster's superior colliculus, *J. Neurosci.*, 12, 2417, 1992.
21. Mooney, R.D., Nikoletseas, M.M., Hess, P.R. et al., The projection from the superficial to the deep layers of the superior colliculus: an intracellular horseradish peroxidase injection study in the hamster, *J. Neurosci.*, 8, 1384, 1988.
22. Rhoades, R.W., Mooney, R.D., Rohrer, W.H. et al., Organization of the projection from the superficial to the deep layers of the hamster's superior colliculus as demonstrated by the anterograde transport of *Phaseolus vulgaris* leucoagglutinin, *J. Comp. Neurol.*, 283, 54, 1989.
23. Behan, M. and Appell, P.P., Intrinsic circuitry in the cat superior colliculus: Projections from the superficial layers, *J. Comp. Neurol.*, 315, 230, 1992.
24. Hall, W.C. and Lee, P., Interlaminar connections of the superior colliculus in the tree shrew. I. The superficial gray layer, *J. Comp. Neurol.*, 332, 213, 1993.
25. Dean, P., Redgrave, P. and Westby, G.W., Event of emergency? Two response systems in the mammalian superior colliculus, *TINS*, 12, 137, 1989.
26. Sparks, D.L., Translation of sensory signals into commands for control of saccadic eye movements: role of primate superior colliculus, *Physiol. Rev.*, 66, 118, 1986.
27. Isa, T., Endo, T., and Saito, Y., The visuo-motor pathway in the local circuit of the rat superior colliculus, *J. Neurosci.*, 15, 8496, 1998.
28. Lee, P.H., Helms, M.C., Augustine, G.J. et al., Role of intrinsic synaptic circuitry in collicular sensorimotor integration, *Proc. Natl. Acad. Sci.*, 94, 13299, 1997.

29. May, P.J. and Hall, W.C., Relationships between the nigrotectal pathway and the cells of origin of the predorsal bundle, *J. Comp. Neurol.*, 226, 357, 1984.
30. Karabelas, A.B. and Moschovakis, A.K., Nigral inhibitory termination on efferent neurons of the superior colliculus. An intracellular horseradish peroxidase study in the cat, *J. Comp. Neurol.*, 239, 309, 1985.
31. Meredith, M.A. and Ramoa, A.S., Intrinsic circuitry of the superior colliculus: Pharmacophysiological identification of horizontally oriented inhibitory interneurons, *J. Neurophysiol.*, 79, 1597, 1998.
32. Munoz, D.P. and Wurtz, R.H., Fixation cells in monkey superior colliculus.I. Characteristics of cell discharge, *J. Neurophysiol.*, 70, 559, 1993.
33. Hikosaka, O. and Wurtz, R.H., Visual and oculomotor functions of monkey substantia nigra pars reticulata. I. Relation of visual and auditory responses to saccades, *J. Neurophysiol.*, 49, 1230, 1983.
34. Joseph, J.P. and Boussaoud, D., Role of the cat substantia nigra pars reticulata in eye and head movements. I. Neural activity, *Exp. Brain Res.*, 57, 286, 1985.
35. Munoz, D.P. and Guitton, D., Control of orienting gaze shifts by the tectoreticulospinal system in the head-free cat. II. Sustained discharges during motor preparation and fixation, *J. Neurophysiol.*, 66, 1624, 1991.
36. Pettit, D.L., Helms, M.C., Lee, P. et al., Local excitatory circuits in the intermediate gray layer of the superior colliculus, *J. Neurophysiol.*, 81, 1424, 1999.
37. Beninato, M. and Spencer, R.F., A cholinergic projection to the rat superior colliculus demonstrated by retrograde transport of horseradish peroxidase and choline acetyltransferase immunohistochemistry, *J. Comp. Neurol.*, 253, 525, 1986.
38. Hall, W.C., Fitzpatrick, D., Klatt, L.L., and Raczkowski, D., Cholinergic innervation pattern of the superior colliculus in the cat, *J. Comp. Neurol.*, 287, 495, 1989.
39. Jeon, C.J., Spencer, R.F., and Mize, R.R., Organization and synaptic connections of cholinergic fibers in the cat superior colliculus, *J. Comp. Neurol.*, 333, 360, 1993.
40. Woolf, N.J. and Butcher, L.L., Cholinergic systems in the rat brain: III. Projections from the pontomesencephalic tegmentum to the thalamus, tectum, basal ganglia, and basal forebrain, *Brain Res. Bull.*, 16, 603, 1986.
41. Aizawa, H., Kobayashi, Y., Yamamoto, M., and Isa, T., Injection of nicotine into the superior colliculus facilitates occurrence of express saccades in monkeys, *J. Neurophysiol.*, 82, 1642, 1999.
42. Jahnsen, H. and Llinás, R., Ionic basis for the electro-responsiveness and oscillatory properties of guinea-pig thalamic neurons *in vitro*, *J. Physiol.* (Lond), 349, 227, 1984.
43. Hounsgaard, J., Hultborn, H., Jespersen, B., and Kiehn, O., Intrinsic membrane properties causing a bistable behavior of alpha-motoneurones, *Exp. Brain Res.*, 55, 391, 1984.
44. Pape, H.C., Queer current and pacemaker: the hyperpolarization-activated cation current in neurons, *Annu. Rev. Physiol.*, 58, 299, 1996.
45. Grantyn, R., Grantyn, A., and Schierwagen, A., passive membrane properties, afterpotentials and repetitive firing of superior colliculus neurons studied in the anesthetized cat, *Exp. Brain Res.*, 50, 377, 1983.
46. Saito, Y. and Isa, T., Electrophysiological and morphological properties of neurons in the rat superior colliculus.I. Neurons in the intermediate layer, *J. Neurophysiol.*, 82, 754, 1999.
47. Lane, R.D., Bennett-Clarke, C.A., Allan, D.M. et al., Immunochemical heterogeneity in the tecto-LP pathway of the rat, *J. Comp. Neurol.*, 333, 210, 1993.
48. Mayer, M.L., Westbrook, G.L., and Guthrie, P.B., Voltage-dependent block by Mg^{2+} of NMDA responses in spinal cord neurons, *Nature*, 309, 261, 1984.

49. Moschovakis, A.K., Karabelas, A.B., and Highstein, S.M., Structure-function relationships in the primate superior colliculus. II. Morphological identity of presaccadic neurons, *J. Neurophysiol.*, 60, 263, 1988.
50. Bozis, A. and Moschovakis, A.K., Neural network simulations of the primate oculomotor system. III. A one-dimensional one-directional model of the superior colliculus, *Biol. Cybern.*, 79, 215, 1998.
51. Istvan, P.J. and Munoz, D.P., Neuronal connectivity in monkey superior colliculus during saccades: evidence from multiple recordings and cross-correlation analysis, *Soc. Neurosci. Abstr.*, 23, 1296, 1997.

8 Development and Use-Dependent Modification of Synaptic Connections in the Visual Layers of the Rodent Superior Colliculus

Rosemarie Grantyn, René Jüttner, and Jochen Meier

CONTENTS

0-8493-0097-5/04/$0.00+$1.50

8.1 INTRODUCTION

Neurons in the superficial layers of the superior colliculus (sSC) are specialized to detect novel visual objects and to provide information on the location of peripheral targets. Figure 8.1 illustrates the main elements of an oligosynaptic pathway, which is responsible for the rapid initiation of saccadic eye movements. The inset of Figure 8.1 illustrates that sSC neurons respond to moving objects and habituate very rapidly.

The position of sSC neurons is critical: it adds a spatial code to the specific visual information that is contained in the action potential sequence. There is general agreement that sSC neurons encode the target position in retinotopic coordinates. Perturbation of the topography reduces the precision of eye and head movements[1] and compromises the integrity of other sensory maps.[2]

Figure 8.2 shows the retinotopic map of the mouse sSC (modified after Reference 3). Information on the retinotopic map of the rat sSC can be found in Reference 4 and, modified, in Reference 5. In mature mice and rats, retinal afferents to the SC mostly derive from the contralateral eye; less than 4% of the retinal input stems from the ipsilateral eye. The medio-lateral (M-L) axis of the retina corresponds to the anterior-posterior (A-P) axis of the sSC, and the dorso-ventral axis of the retina corresponds to the collicular M-L axis. In contrast to the contralateral (crossed) projection, which originates from the entire retina, the ipsilateral (uncrossed) projection is limited to the axons from retinal ganglion cells in the temporal and ventral retina. These fibers terminate in the rostral sSC.

At any given site along the A-P and M-L axis, sSC-neurons receive complementary information about the same visual object. As all the afferent structures share the same retinotopic coordinate system, input representing the same section in the visual field must converge in the sSC and align along the same dorso-ventral (D-V) axis. Therefore, afferents must find the appropriate layers, select the correct target neurons and then occupy the correct sites on the somato-dendritic surface of recipient cells. As a rule of thumb, retinal axons terminate in the most superficial tier of the SGS, while cortical axons from area 17 preferentially occupy the lower parts of the SGS and the SOp. The SOp harbors the terminals from area 18. Horizontally and vertically oriented inhibitory interneurons make synaptic contacts throughout the SGS and SOp.

The retinotopic map in the rodent sSC requires both molecular guidance cues and synaptic activity, and it takes at least 4 weeks to develop. Figure 8.3 summarizes some hallmarks of this process. Here and in the following sections, we shall pref-

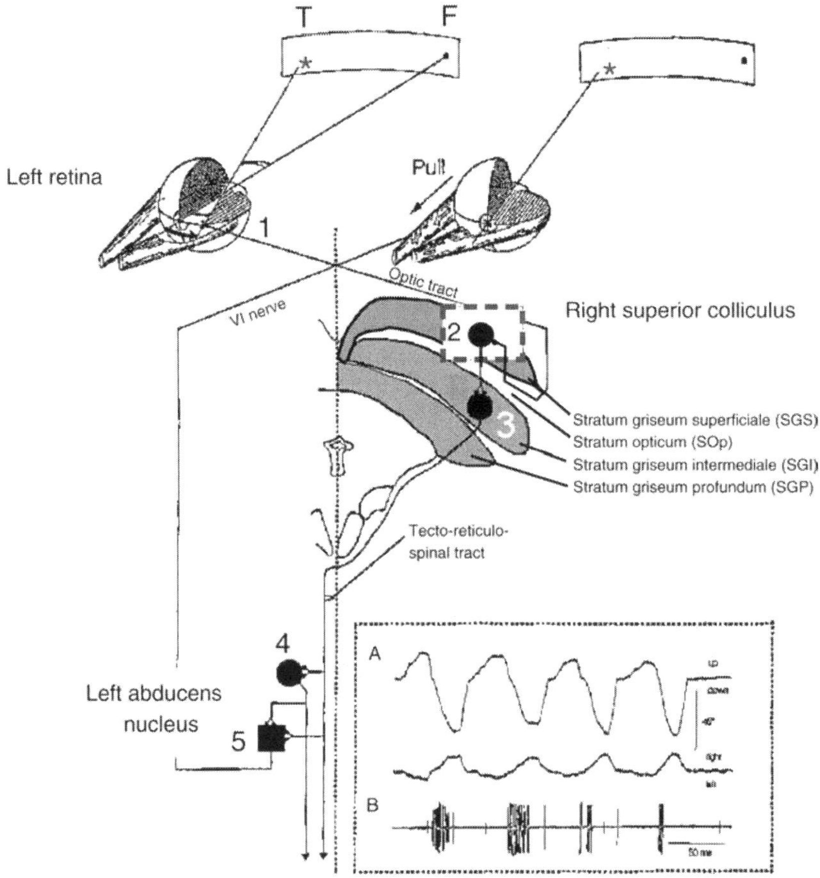

FIGURE 8.1 Scheme of the oligosynaptic pathway underlying orienting eye and head movements towards a visual target. In the center: frontal view on the right dorsal midbrain showing some of the collicular layers and the presence of a direct connection between the superficial and the deeper SC layers. The pathway for horizontal saccadic eye movements includes the following neurons: 1 — retinal ganglion cell, 2 — wide-field vertical neuron, 3 — tectoreticulospinal neuron, 4 — reticulospinal neuron, 5 — abducens motoneuron. Boxed area: visual stimulation (A) and corresponding spike discharge (B) of an sSC neuron. This example was taken from an experiment in an adult awake cat (unpublished experiment by R. Grantyn, A. Berthoz, and A. Grantyn). The scheme also utilizes an original drawing by A. Grantyn.

erentially quote data from mouse and rat, occasionally supplemented with data from hamster and rabbit. It should be noted that the development of the visual system is very similar in mice and rats, although during the 2nd half of embryonic life the mouse is about 1 day ahead. The gestation period of the frequently used pigmented mouse C57BL/6 is 19 days, and eye opening occurs on P13/P14. The gestation period of Wistar rats is 20 days and eye opening occurs on P11. The gestation period of hamsters is 16 days, and eyes are open on day P12.

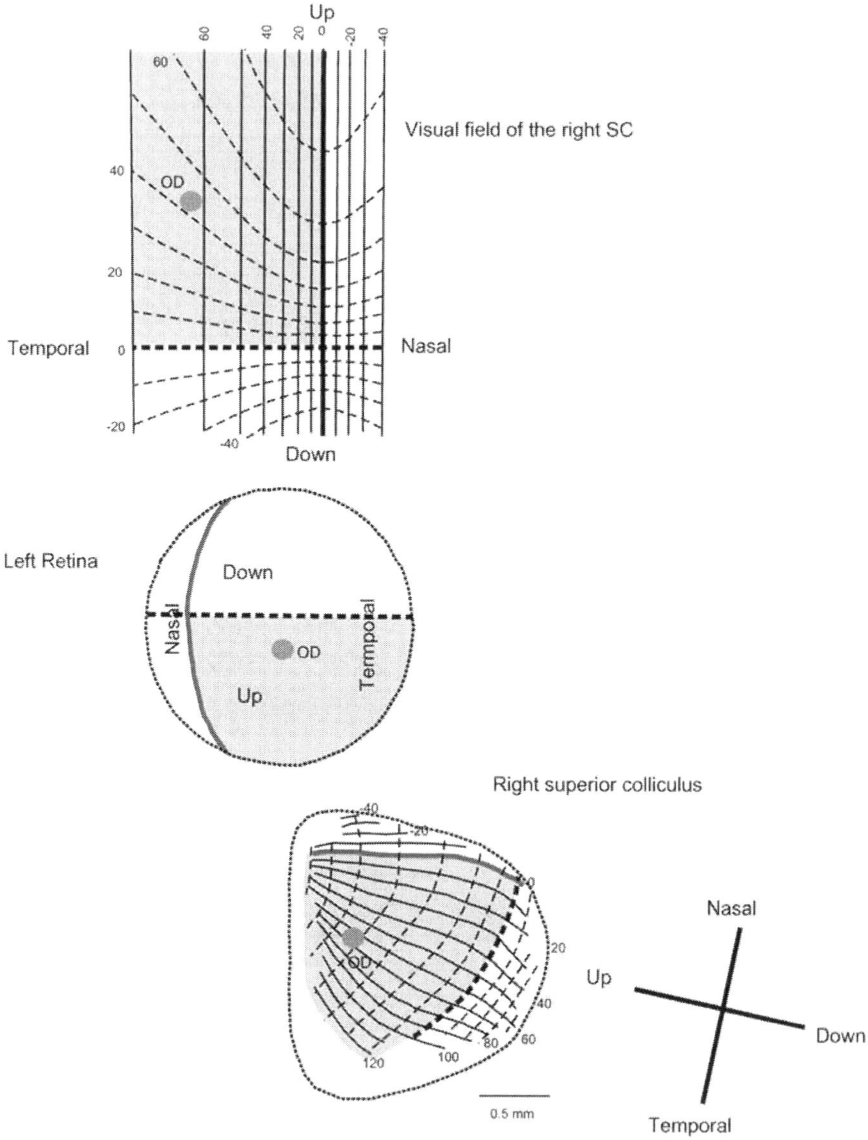

FIGURE 8.2 Relationship between the visual field and its representation in the left retina and right superior colliculus of the adult mouse. (A) Visual field of a mouse as represented in the right sSC. Solid lines: azimuths; dotted lines: elevations. (B) Left retina and areas involved in the initiation of gaze shifts in the nasal, temporal down and up direction. (C) Right SC with superimposed retinotopic map. Solid lines represent azimuths; dotted lines represent elevations. All numbers: degree. Shaded: a section of the visual field and its representation in the left retina and right sSC. (From Dräger, U.C. and Hubel, D.H., Topography of visual and somatosensory projections to mouse superior colliculus, *J. Neurophysiol.*, 39, 91–101, 1976. With permission.)

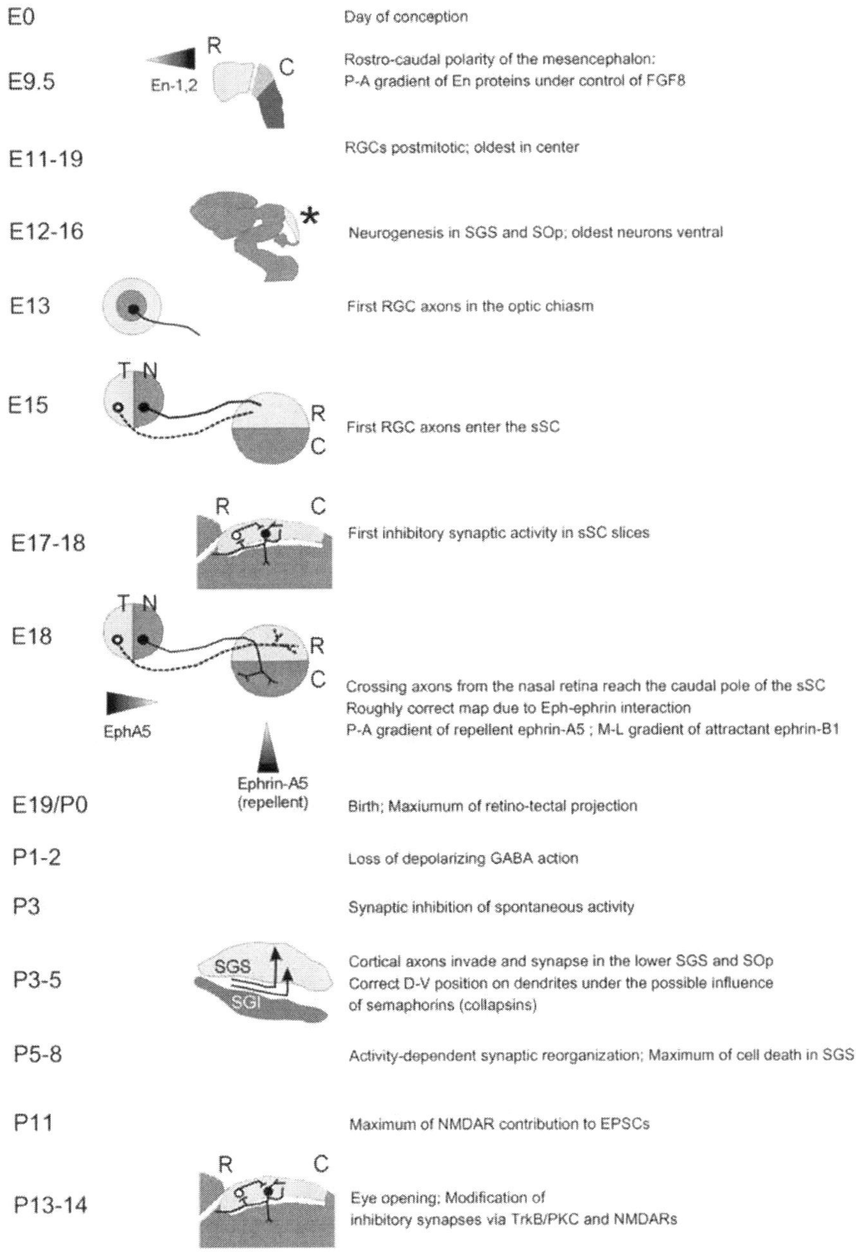

FIGURE 8.3 Steps in the development of the retinotopic map of the sSC before the onset of patterned vision. This scheme summarizes results from several groups, as reviewed in this chapter. Asterisk: mes and tectal plate. The majority of developmental neurobiologists has agreed to refer to the day of conception as embryonic day (E) 0. Accordingly, the day of birth will be postnatal day (P) 0. When a different definition is used in an article quoted, results are redefined in the E0/P0 system.

In mouse, rat, and hamster, the photoreceptors form and mature during the first two weeks after birth. The postnatal development of these rodents is therefore characterized by the existence of a protracted *previsual period* when activity-regulated processes of synapse development presumably rely on intrinsic neuronal activity and/or nonvisual sensory cues.[6,7] It is mostly during this time that inappropriately connected RGCs and sSC neurons vanish. Eye opening marks the onset of patterned vision. While the number of newly formed synapses continues to increase after eye opening,[8,9] their distribution and properties can now be adjusted to the functional requirements of visuomotor behavior. Part of the synapses are eliminated, but in the rat sSC there is no net decrease in synapse number until P30.

One of the major challenges in this research area is to characterize neuron properties and synaptic transmission in their association with developmental, cellular or molecular gradients along the A-P or D-V axes. The work by Cline and colleagues shows how successful this approach can be.[10] Taking advantage of the well-known rostro-caudal gradient of retinal fiber expansion and synapse maturation in the chick optic tectum (OT), they discovered that glutamatergic synapses pass through an initial silent or, rather, activity-dependent stage. At a certain age, the caudal OT preferentially contains immature excitatory glutamatergic synapses characterized by the absence of AMPA receptors. At this stage only NMDA receptors accumulate opposite glutamatergic terminals, and subsequent recruitment of AMPA receptors requires Hebbian-like co-activation of pre- and postsynaptic elements to ensure Ca^{2+} influx as a necessary condition for the insertion of AMPA receptors. This study on the chick OT disclosed a general feature in the development of excitatory glutamatergic synapses in the brain — the activity-dependent transformation from *NMDAR-only* to *NMDA-and-AMPAR* synapses.[11]

The aim of this review is to summarize the available information on A) the molecular gradients underlying axon guidance and synapse formation, and B) the activity-dependent mechanisms of synapse formation and reorganization, which, presumably, underlie the refinement of visual maps. As a starting point, we can assume that this affords a view of earlier and later developmental stages. In reality, both mechanisms are probably acting in parallel and also persist in the mature brain.

8.2 TECTAL POLARITY IS DEFINED BEFORE AXONS ARRIVE

Functional studies in connection with gene mutations in mice have shed light on some of the mechanisms underlying map formation at very early developmental stages, i.e., before SC neurons are born and afferents arrive (see References 12 and 13 for reviews). In mice, the roof of the dorsal midbrain (*mesencephalon*, mes) can be distinguished by day E9.5. Its dorsal half contains the *anlage* of the later SC (see Figure 8.3, asterisk). The mes is separated from the hindbrain by a constriction that is referred to as the mid/hindbrain junction or isthmus. The two most rostral rhombomeres constitute the so called *metencephalon* (met), where rhombomere 1 gives rise to the pons and the cerebellum. The development of the mes/met region is controlled by a group of organizer cells in the isthmus area. A complex set of interactions between genes that encode secreted proteins (first of all members of the FGF and Wnt family) and transcription factors (*Pax*, *Otx*, *Gbx*, and *En* families) initially regulate the formation of mes/met and then control the establishment of rostro-caudal signal gradients, which are important for map formation. The mes/met border represents an interface of *Otx2*- and *Gbx*-expressing cells. Their interaction results in the expression of further genes, first of all *Pax2*, *Wnt1*, and *Fgf8*. The latter two encode secreted proteins with a slightly larger radius of action. FGF8, in turn, organizes the surrounding neural tissue. Mice carrying a hypomorphic *Fgf8* allele suffer from failure in cell migration and lack the posterior midbrain.[14]

FGF8 also controls the expression of the homeodomain transcription factor genes *En-1* and *En-2*, the vertebrate analogues of the *Drosophila* polarity segment gene *engrailed*.[15] *En* proteins distribute in the posterior mes in a caudal to rostral gradient. Of the already identified molecules, which could contribute to the retinotopic topography by forming concentration gradients along the A-P axis, *En-1* and *En-2* are the first to appear.[16] In the mouse, *En* expression starts on E8.[17] At least in the chick, over-expression of *En-1* and *En-2* results in a disturbance of the normal topography of retinal axons in the OT.[18,19] Very likely, this is caused by a misexpression of gradient-forming axon-repelling molecules.

8.3 COMPLEMENTARY GRADIENTS OF GUIDANCE CUES CONTROL INITIAL RETINAL AXON TOPOGRAPHIES

The search for molecules that guide axonal growth in the immature tectum is possibly one of the most exciting stories of current neurobiology (see References 20 to 23 for comprehensive reviews). For decades, this search was driven by the ingenious chemoaffinity hypothesis of Sperry,[24] who proposed that topographically ordered connections are formed between retinal axons and tectal neurons under the influence of chemical tags. Sperry also suggested that the specificity of neural connections underlying the two-dimensional map in the OT is established by complementary distribution of tag gradients in sensory and target structures. This notion was validated by the discovery of spatial signal gradients in the avian OT (for instance References 25 to 28) and in the mouse sSC.[29]

With their famous stripe assays, Bonhoeffer and colleagues showed that neurites from the temporal retina prefer to grow on membranes from the anterior tectum/colliculus, and that they are repelled by a signal provided by the posterior tectum/colliculus. Fibers from the nasal retina did not show such preference. One of the repulsive axon guidance signals (initially termed RAGs) was later identified to be ephrin-A5.[30] Ephrin-A5 is a ligand to three of the seven known EphA receptors.[22] In the chick, a molecular concentration gradient exists also for ephrin-A2 in the OT, and its receptor EphA3 in the retina.[31]

Bi-directional signalings emerging from Eph-ephrin interactions mediate a large number of contact-dependent developmental processes, such as boundary formation, cell migration, axon guidance, and vasculogenesis,[32,33] and they are most significant for the initial topography of retinal axons in the rodent sSC. In the mouse sSC ephrin-A5 is expressed according to an increasing A-P gradient, while ephrin-A2 expression is maximal in the middle of the A-P axis.[34] In the mouse retina, EphA5-expression follows an increasing nasal to temporal gradient.[35] This pattern is compatible with a repellent action of caudal, ephrin-A5-positive, sSC cells on temporal RGC axons from the temporal retina, which carry EphA5 (Figure 8.3). Indeed, comparison of retinal axon trajectories in the sSC of wild-type and ephrin-A5 null mice showed that, in the absence of ephrin-A5, RGC axons produce a large overshoot toward the caudal pole of the SC.[34]

The ephrin-A5/EphA5 gradient is not the only concentration gradient affecting the growth of retinal axons in the developing sSC. Differentially distributed attractants assist the establishment of the initial retinotectal topography.[36] Moreover, true axon guidance molecules may combine with molecular gradients, which regulate axon ramification rather than axon targeting. The available quantitative gradient-based models (see Reference 21 for a discussion) mostly assume that in-growing axons directly target a single destination (with some overshoot). The size of a repellent signal is then derived from the product of the concentrations of Eph receptors and ephrin ligand.[37] However, in rodents and other mammals guidance molecules may regulate branching rather than the targeting of the most advanced growth cone. Tracing of DiI-stained retinal axons in the rat SC between E20 and P19 showed that retinal axons issue numerous collaterals on their way to their most distal destination.[38] Ephrin-A ligands of the SC prevent the formation of branching collaterals in the caudal SC.[39]

The existence of attractant mechanisms and their possible role in the regulation of collateral formation has recently been demonstrated in the amphibian and mouse SC.[40] Since, in mice, EphBx-containing axons from the ventral retina are guided to the ephrin-B1-expressing medial SC, it was suggested that ephrin-B1 acts as attractant. However, in the *Xenopus*, the ephrin-B-ligand is expressed by the RGCs, while the optic tectum contains the EphB-receptor.[41] This is called reverse signaling, in contrast to the forward signaling in the mouse SC, where the EphB receptors are in the growth cones of the arriving axons. Double mutation of EphB receptors (for instance *EphB2-/-* and *EphB3-/-*) results in aberrant termination patterns. Apart from the normal termination zone, mutant mice retain additional ectopic terminations more laterally.[40] Still more effort is needed, however, to clarify the interplay of different guidance molecules and the respective intracellular signal transduction pathways resulting in growth cone attraction or repulsion.

8.4 THE DEVELOPMENT OF THE RETINOTECTAL MAP PRECEDES THE FORMATION AND ALIGNMENT OF CORTICOTECTAL CONNECTIONS

8.4.1 Schedule of Retinotectal Pathway formation

Crossed retinal axons invade the tectal plate by E15 in mouse[42,43] and by E16.5 in rat.[44] The caudal pole will only be reached 2 days later.[42] Thus, during a time window of at least 2 days, the caudal half of the SC lags behind the development of the rostral SC.

In mature rodents the retinal projection arises mostly from the contralateral eye. Less than 4% of retinal ganglion cell axons project ipsilaterally. However, after birth (P3–5) the ipsilateral retinotectal pathway is much stronger,[42,45,46] and the entire sSC is invaded by ipsilateral retinal axons. Most of the ipsilaterally projecting neurons do not form synaptic contacts[47] and die within the first 2 weeks of life. The mature pattern is reached by P10–15.[45]

Initially, fiber bundles occupy the entire D-V axis of the sSC, but later they disappear from the SGS and only remain in the optic layer.[43,44] Between P0 and P4, when the border between SGS and SOp is seen and all other layers are distinct, dorsal collaterals are given off by the main retinal axons, which follow a rostro-caudal and latero-medial trajectory in the SOp. The collaterals arborize in the SGS and terminate at the appropriate D-V level.[43,48] An earlier idea that the SGS and SOp separate from each other because the SGS neurons migrate through the retinal fibers to assume a more dorsal position was later abandoned.[48]

8.4.2 Alignment of the Corticotectal and Other Afferent Pathways

Potentially, all direct and indirect targets of retinal ganglion cells could contribute to the retinotopic map emerging in the sSC. In addition to the contralateral retina, this list includes the ipsilateral visual cortex, the contralateral sSC, the ventral geniculate nucleus, the intergeniculate leaflet, the suprachiasmatic nucleus, the nuclei of the accessory optic tract, and the parabigeminal nucleus. But, apart from the first two, these connections have not been studied from the viewpoint of map formation and therefore will not be discussed in this chapter.

The termination of the corticotectal and all other visual afferents must be organized in such a way that at any given site the receptive fields of the afferents overlap with the receptive fields of the sSC neurons they contact. This *retinotopic fidelity* of afferent and target cells has been extensively studied in the cat, originally by McIlwain and colleagues (see, for instance, Reference 49).

One way to ensure retinotopic fidelity in the sSC is to let the retinal axons establish a coarse map before the axons from other visual structures have a chance to form connections. Indeed, the corticotectal axons enter the sSC much later than retinotectal axons. Layer V neurons from different areas of the neocortex first extend a primary axon toward the brainstem and spinal cord. Until P4 cortical axons bypass the colliculus and *wait* for the correct time to give off collaterals and to invade the sSC.[50–53] All subcortical targets, including the sSC, are contacted by collaterals formed at stereotypic positions, and not by bifurcation of the leading axon pole.[54]

In the adult, the corticotectal pathway originates from the layer V pyramidal neurons of the ipsilateral visual cortex.[53,55,56] However, as in the case of the retinotectal pathways, sSC neurons transiently receive exuberant connections from a wider range of cortical sources. In neonates, connections also arise from nonvisual areas.[56]

In older animals (in the rat: P12), the origin of cortical projection to the sSC is restricted to areas 17 and 18,[56] and termination is confined to the lower half of the SGS and the upper SOp.[57]

Morphological studies at the light and EM level provide ample evidence that not only retinal but also cortical afferents initially branch and terminate throughout the entire depth of the SGS and SOp.[53,58] Later in development, they assume a more defined, dorsoventrally narrow position.[42,43,53,59] In adult rodents, the contralateral retinal projection is heaviest within the upper 100 μm of the sSC.[60,61] In mice, small contralateral RGC (presumably W-cell) axons terminate in the zonal layer of the upper SGS, larger contralateral RGC (presumably Y-cell) axons terminate in the lower SGS.[62] Ipsilateral RGCs terminate in the SOp, forming characteristic patches.[42,45] Corticotectal afferents also segregate. Axons from the striate cortex (area 17) preferentially terminate in the lower SGS and SOp,[57,60] while axons from extrastriate cortical areas terminate deeper in the SC,[55] but necessarily in register with the retinotopic afferents.

It is not yet clear which signals guide the corticotectal axons to the correct destination and limit their termination along the A-P, M-L, and D-V axes of the developing sSC. Most likely, chemical cues operate not only in the selection of a target area but also in the selection of appropriate target cells. In any case, the connections made with collicular or tectal explants are always arising from layer V pyramidal neurons.[63,64] Axonal growth cones of cortical neurons *in vitro* respond to gradients of semaphorins.[65] Sema3A, also known as collapsin-1, was the first member of the semaphorin family to be identified, and since then has been considered the prototypic growth cone collapse factor. Other semaphorins act as attractants. Semaphorins exercise their function through receptor complexes composed of members of the neuropilin (NP) and/or plexin protein families.[66] *In situ* hybridization showed that Sema 3A is expressed in the *Xenopus* OT.[67] Neuropilin-1 (NP-1) is abundantly present in developing retinal ganglion cells.[68] In *Xenopus*, Sema 3A/NP-1 interaction causes collapse of retinal growth cones.[67] Although semaphorins are expected to influence the termination of cortical axons in the sSC, this issue still awaits further experimental analysis.

The stratification of visual afferents implies that in-growing axons are steered by molecular tags forming an *adhesive landscape* along the D-V axis of the sSC.[69] In neurons that span the entire D-V axis, these tags must be targeted and stabilized at specific dendritic compartments. A specific dendritic location was observed with immunostaining of cadherins, a family of Ca^{2+}-dependent cell adhesion molecules (see Reference 70 for a detailed review). N-cadherin is concentrated in the synaptic cleft of retinotectal synapses.[69] N-cadherin becomes concentrated in the retinorecipient laminae upon formation of retinotectal contacts, but distributes diffusely when the eyes are removed. In the chicken optic tectum, which consists of 16 very distinct *laminae*, each layer contains a characteristic set of cadherins. Blocking antibodies against N-cadherin produce exuberant projections or prevent arborization.[71] In the mouse, cadherin-11 is predominantly expressed in the superficial collicular layers,[72] while cadherin-8 was found in patches in the intermediate gray layer.[73] However, the proposed role of cadherins in synaptic specificity and maintenance still awaits verification in the rodent SC.

8.5 BIRTH AND MIGRATION OF NEURONS FOLLOW A VENTRO-DORSAL BUT NOT ANTERIOR-POSTERIOR GRADIENT

A longstanding question is whether the density of collicular neurons obeys any gradients along the A-P or M-L axes.[74,75] In addition, it has been asked to what extent the temporal sequence of neuron birth and migration could establish a D-V gradient and influence the formation of layers.[42,48]

Unfortunately, no map-related gradients have been quantitatively evaluated in the sSC. Moreover, no rostral lead could be detected in the generation of mouse or rat sSC neurons. This contrasts with the early development of the rostral tectum in birds.[76] However, a ventro-dorsal sequence of neurogenesis was reported by several groups.[48,77,78] The birth of SGS-neurons lags significantly behind neurogenesis in the SOp. The birth of very large neurons in the SGI (presumably the tectoreticulospinal neurons, TRSNs) occurs even earlier, with a peak at E12.[78,79] Taking advantage of this developmental schedule, it has been possible to isolate the precursors of deep layer projection neurons and to investigate their pattern of ion channel expression at early stages of neuron differentiation.[80] Unfortunately, such a strategy cannot be applied to neurons of the sSC.

Collicular neurons originate in the thick ventricular zone close to the ventral surface of the tectal plate and then migrate along radial glia to their final position. A quantitative radioactive thymidine study in rat[78] showed that neurons in the SGS complete their last mitotic cycle between E13 and E17 (peak at E15). Neurogenesis in the mouse is roughly one day ahead.[48]

In Nissl-stained sections of the rat or mouse sSC, a layered pattern is first distinguishable at E15. The superficial layers (still without a boundary between SGS and SOp) are defined on day E17. By P0, neuron migration is largely complete, and by P4 all layers can be seen nearly as well as in the adult, i.e., SGZ, SGS, and SOp (see Figure 8.1) are now distinct. Having a thickness of <200 μm, the sSC from a newborn mouse can be well maintained as a whole-mount slice.[81] The sSC of P0 rats is considerably thicker (Figure 8.4A), reaching 350 to 390 μm,[44,82] which is more than 80% of the adult value.[83]

8.6 VENTRO-DORSAL POSITION, DENDRITIC GEOMETRY, AXONAL PROJECTION, AND MOLECULAR MARKERS PROVIDE A BASIS FOR NEURON IDENTIFICATION IN THE MATURE AND DEVELOPING COLLICULUS

Electrophysiological recordings from horizontal sSC slices revealed that by E17 and 18, respectively, neurons in mice and rats are able to generate synaptic currents. This is about the time when RGCs begin to accumulate retrogradely transported markers,[44] but cortical afferents are still absent (see Figure 8.3). Consequently, axon in-growth and ramification, on one side, and neuron migration and dendrite extension, on the other side, should largely occur in parallel.[43,48] This implies that afferent

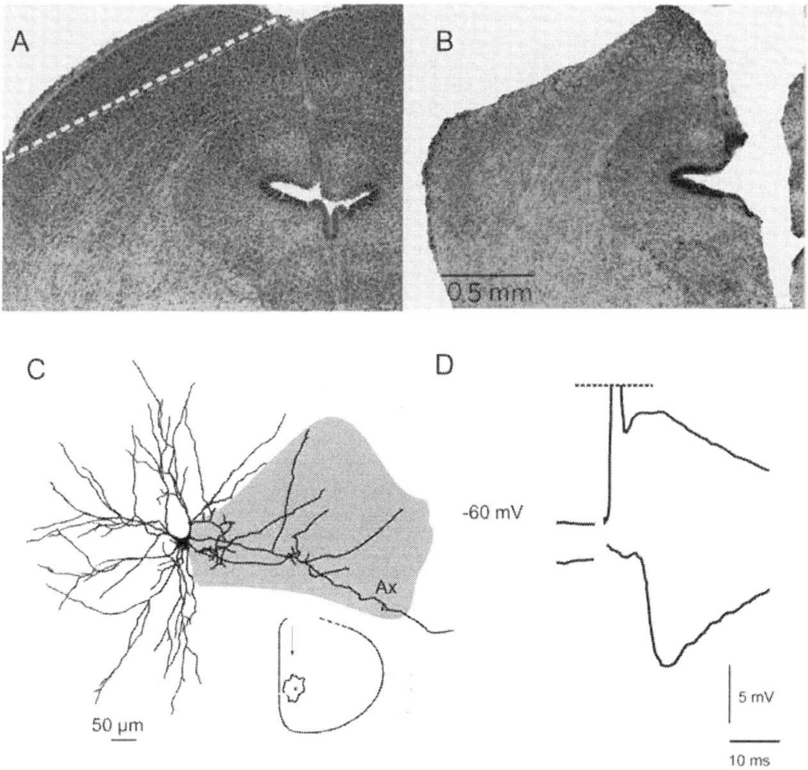

FIGURE 8.4 Section of the late embryonic rat sSC (A, B) and recording of an action potential/IPSP sequence (D) from a wide-field vertical neuron (C). (A, B) Nissl-stained section from an E20 rat. Already by this early age the boundaries of the sSC are discernible at least on the lateral side. Between E18 and P28 the sSC can be maintained as a whole mount preparation with only one ventral cut. (C) Recording from an E21 sSC whole mount. Electrical stimulation rostrally to the position of the soma (arrow head in D) induced a sequence of antidromic action potential (truncated at the level of the dotted line) and an IPSP that is outward due to Cl^- influx. Membrane hyperpolarization due to current injection blocked antidromic invasion and inverts the IPSC. (C) WFVN after reconstruction. The shadowed area indicates the local axonal projection as reconstructed from the partially labeled axon (ax). Inset: Projection of the dendritic area of this WFVN on the surface of the sSC.

input, or its lack at the appropriate time, could influence the dendritic geometry of sSC neurons.[84]

Comparing the dendritic trees of tectogeniculate neurons in normal and neonatally enucleated adult hamsters, Mooney and colleagues discovered that enucleation caused an abnormal orientation of dendrites toward the deeper layers where sSC neurons could find somatosensory afferents as a possible substitute for the missing retinal input.[85] This is an important example of afferent-dependent plasticity of dendrite growth. However, more information is necessary to clarify the mechanisms through which specific neuron types in the sSC establish their characteristic position and dendritic orientation.

The obvious problem for developmental studies is that before neurons reach maturity one cannot easily apply the criteria for neuron identification defined in adult preparations. Only in a few fortunate cases (such as RGCs) is a molecular marker (Thy-1) available, which then allows a search for other distinctive properties (for instance, the selective expression of P2X receptors in RGCs).[86] Although such natural molecular tags are not yet known for sSC neurons, research on the basic properties of sSC neurons and synapses has recently become more active.

A frequently applied classification of sSC neurons in rodents was proposed by Langer and Lund.[87] They described the following neuron types: wide-field vertical neurons (WFVNs), narrow field vertical neurons (NFVNs), piriform or stellate neurons, horizontal neurons (HNs), and marginal cells.

8.6.1 Wide-Field Vertical Neurons

WFVNs are the largest cells of the sSC (soma diameter in rat: >20 to 25 μm). They reside in the lowermost SGS and in the SOp[88–90] and have radial spine-bearing dendrites with a predominantly dorsal orientation.[87,91–94] Therefore, they represent a potential target for all visual afferents. In addition, they may receive some input from fibers projecting to the SGI.[90] Although, WFVNs lack an easily recognizable *electrophysiological signature*,[95] they tend to generate multiple spikes in response to a single short stimulus, which might help to identify WFVNs at least in mature slice preparations.[96] In addition, WFVNs (but also HNs) preferentially express inwardly rectifying, highly Ca^{2+}-permeable AMPA receptors.[97]

From a histochemical viewpoint, WFVNs appear to represent a more or less homogeneous neuron population. The majority of large SOp neurons[5] or cells projecting into the LP stain for calbindin[98] and, *vice versa*, 91% of the calbindin-immunoreactive neurons project to the LP.[99] Immunoreactivity (IR) for calbindin seldom or never co-localizes with GABA[99,100] or parvalbumin.[101,102] Unfortunately, in the rat sSC, calbindin-IR is not an exclusive marker for WFVNs.[99,102] Nevertheless, together with a dendritic reconstruction, calbindin-IR can probably serve as a molecular marker to recognize putative WFVN/LP-projection neurons from other neurons in the deep SGS or SOp.

The function of rodent WFVNs was first investigated by Mooney and colleagues.[90] They classified SGS neurons according to their axonal projection into the lateral posterior (pulvinar) nucleus (LP) or the dorsal LGN (LGNd), characterized the corresponding receptive field properties, labeled the cells with intracellular HRP injections, and reconstructed their dendrites completely and their axons partially. This elegant work provided direct evidence for the following conclusion: most WFVNs are movement-sensitive, and they nearly always project to the LP. By connecting the retina with the LP and then the visual cortex, WFVNs represent a major element of the extrageniculate visual pathway.

In addition to these extracollicular targets, hamster WFVNs establish a link with the deeper SC layers.[92] A projection from the sSC to the deeper SC layers has also been described for other species,[103–107] including primates.[108] An important finding was that this intracollicular visuo-motor pathway is not exclusively formed by a specialized class of *interlayer neurons*, which solely project to the deeper SC layers, but also by projection neurons with a variety of distant targets.[90,108] An interesting question was therefore how the positional code is preserved within the SC. The most detailed information on this issue comes from axon reconstruction in primates.[108] It was found that in individual sSC neurons the axonal projection in the deeper SC layers does not significantly exceed the dendritic territory in the sSC. Our own material from postnatal rat (10 partially reconstructed axons) showed, however, that on their way to rostral destinations WFVNs issue regularly spaced (about 150 μm) *recurrent* collaterals from the SOp back into the SGS. As collicular effects on LP neurons are mostly excitatory,[109] one can imagine that these WFVN collaterals spread electrical activity to other sSC territories. This should blur the positional information provided by the sSC, unless inhibition is active to restrict the activation focus. Indeed, experiments with collicular injections of $GABA_AR$ agonists or antagonists showed decreased precision of saccadic eye movements.[1,110]

Figure 8.4C illustrates a morphologically and electrophysiologically characterized WFVN from a P21 slice of the rat sSC.[111] In the well-preserved collicular whole-mounts used for these tests, WFVNs generated IPSPs in response to coarse electrical activation in the rostrolateral SOp (Figure 8.4D). Short-latency spikes were seen in at least 50% of the cases (>40 neurons tested between P9 and P28). That these short latency spikes in fact represent antidromic action potentials was never rigorously tested. But it would not seem unlikely, considering the course of the main axons of WFVNs in the SWOp.

The presence of local collaterals is consistent with the existence of a circuit for recurrent inhibition. Apart from restricting the visual receptive fields on a functional level, recurrent inhibition could also play a role in activity-dependent modification of retinal or cortical inputs. For instance, experiments in the amphibian OT showed[112] that synaptic responses to retinal stimulation were depressed when occurring during a critical interval *after* a large EPSP/spike (Figure 8.5). This is a condition that is thought to decrease the chances for synapse stabilization.

8.6.2 Horizontal Neurons

In contrast to the projection neurons of the sSC, intrinsic inhibitory neurons of the rodent sSC often possess only two stem dendrites, without spines.[111,113] With a soma size of 12 to 15 μm they are considerably smaller than WFVNs. These bipolar neurons represent another distinct phenotype of the rodent sSC.[87,93,97,105,114,115] As the name implies, the dendritic arbor of HNs orients itself parallel to the sSC surface. In hamsters, the average mediolateral extension of intracellularly labeled HNs is 637 μm.[116] The distant dendrites have unusual geometries, being highly branched, tufted or curly. Cells with bipolar dendritic morphology are also found at an early postnatal age.[82,117]

HNs typically use GABA as a neurotransmitter[118] and may be involved in the recurrent or surround inhibition that limits visual responses in time and space. Interestingly, some HNs also project to the deeper SC layers.[92] A horizontal orientation of dendrites is, of course, only a modest criterion to distinguish an inhibitory interneuron. In fact, some horizontally oriented cells were found to project into the LP[90] and to express calbindin.[5] Therefore, to regard an HN as an inhibitory inter-

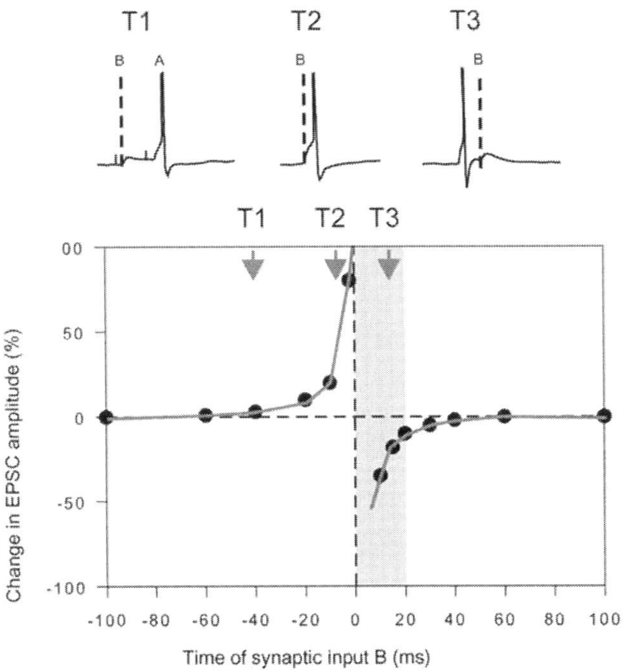

FIGURE 8.5 Interval-dependent potentiation of a weak EPSP induced by electrical stimulation in the amphibian retina before (T1), during (T2), and after (T3) a strong retinal stimulus initiating a larger EPSP able to initiate a spike. Shadowed area indicates period of depression of the weaker EPSC. (Modified from Zhang, L.I., Tao, H.W., Holt, C.E., Harris, W.A., and Poo, M.M., A critical window for cooperation and competition among developing retinotectal synapses, *Nature*, 395, 37–44, 1998. With permission.)

neuron, dendritic reconstruction should be combined with demonstration of GABA- or GAD-IR.

8.7 SYNAPTOGENESIS STARTS WITH GABA AND GLYCINE

If one accepts the possibility that inhibitory connections restrict the spread of excitatory input in the sSC, one is led to the next question: When are inhibitory synapses available for such function?

With EM techniques synapses in the sSC were detected as early as E17.[44] In fact, these were synapse-like membrane appositions that mostly lacked vesicles and, until recently, little could be said about the functional capacity of these terminals. Using patch clamp recording from embryonic collicular slices we have now shown that sSC neurons are competent for synaptic transmission as early as E17.[81,119] Moreover, until P2, spontaneous synaptic activity is exclusively generated by inhibitory synapses. Figure 8.6 shows a specimen record of unitary evoked IPSCs (eIPSCs) from an E17 mouse sSC neuron. Unitary IPSCs are the responses to activation

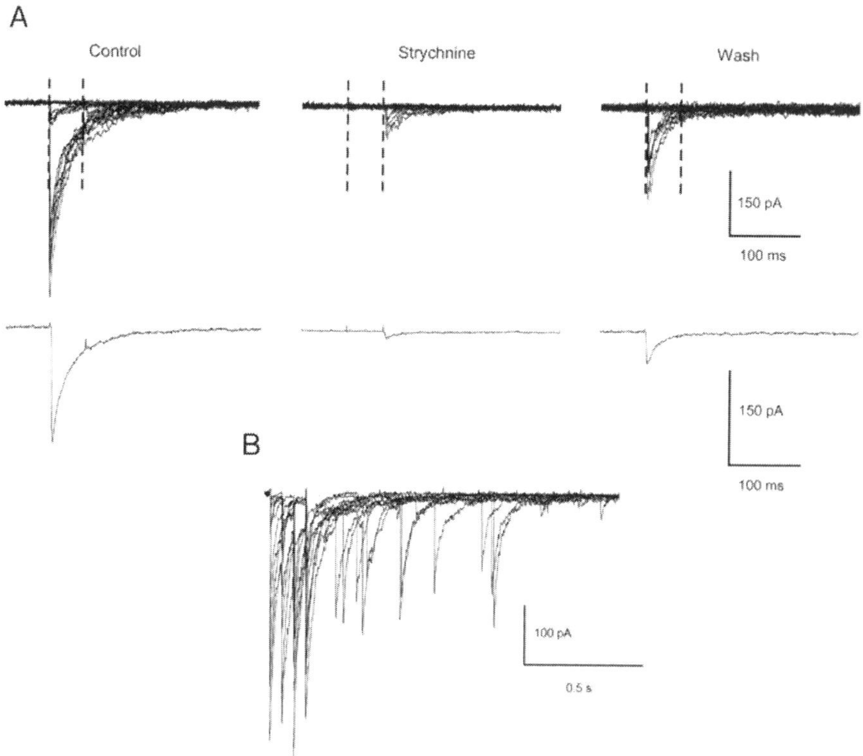

FIGURE 8.6 Specimen recordings of inhibitory synaptic currents in horizontal SC slices from E17 mice. (A) Paired pulse stimulation reveals strong depression at intervals < 5 s, which is presumably due to temporal unavailability of postsynaptic receptors. Strychnine at concentrations between 0.5 and 3 μ*M* is a weak antagonist at these synapses. It completely blocks the response to the first pulse and unmasks a smaller response that was protected from desensitization by strychnine occupancy of the receptor. (B) Repetitive stimulation by at least four stimuli induced delayed and asynchronous release, in addition to the partially failing stimulus-locked eIPSC. The asynchronous response component is a characteristic feature of neonatal synapses and presumably due to the immaturity of the presynaptic buffering systems.

of just one presynaptic cell. Note that at this early stage strychnine acts as a weak antagonist at inhibitory synapses (see Reference 120 on the function of GlyRs in the sSC). In the embryonic and early postnatal rat and mouse sSC amplitudes of unitary eIPSCs were surprisingly large and two to three times longer than in the adult. Evoked IPSCs exceeded the amplitudes of mIPSCs by a factor of 4 to 12, indicating that action potential-mediated transmitter release is already more effective than spontaneous release. Repeated activation at short intervals induces strong synaptic depression, and transmission switches to the asynchronous mode of transmitter release (Figure 8.6B and Reference 121). In summary, synaptic activity in the sSC starts with inhibitory synaptic responses. Immature inhibitory synaptic transmission is characterized by prolonged duration, high quantal content after single stimuli, and depression and asynchronous release during and after repetitive activation.

It should be pointed out, that the presence of inhibitory synaptic activity in slices from embryonic mice or rats need not be interpreted as proof of the existence of classical inhibitory synapses. There are at least two alternative possibilities: (1) the inhibitory synaptic transmitter is released from nonsynaptic structures; (2) inhibitory synaptic contacts are formed in the slice, but as a pathophysiological reaction to tissue isolation and *in vitro* maintenance.

To address this problem, we performed immunostaining in frontal sections prepared from immediately fixed sections of the embryonic rat sSC.[119] Synaptic terminals were identified by immunoreactivity (IR) to synaptophysin (Syp). Figure 8.7A shows that some Syp-immunoreactive puncta were indeed present in the immediately fixed colliculi of E19 rats. In the vast majority of terminals, Syp-IR colocalized with the vesicular inhibitory amino acid transporter VIAAT (Figure 8.7A and 8.7C), an excellent marker of GABAergic and glycinergic terminals at early developmental stages.[122] Only 25% of the Syp-immunoreactive puncta in the E19 sSC were devoid of VIAAT-IR. About the same fraction of terminals showed IR for the glutamate transporter (GluT). Although we observed a significantly larger number of inhibitory terminals in slices maintained *in vitro* (Figures 8.7B to 8.7E), there is no doubt that the E19 rat sSC contains inhibitory synaptic terminals. Similar results were obtained from the embryonic mouse sSC. We therefore concluded that in the rat and mouse sSC synaptogenesis starts before birth, with an initial predominance of inhibitory contacts.

These findings create the need of some corrections in the already established concepts concerning synapse development and map formation in the sSC. For a long time, the role of GABAergic synapses in the formation of topographically ordered retinal and cortical projections received little consideration. On the contrary, it was believed that intrinsic inhibitory circuits in the sSC form under the influence of, and in accordance with, the synaptogenesis of retinal and cortical afferents.[8,123] Although an effect of glutamatergic input on the development of GABAergic inhibition was confirmed in our lab (see Section 8.9.2), it now seems unlikely that inhibitory synaptogenesis is triggered by preexisting glutamatergic synapses because: (1) in late embryonic sSC slices, glutamatergic synaptic activity is virtually absent, while IPSCs can be found in many sSC neurons (>40% at E19 in the rat); (2) a strong majority of synaptophysin-IR terminals displays VIAAT-IR (75%); and (3) the relative number of synaptophysin-IR terminals with GluT-IR remains low (<25%) until day P1.

The unexpectedly dominant role of inhibitory synapses in the immature sSC can be understood better if one considers that at this stage of development GABA and Gly act as depolarizing transmitters. Thus, early in development GABAergic synapses may fulfil a function that later on is carried out by glutamatergic synapses. Synaptic activation of $GABA_A$Rs will cause Ca^{2+} influx through voltage-activated Ca^{2+} channels (Figure 8.8A to 8.8C), with all the well-known consequences for enzyme activation and gene expression. A leading role of GABAergic synapses has already been demonstrated in the development of the postnatal hippocampus where GABAergic synapses appear earlier or in synchrony with glutamatergic synapses.[124,125]

Interestingly, both in a collicular culture model of inhibitory synaptogenesis[120] and in slice experiments,[119,126] GABA was found to lose its depolarizing action coincidental with the appearance of glutamatergic synaptic input. In mice, this

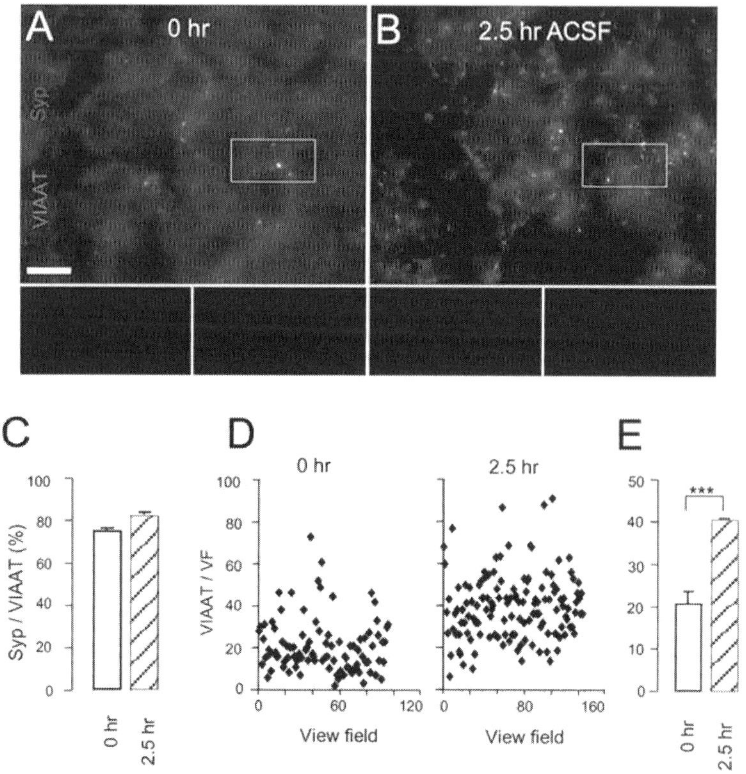

FIGURE 8.7 State of inhibitory synaptogenesis in horizontal sSC slices from E19 rats. (A) Immediately fixed slice. Note presence of synaptophysin-(Syp)- immunoreactive puncta in contact with the somata and proximal dendrites of sSC neurons. Nearly 80% of the Syp-positive terminals stained with the inhibitory synaptic marker VIAAT. Cell counts indicate that the average number of VIAAT-positive terminals per neuron is 0.8. During a 2.5-hr period of slice maintenance in artificial cerebrospinal fluid (ACSF) the number of Syp- and VIAAT-IR terminals doubles (B to E), while the neuron number remains the same. Together with the data from Figure 8.6 this experiment proves the capacity of embryonic rodent sSC neurons to form inhibitory synaptic contacts. VF — view field. Scale bar: 10 μm, boxes 4.8 μm. (Modified from Meier, J., Akyeli, J., Kirischuk, S., and Grantyn, R., GABA(A) receptor activity and PKC control inhibitory synaptogenesis in CNS tissue slices, *Mol. Cell. Neurosci.*, 23, 600–613, 2003. With permission.)

happens between P1 and P2 (see Figure 8.8D). As early as P3, application of $GABA_A$ antagonists caused disinhibition of spike generation.[81,127] This means the time window during which GABA and Gly can act as depolarizing neurotransmitters is narrow (only 5 days), the more so as the total time needed to reach adult levels of synapse numbers is 5 to 6 weeks (see Figure 8.3).

Although short in duration, the period of depolarizing GABA may be critical for further synapse formation. As $GABA_AR$ activation induces Ca^{2+} influx in collicular neurons (Figures 8.8A to 8.8C, 8.9A and 8.9B), formation of additional synapses may depend on the state of $GABA_ARs$ and/or Ca^{2+} channels. We addressed this

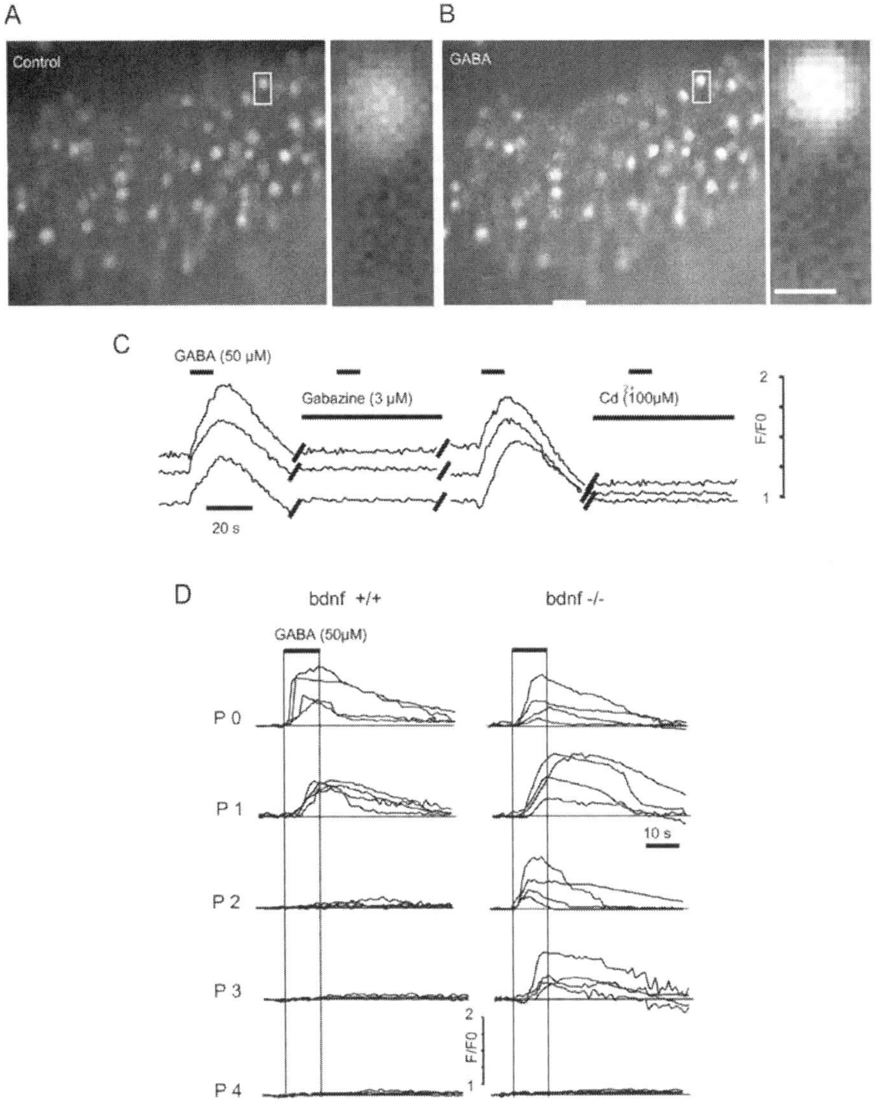

FIGURE 8.8 Ca^{2+} responses of wild-type and BDNF-deficient neurons to GABA in postnatal sSC slices. (A, B) P1 collicular slices after loading with Oregon Green 488 BAPTA-1 AM (control) and during application of GABA (50 μ*M*). A region of interest (ROI, boxed at lower magnification on the left) is shown enlarged. (C) GABA-induced Ca^{2+} elevations were sensitive to the $GABA_A$Rs blocker gabazine and the Ca^{2+} channel blocker Cd^{2+}. (D) Ca^{2+} responses from wild-type and BDNF-deficient slices at different age. Each trace represents one slice and was obtained by averaging the fluorescent signals from at least 100 neurons with responses to KCl. GABA-nonresponsive neurons were included. Fluorescence signals are expressed as relative changes from prestimulus levels ($\Delta F/F_0$).

question in a new slice model of inhibitory synaptogenesis.[119] Figure 8.9C to 8.9E illustrates an experiment with nifedipine, an L-type Ca^{2+} channel blocker. Quantitative evaluation of this and related tests revealed that pharmacological suppression of $GABA_ARs$ (not shown) or low levels of Ca^{2+} influx through L-type Ca^{2+} channels stimulate the formation of additional inhibitory synapses (Figure 8.9F and 8.9G) and the accumulation of $GABA_ARs$ opposite inhibitory terminals (Figure 8.9H). As the other side of the same coin, high levels of $GABA_AR$ activation and Ca^{2+} influx should decrease the likelihood that further contacts are made in collicular slices. This was confirmed.[119] Based on these results, one can predict that $GABA_AR$ potentiating drugs, such as alcohol, barbiturates, and benzodiazepines, should considerably influence the formation of inhibitory circuits at late embryonic stages. In general, these drugs will decrease the number of GABAergic contacts and therefore decrease synaptic inhibition in the neonates, when GABA is no longer a depolarizing transmitter. Apart from causing behavioral disturbances (hyperactivity), the imbalance between excitation and inhibition should increase neuron death. This has already been demonstrated in the lateral thalamus after a single alcohol intoxication.[128]

8.8 AXON TERMINALS ADJUST TO THE NEEDS OF ACTIVE PATHWAYS

In addition to the still enigmatic putative mechanism that uses inhibition to restrict glutamatergic projections, there are several additional factors that could influence the use-dependent reorganization of the retinotopic map. The principal ones are: (1) matching of afferent and target cell populations, (2) axon remodeling, (3) dendrite remodeling and (4) activity-dependent synapse stabilization. In each case, there may be selective up- or down-regulation, i.e., neurogenesis or cell death (1), outgrowth or loss of additional collaterals (2), outgrowth or retraction of dendritic branches (3), and increase or decrease of synapse numbers per unit area (4), not to mention *metaplasticity*, i.e., modification of structural patterns without net changes in numbers. All these mechanisms would serve one main purpose: the adjustment of afferent convergence at any given sSC neuron.[129] Here we examine the effect of lesion and activity blockade on the reorganization of retinotopic maps.

8.8.1 Removal of One Eye or Retinal Lesion Before or After Birth

This manipulation strengthens the uncrossed ipsilateral projection to the rat,[130–132] mouse,[133] and hamster sSC.[134] However, the coordinate system established by the aberrant uncrossed retinotectal pathway becomes ambiguous,[135] and retinal activation is less effective than in control animals.[136,137]

Normally, the initially exuberant ipsilateral retinotectal pathway declines within the first 2 weeks after birth,[45] and is associated with substantial cell death in the retina and, consequently, with a reduction in the number of retinal axons in the optic nerve.[132,138] By the middle of the second postnatal week, the number of RGC axons in the optic nerves of albino rats is only 32.6% of the maximal number at E20.[139]

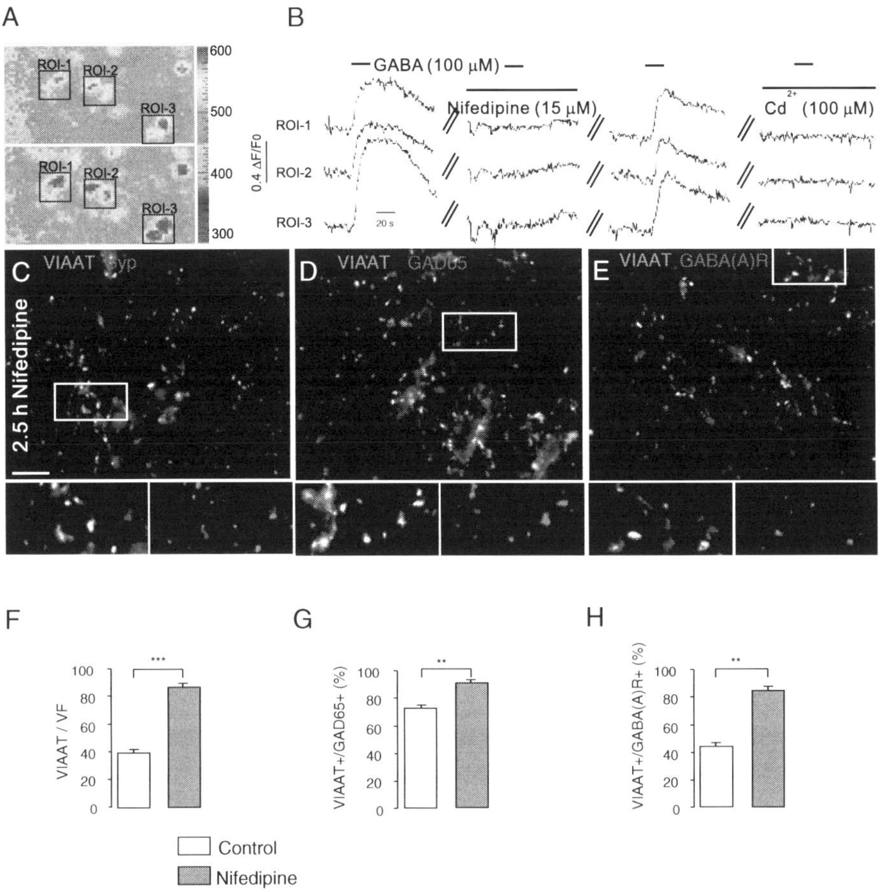

FIGURE 8.9 L-type Ca^{2+} channels play a pivotal role in the generation of GABA-induced Ca^{2+} transients (A, B) and in inhibitory synaptogenesis (C to H). (A) Ca^{2+} transients in response to GABA application. Slices were loaded with Oregon Green 488 BAPTA-1-AM. Pseudo-color image showing three regions of interest (ROI) before (upper panel) and during (lower panel) GABA application. (B) Corresponding traces of the Oregon Green fluorescence signal ($\Delta F/F_0$). Nifedipine or Cd^{2+} were applied before (2 min), during (20 s), and after GABA application (2 min), and led to complete suppression of the GABA-induced fluorescence increase. Scale bar: 10 μm, boxes 4.8 μm, for Ca^{2+} imaging: 30 μm. (C to H) Effects of L-type Ca^{2+} channel activity on the formation of inhibitory synapses. Effect of 2.5-hr incubation in ACSF or nifedipine (15 μ*M*) on the number of VIAAT-positive synaptic terminals (C, F), the co-localization of VIAAT with GAD65 (D, G) and co-localization of VIAAT with $GABA_AR$ α2 (E, H). Note Ca^{2+}-dependent up-regulation of terminal number, GAD65 expression and $GABA_AR$ accumulation. (Modified from Meier, J., Akyeli, J., Kirischuk, S., and Grantyn, R., GABA(A) receptor activity and PKC control inhibitory synaptogenesis in CNS tissue slices, *Mol. Cell. Neurosci.*, 23, 600–613, 2003. With permission.)

It is now widely accepted that RGCs with misplaced collicular projections are more likely to die during the first 1 to 2 postnatal weeks, and that this massive cell death contributes to the increasing precision of the retinotectal map.[140] Neurotrophins are implicated in the regulation of this second apoptotic wave, which regulates the number of RGCs in accordance with the demands of their targets (see References 141 and 142 for reviews).

8.8.2 Ablation or Lesion of the Visual Cortex

This lesion increases the fraction of layer V pyramidal neurons from the remaining visual cortex which project bilaterally. This results in an aberrant crossed cortico-tectal projection with less dense and more confined termination in the sSC.[143,144]

8.8.3 Unilateral Superior Colliculus Lesion

This results in abnormal innervation of the remaining sSC from both retinas.[145] This resembles the situation in structures where visual inputs from both retinas arrive, but segregate into different layers (LGN) or columns (visual cortex). If the density of recrossing axons is sufficiently high, terminations become patchy and may even form stripes.[146] Some axons invade the deeper layers.[145]

8.8.4 Partial Unilateral Collicular Lesion

This causes a compression of the retinotectal projection,[147] but leaves the receptive fields of individual sSC neurons largely unchanged.[148] The in-growing axons find a smaller number of target neurons in the sSC. Therefore, the afferents will be redistributed to the remaining population such that the entire visual field is represented again. Individual retinal axons occupy a smaller area, with fewer branches and fewer terminal boutons, while bouton density remains the same.[149] This was taken as evidence that retinal axons compete for a limited amount of target space.

8.8.5 Activity Blockade with TTX

As in studies of other parts of the brain (see References 150 and 151 for reviews), TTX was applied to investigate whether electrical activity is important for the formation and refinement of the retinal representation.[140,152] Although TTX had little effect on the rate of axon loss from the optic nerve,[153] it increased the survival of RGCs with inappropriate projection in the sSC.[140] In the neonate hamster, Na^+ channel block increased the fraction of surviving RGCs with uncrossed projection and abolished the normally patchy termination pattern in the rostral sSC.[154] Block of retinal activity affected the survival of collicular neurons. In the rat sSC, intraocular TTX injection increased the naturally occurring cell death in the sSC[155,156] by about 50%.[157] This was taken as evidence for a competitive relationship between retinal afferents: the strongly active input wins by suppressing the stabilization of the less active input.[112] Directly or indirectly, this leads to an activity-dependent decision of which RGC or sSC neurons have to die.

Before maturation of the photoreceptors and eye opening, animals lack patterned vision and the capacity to perform orienting movements toward visual targets. This does not mean, however, that network activity is absent. Even sSC slices from P2 mice contain neurons displaying spontaneous spike discharge.[81] Thus, in concert with the signals emerging from the spontaneously active inner retina[158] *previsual* activity of the sSC can influence the properties and distribution of preformed synaptic connections.

The origin of spontaneous synaptic activity in the early postnatal sSC is not yet clear. As already mentioned, glutamatergic spontaneous synaptic activity is scarce at this age, at least in slice preparations. Until P2, GABA-induced depolarization or GABA-modulated Ca^{2+} waves should be the predominant drive for spontaneous activity in the sSC. Then glutamatergic synapses take over. However, the switch from depolarizing to hyperpolarizing GABA action exhibits some degree of plasticity. For instance, studies on *bdnf* knockout animals[126] showed that chronic absence of brain-derived neurotrophic factor delays the loss of $GABA_A$R-mediated depolarization by at least 2 days (Figure 8.8D).

8.9 THE ONSET OF PATTERNED VISION DRIVES SYNAPSES TO MATURE PERFORMANCE

Activity-regulated reorganization of synaptic connections in the rodent sSC is a prolonged process, which probably covers more than 3 to 4 weeks. EM studies in rats showed that the synapse numbers in the SGS reach a maximum (100%) at P24–30 [9] and then decline to about 50% in the adult. The most dramatic increase in the number of synaptic terminals occurs shortly after eye opening, between days P14 and P15. Interestingly, in slices from the mouse sSC, the frequency of spontaneous spike discharge has a similar age dependency (Figure 8.10A and 8.10B). The steepest increment occurs between P15 and P20.[159]

8.9.1 The Role of TrkB Activation

Studies in other brain structures have shown that changes in electrical activity alter the level of neurotrophins and this may modify the location, number, and properties of already existing synapses and also change the speed at which new synapses are formed. The best characterized neurotrophin in the visual system is brain-derived neurotrophic factor (BDNF). In the hamster SC, the amount of BDNF and BDNF mRNA increased due to continuous light rearing, and there were marked diurnal fluctuations in the level of BDNF.[160] BDNF concentrations had a maximum on P12 (day of eye opening in the hamster) and then decreased to reach 63% of the maximum at P18.[161] At all ages studied, the concentration of BDNF was much higher in the SC than in the retina.

The classical theory of neurotrophism stipulates that the survival rate of afferent neurons depends on neurotrophin levels in the target area. At P4, i.e., during the maximal rate of cell death in hamster RGC population, a 13- to 15-fold reduction of RGC pyknosis was obtained by injecting BDNF into the sSC.[162] Similar results were obtained in rat.[163,164] Chronic absence of BDNF due to *bdnf* inactivation had no effect on the number of RGC axons in the optic nerve,[165] but reduced the ability of postnatal RGCs to maintain repetitive discharge.[166] Cell death in layer V of the visual cortex is also not increased by tectal ablation.[167] Whether or not the rate of cell death in the

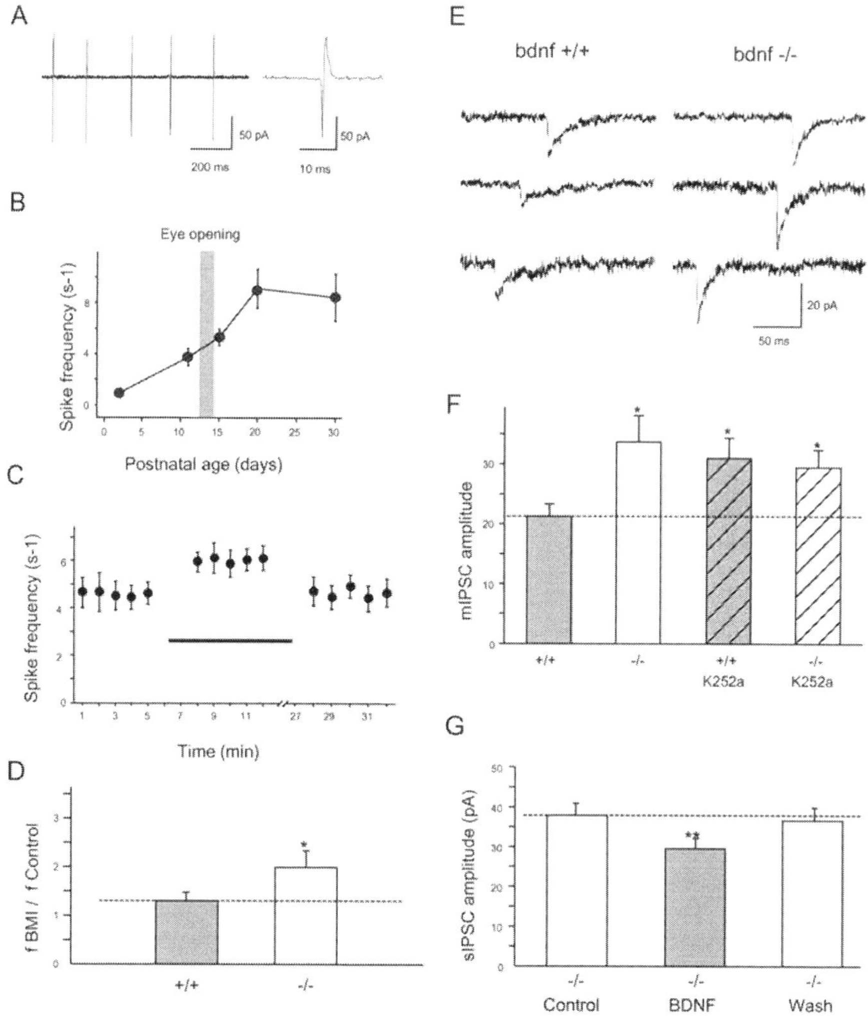

FIGURE 8.10 Effects of BDNF on spontaneous spike discharge and GABAergic synaptic activity in the sSC of P15 mice. (A) Cell-attached recording from a neuron in the sSC of a wild-type mouse. (B) Developmental changes in the average frequency of spontaneous spike discharge in wild-type mice. (C) Disinhibition of spontaneous spike discharge in the presence of bicuculline methiodide (50 μ*M*, horizontal bar). (D) Stronger disinhibition in *bdnf-/-* mice. (E) mIPSC activity (recordings in 1 μ*M* TTX) in wild-type and BDNF-deficient mice. (F) Up-regulation of mIPSC amplitudes in mice lacking BDNF. Same effect caused by a 1.5 hr incubation in the tyrosine kinase inhibitor K252a (200 n*M*). (G) Acute application of BDNF (100 ng/ml for 10 min) reverses the IPSC amplitude increase caused by *bdnf* inactivation. (Modified from Henneberger, C., Jüttner, R., Rothe, T., and Grantyn, R., Postsynaptic action of BDNF on GABAergic synaptic transmission in the superficial layers of the mouse superior colliculus, *J. Neurophysiol.*, 88, 595–603, 2002. With permission.)

visual cortex can be reduced by BDNF injection into the sSC is not yet known. Finally, endogenous BDNF reduces the natural death of neurons in the rat SC[168], which is maximal around P5 to P8.[157] So, by controlling cell death in the visual system during the first 1 to 2 weeks of life, BDNF could influence the process of synaptic refinement.

Apart from its *classical* neurotrophic function, BDNF might play a role in the activity-dependent adjustment of specific synaptic connections.[169] These *modulatory* actions of BDNF have already been examined in the rodent sSC.[126,159] Figures 8.10D, 8.10F, and 8.10G summarize results from experiments with *bdnf-/-* animals. In Figure 8.10C the influence of $GABA_AR$ activity on the spontaneous spike discharge was tested in P15 slices. Application of a $GABA_AR$ blocker induced disinhibition. The degree of disinhibition is significantly higher in the absence of BDNF (Figure 8.10D). These results point to a suppressive effect of BDNF on GABAergic inhibition. Recording eIPSCs and mIPSCs revealed that the chronic absence of BDNF is indeed followed by an up-regulation of inhibitory synaptic responses (Figures 8.10E and 8.10F). This effect could be reproduced by slice incubation in the TrkB antagonist K252a (Figure 8.10F) and reversed by acute application of BDNF (Figure 8.10G). It should be noted that the down-regulation of GABAergic inhibition by BDNF was age-dependent. No difference between *bdnf+/+* and *bdnf-/-* mice could be detected in E13 or younger slices.

These findings led us to the conclusion that the up-regulation of synaptic inhibition in *bdnf-/-* mice was not caused by a slow process of synaptic reorganization, but by the acute absence of BDNF. The suppression of GABAergic inhibition was most likely the result of TrkB-mediated activation of PKC.[159] This is a new aspect of BDNF action in the developing brain and supports the idea of Fritschy and colleagues,[170] who suggested that BDNF down-regulates synaptic inhibition by reducing the $GABA_AR$ surface expression at synaptic sites.

Finally, there is one more remarkable facet in the already big variety of BDNF actions in the developing visual system: it affects the translocation of the Cl^- transporter KCC2 into the neuronal plasma membrane and this is, most likely, the basis for the BDNF-dependent switch from depolarizing to hyperpolarizing GABA action.[126] All these results support the idea that BDNF is one of the major factors that regulate the level of network activity in the sSC around eye opening.

8.9.2 The Role of NMDA Receptor Activation

The onset of patterned vision after eye opening is a critical event in the life of sSC neurons. At least as far as inhibitory synapses are concerned, there are substantial changes in the following properties[81]: (1) After eye opening eIPSCs decay much faster; (2) The expression of the $\alpha 1$ subunit of the $GABA_AR$ increases substantially between P11 and P15; (3) Paired pulse depression is replaced by paired pulse facilitation; (4) Repetitive activation is followed by stimulus-locked IPSCs, not asynchronous responses. After eye opening, (5) IPSCs acquire a stereotyped time course, and (6) rarely exhibit signs of postsynaptic receptor saturation.

The first and second items on this list are illustrated in Figures 8.11A to 8.11C. A change in the subunit composition and shortening of postsynaptic responses is a very common sign of synapse maturation. It also occurs in cholinergic and glutamatergic synapses. The new and interesting finding is that the shortening of GABAergic synaptic currents is caused or influenced by NMDAR activation (Figures 8.11D to 8.11F). This idea is supported by the following results.[171] The change in IPSC duration in sSC neurons was maximal just after the amplitude of NMDAR-mediated EPSC reached a maximal value (Figure 8.11D). Injections of the NMDAR blocker MK801 between P0 and P11 prevented both the shortening of eIPSCs and the developmental up-regulation of the $GABA_AR$ $\alpha 1$ expression (Figures 8.11E and 8.11F).

Thus, our data lend indirect support to the hypothesis of Constantine-Paton and colleagues, who suggested that the development of intrinsic inhibitory circuits in the sSC requires maturation of glutamatergic synaptic activity. They observed that animals receiving NMDAR antagonists during the first 2 postnatal weeks exhibited aberrant retinotectal axons.[172] Lesion-induced sprouting of ipsilateral retinotectal axons was enhanced.[173] Chronic treatment with NMDA beginning from day P8 resulted in a massive decrease in the frequency of spontaneous EPSCs and a reduction in the AMPA/KA-fraction of synaptic currents.[174] Finally, with regard to inhibitory synaptic transmission, this group found that the expression of GAD65 and GAT-3, as well as the appearance of bicuculline-sensitive synaptic activity roughly coincided with the up-regulation and shortening of NMDAR-mediated EPSCs.[123,175] After eye opening, between P17 and P18, IPSCs normally increase in amplitude,[175] but chronic treatment with NMDA from day P8 shifted this increase to an earlier age (P12 to 13).

Although Constantine-Paton and colleagues probably missed the beginning of inhibitory synaptogenesis in the sSC, they were right in suggesting that NMDAR activity can shape the properties of already existing or newly formed inhibitory synapses.[123,175]

8.10 CONCLUSION

During the past decade, the mechanisms enabling the ordered representation of visual space in the rodent sSC have received much attention. While considerable progress has been made in the field of axonal guidance, to date little is known about the development of synaptic connections between individual identified neurons. Therefore, more effort is needed to bridge the gaps between synapses, maps, and sensorimotor functions of the superior colliculus. In short: the mechanisms of map formation in the SC remain an attractive subject for further research.

ACKNOWLEDGMENT

Unfortunately, this review cannot be complete. We are very sorry for not having mentioned a number of outstanding contributions. The questions addressed here represent a selection made in accordance with the FAQs of our current Ph.D. students Jan Akyeli, Christian Henneberger, Sonja Schmidt, and Jan Walter, to whom we would like to dedicate this chapter. The authors thank the editors of this volume for many helpful suggestions on this manuscript. We also wish to thank our colleagues, Alfredo Rodríguez-Tébar and Hannes Schmidt, for their critical reading, David

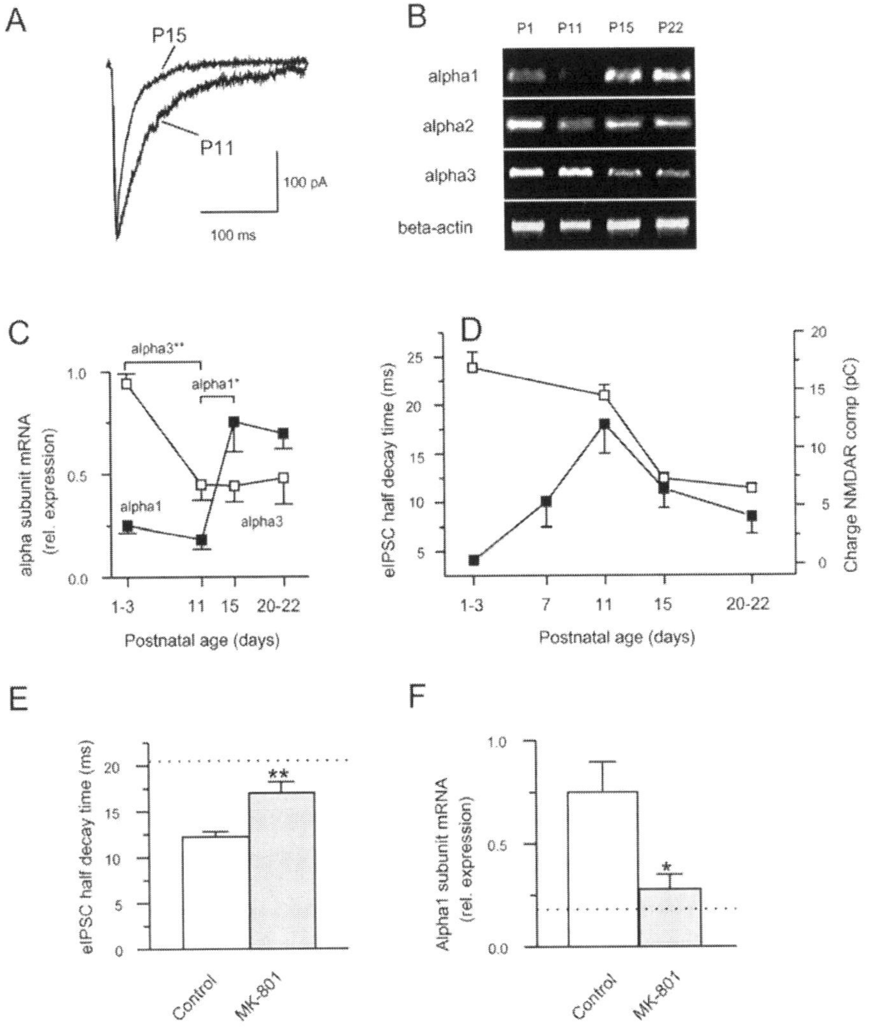

FIGURE 8.11 Developmental plasticity of GABAergic synaptic transmission and its dependence on NMDAR activity. (A) Decrease of eIPSC duration after eye opening (P15), for quantitative results see (D). (B) Semiquantitative RT-PCR of the $GABA_AR$ subunits $\alpha1$, $\alpha2$, and $\alpha3$ in tissue homogenates of the sSC from mice of indicated age. (C) Quantification of results as shown in (B). Mean and SE from three independent PCR reactions for each animal, subunit, time point. (D) Developmental changes in the half-decay time of eIPSCs (open squares) and NMDAR-mediated component of eEPSCs (filled squares). (E) Half-decay time of eIPSCs on P15 in controls and after repeated injection of MK-801 from day P1. (F) relative expression of the $\alpha1$ subunit at P15 in controls and after treatment with MK-801. Dotted line corresponds to level at P11. Asterisks denote significance levels. *: $p<.05$; **: $p<.01$.

Betances for editing the language, and Andrea Schütz for being very helpful in rebuilding our literature database. This work was supported by the DFG-grants SFB515 B2, GRK238, and Ro925 to RG.

LIST OF ABBREVIATIONS

AMPA	alpha-amino-3-hydroxyl-5-methyl-isoxazole-4-propionate
ACSF	artificial cerebrospinal fluid
A-P	anterior-posterior
BDNF	brain-derived neurotrophic factor
D-V	dorso-ventral
FAQs	frequently asked questions
FGF	fibroblast growth factor
GABA	gamma-aminobutyric acid
$GABA_AR$	$GABA_A$ receptor
GAD	glutamate decarboxylase
GAT	GABA transporter
Glu	glutamate
GluT	glutamate transporter
Gly	glycine
GlyT	glycine transporter
HN	horizontal neuron
IR	immunoreactivity
KA	kainate
KCC2	K^+/Cl^- cotransporter
LGNd	dorsal lateral geniculate nucleus
M-L	medio-lateral
LP	lateral posterior nucleus
NMDA	N-methyl-D-aspartate
NMDAR	NMDA receptor
OD	optic disc
OT	optic tectum
PKC	protein kinase C
RACK1	receptor for activated C kinase1
SC	superior colliculus
SGS	stratum griseum superficiale
S-MCPG	(S)-alpha-methyl-4-carboxyphenylglycine
SOp	stratum opticum
SSC	superficial layers of the SC
Syp	synaptophysin
TrkB	tyrosine kinase B
TRSN	tectoreticulospinal neuron

TTX tetrodotoxin
VIAAT vesicular inhibitory amino acid transporter
WFVN wide-field vertical neuron
RGC retinal ganglion cell

REFERENCES

1. Schiller, P.H., Sandell, J.H., and Maunsell, J.H., The effect of frontal eye field and superior colliculus lesions on saccadic latencies in the rhesus monkey, *J. Neurophysiol.*, 57, 1033–1049, 1987.
2. Gutfreund, Y., Zheng, W., and Knudsen, E.I., Gated visual input to the central auditory system, *Science*, 297, 1556–1559, 2002.
3. Dräger, U.C. and Hubel, D.H., Topography of visual and somatosensory projections to mouse superior colliculus, *J. Neurophysiol.*, 39, 91–101, 1976.
4. Siminoff, R., Schwassmann, H.O., and Kruger, L., An electrophysiological study of the visual projection to the superior colliculus of the rat, *J. Comp. Neurol.*, 127, 435–444, 1966.
5. Dreher, B., Barker, D.A., Bath, M.R., and Keay, K.A., Spatiotemporal pattern of ontogenetic expression of calbindin- 28/kD in the retinorecipient layers of rat superior colliculus, *J. Comp. Neurol.*, 376, 223–240, 1996.
6. Stein, B.E., Development of the superior colliculus, *Annu. Rev. Neurosci.*, 7, 95–125, 1984.
7. Mooney, R.D., Fish, S.E., Figley, B.A., and Rhoades, R.W., A transient projection from the trigeminal brainstem complex to the superficial layers of the hamster's superior colliculus, *Exp. Brain Res.*, 86, 367–372, 1991.
8. Lund, R.D. and Lund, J.S., Development of synaptic patterns in the superior colliculus of the rat, *Brain Res.*, 42, 1–20, 1972.
9. Warton, S.S. and McCart, R., Synaptogenesis in the stratum griseum superficiale of the rat superior colliculus, *Synapse*, 3, 136–148, 1989.
10. Wu, G.Y., Malinow, R., and Cline, H.T., Maturation of a central glutamatergic synapse, *Science*, 274, 972–976, 1996.
11. Cline, H.T., Wu, G.Y., and Malinow, R., *In vivo* development of neuronal structure and function, *Cold Spring Harbor Symp. Quant. Biol.*, 61, 95–104, 1996.
12. Nakamura, H., Regionalisation and acquisition of polarity in the optic tectum, *Prog. Neurobiol.*, 65, 473–488, 2001.
13. Rhinn, M. and Brand, M., The midbrain — hindbrain boundary organizer, *Curr. Opin. Neurobiol.*, 11, 34–42, 2001.
14. Meyers, E.N., Lewandoski, M., and Martin, G.R., An Fgf8 mutant allelic series generated by Cre- and Flp-mediated recombination, *Nat. Genet.*, 18, 136–141, 1998.
15. Liu, A. and Joyner, A.L., Early anterior/posterior patterning of the midbrain and cerebellum, *Annu. Rev. Neurosci.*, 24, 869–896, 2001.
16. Itasaki, N. and Nakamura, H., Rostrocaudal polarity of the tectum in birds: correlation of en gradient and topographic order in retinotectal projection, *Neuron*, 8, 787–798, 1992.
17. Rowitch, D.H. and McMahon, A.P., Pax-2 expression in the murine neural plate precedes and encompasses the expression domains of Wnt-1 and En-1, *Mech. Dev.*, 52, 3–8, 1995.

18. Itasaki, N. and Nakamura, H., A role for gradient en expression in positional specification on the optic tectum, *Neuron*, 16, 55–62, 1996.
19. Friedman, G.C. and O'Leary, D.D., Retroviral misexpression of engrailed genes in the chick optic tectum perturbs the topographic targeting of retinal axons, *J. Neurosci.*, 16, 5498–5509, 1996.
20. Holt, C.E. and Harris, W.A., Target selection: invasion, mapping and cell choice, *Curr. Opin. Neurobiol.*, 8, 98–105, 1998.
21. Goodhill, G.J. and Richards, L.J., Retinotectal maps: molecules, models and misplaced data, *Trends Neurosci.*, 22, 529–534, 1999.
22. O'Leary, D.D., Yates, P.A., and McLaughlin, T., Molecular development of sensory maps: representing sights and smells in the brain, *Cell*, 96, 255–269, 1999.
23. Benson, D.L., Colman, D.R., and Huntley, G.W., Molecules, maps and synapse specificity, *Nat. Rev. Neurosci.*, 2, 899–909, 2001.
24. Sperry, R.W., Chemoaffinity and the orderly growth of nerve fiber patterns and connections, *Proc. Natl. Acad. Sci. U.S.A.*, 50, 703–710,
25. Walter, J., Henke-Fahle, S., and Bonhoeffer, F., Avoidance of posterior tectal membranes by temporal retinal axons, *Development*, 101, 909–913, 1987.
26. Trisler, D. and Collins, F., Corresponding spatial gradients of TOP molecules in the developing retina and optic tectum, *Science*, 237, 1208–1209, 1987.
27. Stahl, B., Müller, B., von Boxberg, Y., Cox, E.C., and Bonhoeffer, F., Biochemical characterization of a putative axonal guidance molecule of the chick visual system, *Neuron*, 5, 735–743, 1990.
28. McLoon, S.C., A monoclonal antibody that distinguishes between temporal and nasal retinal axons, *J. Neurosci.*, 11, 1470–1477, 1991.
29. Godement, P. and Bonhoeffer, F., Cross-species recognition of tectal cues by retinal fibers *in vitro*, *Development*, 106, 313–320, 1989.
30. Drescher, U., Kremoser, C., Handwerker, C., Loschinger, J., Noda, M., and Bonhoeffer, F., In vitro guidance of retinal ganglion cell axons by RAGS, a 25 kDa tectal protein related to ligands for Eph receptor tyrosine kinases, *Cell*, 82, 359–370, 1995.
31. Cheng, H.J., Nakamoto, M., Bergemann, A.D., and Flanagan, J.G., Complementary gradients in expression and binding of ELF- 1 and Mek4 in development of the topographic retinotectal projection map, *Cell*, 82, 371–381, 1995.
32. Knoll, B. and Drescher, U., Ephrin-As as receptors in topographic projections, *Trends Neurosci*, 25, 145–149, 2002.
33. Kullander, K. and Klein, R., Mechanisms and functions of Eph and ephrin signalling, *Nat. Rev. Mol. Cell Biol.*, 3, 475–486, 2002.
34. Frisén, J., Yates, P.A., McLaughlin, T., Friedman, G.C., O'Leary, D.D., and Barbacid, M., Ephrin-A5 (AL-1/RAGS) is essential for proper retinal axon guidance and topographic mapping in the mammalian visual system, *Neuron*, 20, 235–243, 1998.
35. Feldheim, D.A., Vanderhaeghen, P., Hansen, M.J., Frisén, J., Lu, Q., Barbacid, M., and Flanagan, J.G., Topographic guidance labels in a sensory projection to the forebrain, *Neuron*, 21, 1303–1313, 1998.
36. Pittman, A.J. and Chien, C.B., Understanding dorsoventral topography: backwards and forwards, *Neuron*, 35, 409–411, 2002.
37. Nakamoto, M., Cheng, H.J., Friedman, G.C., McLaughlin, T., Hansen, M.J., Yoon, C.H., O'Leary, D.D., and Flanagan, J.G., Topographically specific effects of ELF-1 on retinal axon guidance *in vitro* and retinal axon mapping in vivo, *Cell*, 86, 755–766, 1996.
38. Simon, D.K. and O'Leary, D.D., Limited topographic specificity in the targeting and branching of mammalian retinal axons, *Dev. Biol.*, 137, 125–134, 1990.

39. Simon, D.K. and O'Leary, D.D., Responses of retinal axons in vivo and *in vitro* to position-encoding molecules in the embryonic superior colliculus, *Neuron*, 9, 977–989, 1992.
40. Hindges, R., McLaughlin, T., Genoud, N., Henkemeyer, M., and O'Leary, D.D., EphB forward signaling controls directional branch extension and arborization required for dorsal-ventral retinotopic mapping, *Neuron*, 35, 475–487, 2002.
41. Mann, F., Ray, S., Harris, W., and Holt, C., Topographic mapping in dorsoventral axis of the Xenopus retinotectal system depends on signaling through ephrin-B ligands, *Neuron*, 35, 461–473, 2002.
42. Godement, P., Salaün, J., and Imbert, M., Prenatal and postnatal development of retinogeniculate and retinocollicular projections in the mouse, *J. Comp. Neurol.*, 230, 552–575, 1984.
43. Edwards, M.A. and Schneider, G.E., Development of the crossed retinocollicular projection in the mouse, *J. Comp. Neurol.*, 248, 410–421, 1986.
44. Lund, R.D. and Bunt, A.H., Prenatal development of central optic pathways in albino rats, *J. Comp. Neurol.*, 165, 247–264, 1976.
45. Laemle, L.K. and Labriola, A.R., Retinocollicular projections in the neonatal rat: an anatomical basis for plasticity, *Brain Res.*, 255, 317–322, 1982.
46. Robinson, S.R., Horsburgh, G.M., Dreher, B., and McCall, M.J., Changes in the numbers of retinal ganglion cells and optic nerve axons in the developing albino rat, *Dev. Brain Res.*, 35, 161–174, 1987.
47. Jeffery, G., Arzymanow, B.J., and Lieberman, A.R., Does the early exuberant retinal projection to the superior colliculus in the neonatal rat develop synaptic connections?, *Dev. Brain Res.*, 14, 135–138, 1984.
48. Edwards, M.A., Caviness, V.S.J., and Schneider, G.E., Development of cell and fiber lamination in the mouse superior colliculus, *J. Comp. Neurol.*, 248, 395–409, 1986.
49. McIlwain, J.T., Visual receptive fields and their images in superior colliculus of the cat, *J. Neurophysiol.*, 38, 219–230, 1975.
50. Distel, H. and Hollander, H., Autoradiographic tracing of developing subcortical projections of the occipital region in fetal rabbits, *J. Comp. Neurol.*, 192, 505–518, 1980.
51. Reh, T. and Kalil, K., Development of the pyramidal tract in the hamster. I. A light microscopic study, *J. Comp. Neurol.*, 200, 55–67, 1981.
52. Ramirez, J.J., Jhaveri, S., Hahm, J.O., and Schneider, G.E., Maturation of projections from occipital cortex to the ventrolateral geniculate and superior colliculus in postnatal hamsters, *Dev. Brain Res.*, 55, 1–9, 1990.
53. Rhoades, R.W., Figley, B., Mooney, R.D. and Fish, S.E., Development of the occipital corticotectal projection in the hamster, *Exp. Brain Res.*, 86, 373–383, 1991.
54. O'Leary, D.D., Bicknese, A.R., De Carlos, J.A., Heffner, C.D., Koester, S.E., Kutka, L.J., and Terashima, T., Target selection by cortical axons: alternative mechanisms to establish axonal connections in the developing brain, *Cold Spring Harb. Symp. Quant. Biol.*, 55, 453–468, 1990.
55. Olavarria, J. and Van Sluyters, R.C., The projection from striate and extrastriate cortical areas to the superior colliculus in the rat, *Brain Res.*, 242, 332–336, 1982.
56. Thong, I.G. and Dreher, B., The development of the corticotectal pathway in the albino rat, *Brain Res.*, 390, 227–238, 1986.
57. Harvey, A.R. and Worthington, D.R., The projection from different visual cortical areas to the rat superior colliculus, *J. Comp. Neurol.*, 298, 281–292, 1990.
58. Thong, I.G. and Dreher, B., The development of the corticotectal pathway in the albino rat: transient projections from the visual and motor cortices, *Neurosci. Lett.*, 80, 275–282, 1987.

59. Sachs, G.M., Jacobson, M., and Caviness, V.S.J., Postnatal changes in arborization patterns of murine retinocollicular axons, *J. Comp. Neurol.*, 246, 395–408, 1986.
60. Lund, R.D., Synaptic patterns of the superficial layers of the superior colliculus of the rat, *J. Comp. Neurol.*, 135, 179–208, 1969.
61. Schönitzer, K. and Holländer, H., Retinotectal terminals in the superior colliculus of the rabbit: a light and electron microscopic analysis, *J. Comp. Neurol.*, 223, 153–162, 1984.
62. Hofbauer, A., Depth segregation of retinal ganglion cells projecting to mouse superior colliculus, *J. Comp. Neurol.*, 234, 465–474, 1985.
63. Yamamoto, N., Yamada, K., Kurotani, T., and Toyama, K., Laminar specificity of extrinsic cortical connections studied in coculture preparations, *Neuron*, 9, 217–228, 1992.
64. Novak, N. and Bolz, J., Formation of specific efferent connections in organotypic slice cultures from rat visual cortex cocultured with lateral geniculate nucleus and superior colliculus, *Eur. J. Biochem.*, 5, 15–24, 1993.
65. Bagnard, D., Lohrum, M., Uziel, D., Puschel, A.W., and Bolz, J., Semaphorins act as attractive and repulsive guidance signals during the development of cortical projections, *Development*, 125, 5043–5053, 1998.
66. Liu, B.P. and Strittmatter, S.M., Semaphorin-mediated axonal guidance via Rho-related G proteins, *Curr. Opin. Cell Biol.*, 13, 619–626, 2001.
67. Campbell, D.S., Regan, A.G., Lopez, J.S., Tannahill, D., Harris, W.A., and Holt, C.E., Semaphorin 3A elicits stage-dependent collapse, turning, and branching in Xenopus retinal growth cones, *J. Neurosci.*, 21, 8538–8547, 2001.
68. Kawakami, A., Kitsukawa, T., Takagi, S., and Fujisawa, H., Developmentally regulated expression of a cell surface protein, neuropilin, in the mouse nervous system, *J. Neurobiol.*, 29, 1–17, 1996.
69. Yamagata, M., Herman, J.-P., and Sanes, J.R., Lamina-specific expression of adhesion molecules in developing chick optic tectum, *J. Neurosci.*, 15, 4556–4571, 1995.
70. Redies, C., Cadherins in the central nervous system, *Prog. Neurobiol.*, 61, 611–648, 2000.
71. Inoue, A. and Sanes, J.R., Lamina-specific connectivity in the brain: regulation by N-cadherin, neurotrophins, and glycoconjugates, *Science*, 276, 1428–1431, 1997.
72. Kimura, Y., Matsunami, H., and Takeichi, M., Expression of cadherin-11 delineates boundaries, neuromeres, and nuclei in the developing mouse brain, *Dev. Dyn.*, 206, 455–462, 1996.
73. Korematsu, K., Nishi, T., Okamura, A., Goto, S., Morioka, M., Hamada, J., and Ushio, Y., Cadherin-8 protein expression in gray matter structures and nerve fibers of the neonatal and adult mouse brain, *Neuroscience*, 87, 303–315, 1998.
74. Olivier, E., Chat, M., and Grantyn, A., Rostrocaudal and lateromedial density distributions of superior colliculus neurons projecting in the predorsal bundle and to the spinal cord: a retrograde HRP study in the cat, *Exp. Brain Res.*, 87, 268–282, 1991.
75. Grantyn, A., Brandi, A.M., Dubayle, D., Graf, W., Ugolini, G., Hadjidimitrakis, K., and Moschovakis, A., Density gradients of trans-synaptically labeled collicular neurons after injections of rabies virus in the lateral rectus muscle of the rhesus monkey, *J. Comp Neurol.*, 451, 346–361, 2002.
76. Garcia-Calero, E., Martinez-de-la-Torre, M., and Puelles, L., The avian griseum tectale: cytoarchitecture, NOS expression and neurogenesis, *Brain Res. Bull.*, 57, 353–357, 2002.
77. Brückner, G., Mares, V., and Biesold, D., Neurogenesis in the visual system of the rat. An autoradiographic investigation, *J. Comp. Neurol.*, 166, 245–256, 1976.
78. Altman, J. and Bayer, S.A., Time of origin of neurons of the rat superior colliculus in relation to other components of the visual and visuomotor pathway, *Exp. Brain Res.*, 42, 424–434, 1981.

79. Mustari, M.J., Lund, R.D., and Graubard, K., Histogenesis of the superior colliculus of the albino rat: a tritiated thymidine study, *Brain Res.*, 164, 39–52, 1979.
80. Grantyn, R., Perouansky, M., Rodríguez-Tébar, A., and Lux, H.-D., Expression of depolarizing voltage- and transmitter-activated currents in neuronal precursor cells from the rat brain is preceded by a proton-activated sodium current, *Dev. Brain Res.*, 49, 150–155, 1989.
81. Jüttner, R., Meier, J., and Grantyn, R., Slow IPSC kinetics, low levels of alpha1 subunit expression and paired-pulse depression are distinct properties of neonatal inhibitory GABAergic synaptic connections in the mouse superior colliculus, *Eur. J. Neurosci.*, 13, 2088–2098, 2001.
82. Labriola, A.R. and Laemle, L.K., Cellular morphology in the visual layer of the developing rat superior colliculus, *Exp. Neurol.*, 55, 247–268, 1977.
83. Albers, F.J., Meek, J., and Nieuwenhuys, R., Morphometric parameters of the superior colliculus of albino and pigmented rats, *J. Comp. Neurol.*, 274, 357–370, 1988.
84. Mooney, R.D. and Rhoades, R.W., Determinants of axonal and dendritic structure in the superior colliculus, *Prog. Brain Res.*, 95, 57–67, 1993.
85. Mooney, R.D., Savage, S.V., Hobler, S., King, T.D. and Rhoades, R.W., Normal development and effects of deafferentation on the morphology of superior collicular neurons projecting to the lateral posterior nucleus in hamster, *J. Comp. Neurol.*, 315, 413–430, 1992.
86. Taschenberger, H., Juttner, R., and Grantyn, R., Ca2+-permeable P2X receptor channels in cultured rat retinal ganglion cells, *J. Neurosci.*, 19, 3353–3366, 1999.
87. Langer, T.P. and Lund, R.D., The upper layers of the superior colliculus of the rat: A Golgi study, *J. Comp. Neurol.*, 158, 405–436, 1974.
88. Donnelly, J.F., Thompson, S.M., and Robertson, R.T., Organization of projections from the superior colliculus to the thalamic lateral posterior nucleus in the rat, *Brain Res.*, 288, 315–319, 1983.
89. Sugita, S., Otani, K., Tokunaga, A., and Terasawa, K., Laminar origin of the tecto-thalamic projections in the albino rat, *Neurosci. Lett.*, 43, 143–147, 1983.
90. Mooney, R.D., Nikoletseas, M.M., Ruiz, S.A., and Rhoades, R.W., Receptive-field properties and morphological characteristics of the superior collicular neurons that project to the lateral posterior and dorsal lateral geniculate nuclei in the hamster, *J. Neurophysiol.*, 59, 1333–1351, 1988.
91. Grantyn, R., Gaze control through superior colliculus: Structure and function, in *Anatomy of the Oculomotor System, Review of Oculomotor Research,* Vol. 2, Buettner-Ennever, J., Ed., Elsevier, Amsterdam, 1988, pp. 273–331.
92. Mooney, R.D., Nikoletseas, M.M., Hess, P.R., Allen, Z., Lewin, A.C., and Rhoades, R.W., The projection from the superficial to the deep layers of the superior colliculus: an intracellular horseradish peroxidase injection study in the hamster, *J. Neurosci.*, 8, 1384–1399, 1988.
93. Hall, W.C. and Lee, P., Interlaminar connections of the superior colliculus in the tree shrew. III: The optic layer, *Vis. Neurosci.*, 14, 647–661, 1997.
94. Hilbig, H., Merbach, M., Krause, J., Gartner, U., and Stubbe, A., Dendritic organization of neurons of the superior colliculus in animals with different visual capability, *Brain Res. Bull.*, 51, 255–265, 2000.
95. Lo, F.S., Cork, R.J., and Mize, R.R., Physiological properties of neurons in the optic layer of the rat's superior colliculus, *J. Neurophysiol.*, 80, 331–343, 1998.
96. Ozen, G., Augustine, G.J., and Hall, W.C., Contribution of superficial layer neurons to premotor bursts in the superior colliculus, *J. Neurophysiol.*, 84, 460–471, 2000.

97. Endo, T. and Isa, T., Functionally different AMPA-type glutamate receptors in morphologically identified neurons in rat superficial superior colliculus, *Neuroscience*, 108, 129–141, 2001.
98. Lane, R.D., Bennett-Clarke, C.A., Allan, D.M., and Mooney, R.D., Immunochemical heterogeneity in the tecto-LP pathway of the rat, *J. Comp. Neurol.*, 333, 210–222, 1993.
99. Lane, R.D., Allan, D.M., Bennett-Clarke, C.A., Howell, D.L., and Rhoades, R.W., Projection status of calbindin- and parvalbumin-immunoreactive neurons in the superficial layers of the rat's superior colliculus, *Vis. Neurosci.*, 14, 277–286, 1997.
100. Behan, M., Jourdain, A., and Bray, G.M., Calcium binding protein (calbindin D28k) immunoreactivity in the hamster superior colliculus: ultrastructure and lack of colocalization with GABA, *Exp. Brain Res.*, 89, 115–124, 1992.
101. Barker, D.A. and Dreher, B., Spatiotemporal patterns of ontogenetic expression of parvalbumin in the superior colliculi of rats and rabbits, *J. Comp. Neurol.*, 393, 210–230, 1998.
102. Cork, R.J., Baber, S.Z., and Mize, R.R., $\text{Calbindin}_{\text{D28k}}$- and parvalbumin-immunoreactive neurons form complementary sublaminae in the rat superior colliculus, *J. Comp. Neurol.*, 394, 205–217, 1998.
103. Grantyn, R., Ludwig, R., and Eberhardt, W., Neurons of the superficial tectal grey. An intracellular HRP-study on the kitten superior colliculus *in vitro*, *Exp. Brain Res.*, 55, 172–176, 1984.
104. Behan, M. and Appell, P.P., Intrinsic circuitry in the cat superior colliculus: Projections from the superficial layers, *J. Comp. Neurol.*, 315, 230–243, 1992.
105. Lee, P. and Hall, W.C., Interlaminar connections of the superior colliculus in the tree shrew. II: Projections from the superficial gray to the optic layer, *Vis. Neurosci.*, 12, 573–588, 1995.
106. Isa, T., Endo, T., and Saito, Y., The visuo-motor pathway in the local circuit of the rat superior colliculus, *J. Neurosci.*, 18, 8496–8504, 1998.
107. Hilbig, H. and Schierwagen, A., Interlayer neurones in the rat superior colliculus: A tracer study using DiI/Di-ASP, *Neuroreport*, 5, 477–480, 1994.
108. Moschovakis, A.K., Karabelas, A.B., and Highstein, S.M., Structure-function relationships in the primate superior colliculus. I. Morphological classification of efferent neurons, *J. Neurophysiol.*, 60, 232–262, 1988.
109. Casanova, C. and Molotchnikoff, S., Influence of the superior colliculus on visual responses of cells in the rabbit's lateral posterior nucleus, *Exp. Brain Res.*, 80, 387–396, 1990.
110. Hikosaka, O. and Wurtz, R.H., Effects on eye movements of a GABA agonist and antagonist injected into monkey superior colliculus, *Brain Res.*, 272, 368–372, 1983.
111. Warton, S.S., Perouansky, M., and Grantyn, R., Development of GABAergic synaptic connections in vivo and in cultures from the rat superior colliculus, *Dev. Brain Res.*, 52, 95–111, 1990.
112. Zhang, L.I., Tao, H.W., Holt, C.E., Harris, W.A., and Poo, M.M., A critical window for cooperation and competition among developing retinotectal synapses, *Nature*, 395, 37–44, 1998.
113. Kraszewski, K. and Grantyn, R., Development of GABAergic connections *in vitro*: Increasing efficacy of synaptic transmission is not accompanied by changes in miniature currents, *J. Neurobiol.*, 23, 766–781, 1992.
114. Tokunaga, A. and Otani, K., Dendritic patterns of neurons in the rat superior colliculus, *Exp. Neurol.*, 52, 189–205, 1976.

115. Mooney, R.D., Nikoletseas, M.M., King, T.D., Savage, S.V., Weaver, M.T., and Rhoades, R.W., Structural and functional consequences of neonatal deafferentation in the superficial layers of the hamster's superior colliculus, *J. Comp. Neurol.*, 315, 398–412, 1992.
116. Rhoades, R.W., Rohrer, W.H., Mooney, R.D., and Ruiz, S., The orientation of horizontal cell dendrites in the superior colliculus of the hamster: an analysis based on three-dimensional reconstruction of intracellularly injected neurons, *Exp. Brain Res.*, 76, 229–238, 1989.
117. Mathers, L.H.J., Postnatal maturation of neurons in the rabbit superior colliculus, *J. Comp. Neurol.*, 173, 439–456, 1977.
118. Mize, R.R., Spencer, R.F., and Sterling, P., Two types of GABA-accumulating neurons in the superficial gray layer of the cat superior colliculus, *J. Comp. Neurol.*, 206, 180–192, 1982.
119. Meier, J., Akyeli, J., Kirischuk, S., and Grantyn, R., GABA(A) receptor activity and PKC control inhibitory synaptogenesis in CNS tissue slices, *Mol. Cell. Neurosci.,* 23, 600–613, 2003.
120. Meier, J., Jüttner, R., Kirischuk, S., and Grantyn, R., Synaptic anchoring of glycine receptors in developing collicular neurons under control of metabotropic glutamate receptor activity, *Mol. Cell. Neurosci.*, 324–340, 2002.
121. Kirischuk, S. and Grantyn, R., Intraterminal Ca2+ concentration and asynchronous transmitter release at single GABAergic boutons in rat collicular cultures, *J. Physiol.*, 548, 753–764, 2003.
122. Sagné, C., El Mestikawy, S., Isambert, M.F., Hamon, M., Henry, J.P., Giros, B., and Gasnier, B., Cloning of a functional vesicular GABA and glycine transporter by screening of genome databases, *FEBS Lett.*, 417, 177–183, 1997.
123. Shi, J., Aamodt, S.M., and Constantine-Paton, M., Temporal correlations between functional and molecular changes in NMDA receptors and GABA neurotransmission in the superior colliculus, *J. Neurosci.*, 17, 6264–6276, 1997.
124. Davies, P., Anderton, B., Kirsch, J., Konnerth, A., Nitsch, R., and Sheetz, M., First one in, last one out: The role of GABAergic transmission in generation and degeneration, *Prog. Neurobiol.*, 55, 651–658, 1998.
125. Ben-Ari, Y., Khazipov, R., Leinekugel, X., Caillard, O., and Gaiarsa, J.-L., GABAA, NMDA and AMPA receptors: a developmentally regulated 'mènage à trois', *Trends Neurosci.*, 20, 523–529, 1997.
126. Akyeli, J., Meier, J., Kirischuk, S., and Grantyn, R., BDNF facilitates the translocation of KCC2 into the neuronal plasma membrane and regulates the switch from depolarizing to hyperpolarizing GABA action, Abstr., 6th IBRO Congr. Neurosci., 2019, 2003.
127. Lo, F.S. and Mize, R.R., Synaptic regulation of L-type Ca(2+) channel activity and long-term depression during refinement of the retinocollicular pathway in developing rodent superior colliculus, *J. Neurosci. (Online)*, 20, RC58, 2000.
128. Ikonomidou, C., Bittigau, P., Ishimaru, M.J., Wozniak, D.F., Koch, C., Genz, K., Price, M.T., Stefovska, V., Horster, F., Tenkova, T., Dikranian, K., and Olney, J.W., Ethanol-induced apoptotic neurodegeneration and fetal alcohol syndrome, *Science*, 287, 1056–1060, 2000.
129. Xiong, M. and Finlay, B.L., What do developmental mapping rules optimize?, *Prog. Brain Res.*, 112, 351–361, 1996.
130. Lund, R.D. and Lund, J.S., Plasticity in the developing visual system: the effects of retinal lesions made in young rats, *J. Comp. Neurol.*, 169, 133–154, 1976.

131. Land, P.W. and Lund, R.D., Development of the rat's uncrossed retinotectal pathway and its relation to plasticity studies, *Science*, 205, 698–700, 1979.
132. Jeffery, G. and Perry, V.H., Evidence for ganglion cell death during development of the ipsilateral retinal projection in the rat, *Dev. Brain Res.*, 2, 176–180, 1982.
133. Godement, P., Saillour, P., and Imbert, M., The ipsilateral optic pathway to the dorsal lateral geniculate nucleus and superior colliculus in mice with prenatal or postnatal loss of one eye, *J. Comp. Neurol.*, 190, 611–626, 1980.
134. Frost, D.O. and Schneider, G.E., Plasticity of retinofugal projections after partial lesions of the retina in newborn Syrian hamsters, *J. Comp. Neurol.*, 185, 517–567, 1979.
135. Cunningham, T.J. and Speas, G., Inversion of anomalous uncrossed projections along the mediolateral axis of the superior colliculus: implications for retinocollicular, *Brain Res.*, 88, 73–79, 1975.
136. Chow, K.L., Ostrach, L.H., Crabtree, J.W., Bernegger, O., Baumbach, H.D., and Lawson, R., Anomalous uncrossed retinal projections fail to activate superior colliculus neurons in rabbits unilaterally enucleated by fetal surgery. PG, *J. Comp. Neurol.*, 196, 189–204, 1981.
137. Takeuchi, H., Fukuda, Y., Hara, Y., and Hsiao, C., Physiological properties of expanded ipsilateral retinocollicular projection in neonatally enucleated albino rats, *Brain Res.*, 231, 191–196, 1982.
138. Martin, P.R., Sefton, A.J., and Dreher, B., The retinal location and fate of ganglion cells which project to the ipsilateral superior colliculus in neonatal albino and hooded rats, *Neurosci. Lett.*, 41, 219–226, 1983.
139. Crespo, D., O'Leary, D.D., and Cowan, W.M., Changes in the numbers of optic nerve fibers during late prenatal and postnatal development in the albino rat, *Dev. Brain Res.*, 19, 129–134, 1985.
140. O'Leary, D.D., Fawcett, J.W., and Cowan, W.M., Topographic targeting errors in the retinocollicular projection and their elimination by selective ganglion cell death, *J. Neurosci.*, 6, 3692–3705, 1986.
141. Von Bartheld, C.S., Neurotrophins in the developing and regenerating visual system, *Histol. Histopathol.*, 13, 437–459, 1998.
142. Frost, D.O., BDNF/trkB signaling in the developmental sculpting of visual connections, *Prog. Brain Res.*, 134, 35–49, 2001.
143. Mustari, M.J. and Lund, R.D., An aberrant crossed visual corticotectal pathway in albino rats, *Brain Res.*, 112, 37–44, 1976.
144. Rhoades, R.W. and Chalupa, L.M., Functional and anatomical consequences of neonatal visual cortical damage in superior colliculus of the golden hamster, *J. Neurophysiol.*, 41, 1466–1494, 1978.
145. So, K.F., Development of abnormal recrossing retinotectal projections after superior colliculus lesions in newborn Syrian hamsters, *J. Comp. Neurol.*, 186, 241–257, 1979.
146. Mooney, R.D., Klein, B.G., Szczepanik, A.M., and Rhoades, R.W., Extensive recrossing of retinotectal axons after neonatal unilateral superior collicular lesions in hamster, *Brain Res.*, 351, 297–313, 1985.
147. Udin, S.B. and Schneider, G.E., Compressed retinotectal projection in hamsters: fewer ganglion cells project to tectum after neonatal tectal lesions, *Exp. Brain Res.*, 43, 261–269, 1981.
148. Pallas, S.L. and Finlay, B.L., Conservation of receptive-field properties of superior colliculus cells after developmental rearrangements of retinal input, *Vis. Neurosci.*, 2, 121–135, 1989.

149. Xiong, M., Pallas, S.L., Lim, S., and Finlay, B.L., Regulation of retinal ganglion cell axon arbor size by target availability: mechanisms of compression and expansion of the retinotectal projection, *J. Comp. Neurol.*, 344, 581–597, 1994.
150. Kalil, R.E., The influence of action potentials on the development of the central visual pathway in mammals, *J. Exp. Biol.*, 153, 261–276, 1990.
151. Goodman, C.S. and Shatz, C.J., Developmental mechanisms that generate precise patterns of neuronal connectivity, *Cell*, 72 Suppl., 77–98, 1993.
152. Fawcett, J.W., O'Leary, D.D., and Cowan, W.M., Activity and the control of ganglion cell death in the rat retina, *Proc. Natl. Acad. Sci. U.S.A.*, 81, 5589–5593, 1984.
153. O'Leary, D.D., Crespo, D., Fawcett, J.W., and Cowan, W.M., The effect of intraocular tetrodotoxin on the postnatal reduction in the numbers of optic nerve axons in the rat, *Brain Res.*, 395, 96–103, 1986.
154. Thompson, I. and Holt, C., Effects of intraocular tetrodotoxin on the development of the retinocollicular pathway in the Syrian hamster, *J. Comp. Neurol.*, 282, 371–388, 1989.
155. Finlay, B.L., Berg, A.T., and Sengelaub, D.R., Cell death in the mammalian visual system during normal development: II. Superior colliculus, *J. Comp. Neurol.*, 204, 318–324, 1982.
156. Giordano, D.L., Murray, M., and Cunningham, T.J., Naturally occurring neuron death in the optic layers of superior colliculus of the postnatal rat, *J. Neurocytol.*, 9, 603–614, 1980.
157. Galli-Resta, L., Ensini, M., Fusco, E., Gravina, A., and Margheritti, B., Afferent spontaneous electrical activity promotes the survival of target cells in the developing retinotectal system of the rat, *J. Neurosci.*, 13, 243–250, 1993.
158. Meister, M., Wong, R.O.L., Baylor, D.A., and Shatz, C.J., Synchronous bursts of action potentials in ganglion cells of the developing mammalian retina, *Science*, 252, 939–943, 1991.
159. Henneberger, C., Jüttner, R., Rothe, T., and Grantyn, R., Postsynaptic action of BDNF on GABAergic synaptic transmission in the superficial layers of the mouse superior colliculus, *J. Neurophysiol.*, 88, 595–603, 2002.
160. Pollock, G.S., Vernon, E., Forbes, M.E., Yan, Q., Ma, Y.T., Hsieh, T., Robichon, R., Frost, D.O., and Johnson, J.E., Effects of early visual experience and diurnal rhythms on BDNF mRNA and protein levels in the visual system, hippocampus, and cerebellum, *J. Neurosci.*, 21, 3923–3931, 2001.
161. Frost, D.O., Ma, Y.T., Hsieh, T., Forbes, M.E., and Johnson, J.E., Developmental changes in BDNF protein levels in the hamster retina and superior colliculus, *J. Neurobiol.*, 49, 173–187, 2001.
162. Ma, Y.T., Hsieh, T., Forbes, M.E., Johnson, J.E., and Frost, D.O., BDNF injected into the superior colliculus reduces developmental retinal ganglion cell death, *J. Neurosci.*, 18, 2097–2107, 1998.
163. Spalding, K.L., Tan, M.M., Hendry, I.A., and Harvey, A.R., Anterograde transport and trophic actions of BDNF and NT-4/5 in the developing rat visual system, *Mol. Cell. Neurosci.*, 19, 485–500, 2002.
164. Isenmann, S., Cellerino, A., Gravel, C., and Bahr, M., Excess target-derived brain-derived neurotrophic factor preserves the transient uncrossed retinal projection to the superior colliculus, *Mol. Cell. Neurosci.*, 14, 52–65, 1999.
165. Cellerino, A., Carroll, P., Thoenen, H., and Barde, Y.-A., Reduced size of retinal ganglion cell axons and CNS-specific hypomyelination in mice lacking brain-derived neurotrophic factor, *Mol. Cell. Neurosci.*, 9, 397–408, 1997.

166. Rothe, T., Bähring, R., Carroll, P., and Grantyn, R., Repetitive firing deficits and reduced sodium current density in retinal ganglion cells developing in the absence of BDNF, *J. Neurobiol.*, 40, 407–419, 1999.
167. Pallas, S.L., Gilmour, S.M., and Finlay, B.L., Control of cell number in the developing neocortex. I. Effects of early tectal ablation, *Brain Res.*, 471, 1–11, 1988.
168. Caleo, M., Menna, E., Chierzi, S., Cenni, M.C., and Maffei, L., Brain-derived neurotrophic factor is an anterograde survival factor in the rat visual system, *Curr. Biol.*, 10, 1155–1161, 2000.
169. Poo, M.M., Neurotrophins as synaptic modulators, *Nat Rev Neurosci*, 2, 24–32, 2001.
170. Brünig, I., Penschuck, S., Berninger, B., Benson, J., and Fritschy, J.M., BDNF reduces miniature inhibitory postsynaptic currents by rapid downregulation of GABA(A) receptor surface expression, *Eur. J. Neurosci.*, 13, 1320–1328, 2001.
171. Jüttner, R., Schmidt, S., Meier, J., Wagner, K.D., Rothe, T., and Grantyn, R., NMDA receptor dependent changes in the characteristics of GABAergic inhibition in the mouse superior colliculus before and after eye opening, unpublished data.
172. Simon, D.K., Prusky, G.T., O'Leary, D.D., and Constantine-Paton, M., *N*-Methyl-D-aspartate receptor antagonists disrupt the formation of a mammalian neural map, *Proc. Natl. Acad. Sci. U.S.A.*, 89, 10593–10597, 1992.
173. Colonnese, M.T. and Constantine-Paton, M., Chronic NMDA receptor blockade from birth increases the sprouting capacity of ipsilateral retinocollicular axons without disrupting their early segregation, *J. Neurosci.*, 21, 1557–1568, 2001.
174. Shi, J., Aamodt, S.M., Townsend, M., and Constantine-Paton, M., Developmental depression of glutamate neurotransmission by chronic low-level activation of NMDA receptors, *J. Neurosci.*, 21, 6233–6244, 2001.
175. Aamodt, S.M., Shi, J., Colonnese, M.T., Veras, W., and Constantine-Paton, M., Chronic NMDA exposure accelerates development of GABAergic inhibition in the superior colliculus, *J. Neurophysiol.*, 83, 1580–1591, 2000.

9 Mechanisms Underlying Development of the Retinocollicular Pathway

R. Ranney Mize and Thomas E. Salt

CONTENTS

9.1 INTRODUCTION

The mammalian superior colliculus (SC) and its vertebrate homologue, the optic tectum, have long been favorite models for study of developmental processes in brain. These studies range from those focusing upon the initial stages of axon outgrowth of retinal ganglion cells to those examining the final formation of topographically correct synaptic connections in the SC. The early stages in the formation of this pathway are directed by chemotaxic molecules, both attractants and repellents,

0-8493-0097-5/04/$0.00+$1.50

which operate largely independently of activity or visual experience. The later stages of development, particularly the refinement of axonal connections that result in a precise topographic map of the retina onto SC, are dependent upon both impulse activity and visual experience, as well as upon a variety of signaling molecules operating at both the presynaptic terminal and within the postsynaptic target. This chapter reviews these later stages of development, including evidence for the involvement of various receptors and channels, and the changes in physiology of the synapse that accompany this refinement.

9.2 THE RODENT SUPERIOR COLLICULUS AS A MODEL FOR PATHWAY DEVELOPMENT

The rodent superior colliculus is an excellent model in which to study processes of synaptic refinement because the afferent pathways to SC from both the retina and visual cortex undergo substantial refinement after birth. Thus, like the ferret lateral geniculate nucleus and visual cortex, the later stages of refinement of the pathways to the rodent superior colliculus can be examined after birth has already occurred, at a time when it is relatively easy to trace the pathways using anatomical techniques, to make extra- or intracellular recordings from the SC, and to manipulate the pathways experimentally using pharmacological tools or lesion techniques. The rodent SC is also an increasingly valuable model for these types of studies because it is possible to examine genetically modified mice in which genes for a variety of transmitter receptors, channels, and signaling molecules have been modified or deleted.

9.3 ANATOMICAL DEVELOPMENT OF THE AFFERENT PATHWAYS TO SUPERIOR COLLICULUS

The development of the retinocollicular pathways in rodents is robust during the first few days after birth. The rodent is a lateral eyed species, and so each SC receives a massive input from the contralateral eye, where all but a small binocular portion of the visual field is represented in the SC only via that pathway. This pathway has a precise topographic organization such that the superior visual field is represented in the medial SC, the inferior field in the lateral SC, the nasal visual field in the rostral SC and the temporal visual field in the caudal SC.[1,2] In addition, each SC receives a much less extensive input from the ipsilateral eye, which represents a small region of visual space within the temporal retina (nasal field) in which there is binocular vision.[3–5] This pathway is also topographically organized. In the normal adult, the contralateral pathway fills the entire rostral-caudal extent of SC, but its fibers for the most part are confined to the dorsal two-thirds of the superficial gray layer. The domain of the ipsilateral pathway is much more restricted. The ipsilateral pathway consists of patches of terminal fibers within the rostral most 200 to 300 μm of the SC and also a continuous band of fibers at the medial edge of SC, which extends through almost the entire rostral-caudal length of SC. These ipsilateral fibers form a ventral tier of terminations that lie within the ventral one-third of the superficial gray layer (SGL) and uppermost optic layer (OL), below and largely segregated from those of the contralateral pathway in both rats and mice.[3,6,7]

This topographic organization develops only gradually during early postnatal life in both rats and mice. Initially, both the contralateral and ipsilateral retinocollicular pathways innervate the entire superficial SC, and there is substantial if not complete overlap in the territories normally occupied by only one or the other pathway in adult animals. This overlap has been recognized for a number of years in a variety of rodents,[1,8–11] but recent studies using the very sensitive tracer cholera toxin subunit B show that the two pathways may overlap even more extensively than previously realized, at least in mouse.[4–6] The process responsible for refinement appears also to be somewhat more protracted in time. The overlap occurs not only in fibers being misdirected to topographically incorrect targets within the more caudal and lateral SC, but also in the dorsal-ventral plane where many ipsilateral fibers innervate the upper SGL, which is normally the exclusive domain of the contralateral pathway. The pattern is similar in mouse, where both pathways extensively overlap between P1 and P4, and then gradually retract during the next 2 to 4 postnatal weeks. This process includes the pruning or elimination of misdirected axons and the expansion of appropriately targeted axons and their arbors of boutons, resulting in an appropriate topographic map and normal vision.

The development of the ipsilateral pathway is the easiest to study experimentally because it retracts to such a small region of SC in the adult. This pathway is refined in stages. In mouse, studies show that at P4, it covers much of the SC in the rostral caudal plane and the entire SGL in the dorsal ventral plane (Figure 9.1). By P9, it has retracted so that it occupies mostly the ventral SGL, forming a distinctive tier below the more dorsal contralateral input. However, many inappropriately targeted fibers remain within the more caudal and lateral regions of SC (Figure 9.1). By P15, the pathway has further retracted so that the fibers are now located only in the rostral and medial SC where they come into retinotopic register with the contralateral eye.[4–6] The arbors of correctly targeted fibers also increase in complexity with expanded branching and an increase in the density of boutons.[1,12] This process involves normal retinal ganglion cell death,[1] and, as we discuss below, a variety of receptors, channels, and signaling molecules.

9.4 THE ROLE OF THE *N*-METHYL-D-ASPARTATE RECEPTOR

The *N*-methyl-D-aspartate (NMDA) receptor is one of the ionotropic glutamate receptors. It is constructed from several subunits, the NR1, and the NR2a, b, c, d subunits. Incorporation of different subunits allows the construction of receptor complexes with different pharmacological and biophysical properties.[13,14] The NMDA receptor ion channel has been shown to be more calcium-permeable than other ionotropic glutamate receptors. Furthermore, the ion channel is blocked by magnesium ions at membrane potentials close to the resting one, thus effectively making NMDA receptor mediated responses voltage dependent under physiological conditions. Both of

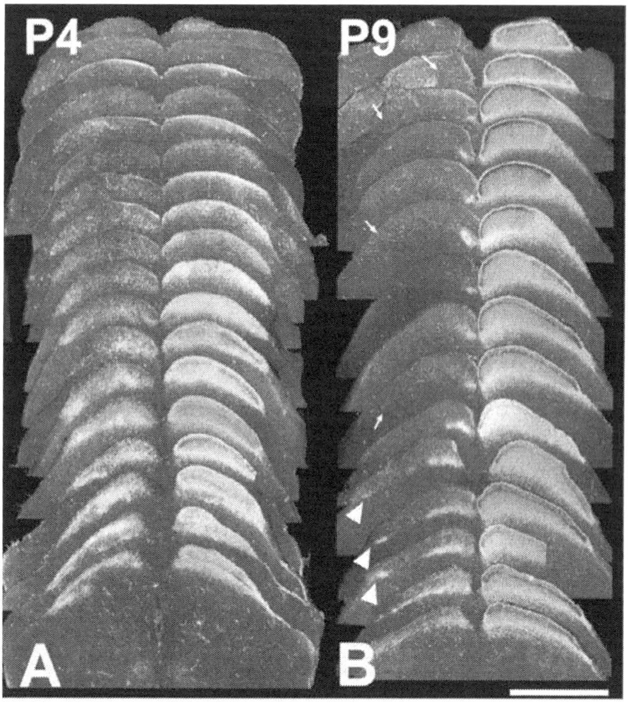

FIGURE 9.1 Sequential series of sections from the superior colliculus (SC) of normal C57/BL6 mice in which the retinocollicular pathway has been labeled after injection into one eye of the cholera toxin subunit B conjugated to horseradish peroxidase (HRP). The label is visualized by the chromagen diaminobenzidine (DAB) and illustrated using dark field optics. The right SC is contralateral, the left SC ipsilateral to the injected eye. Sections of the SC are stacked from rostral (bottom) to caudal (top). Note that the ipsilateral label at P4 extends throughout almost the entire SC, while it is more concentrated within the rostro-medial SC by P9. Scale bar = 1 mm. (Modified from Figure 1 of Wu, H.H., Cork, R.J., and Mize R.R., Development of the ipsilateral retinocollicular pathway in normal and NOS gene knockout mice, *J. Comp. Neurol.*, 426, 651–665, 2000. With permission.)

these properties are likely to be important factors in mechanisms of learning and memory and synaptic plasticity (see References 15 and 16 for reviews).

Evidence that the NMDA receptor is important for the development of the retinocollicular pathway in rodents was first published in 1992. The NMDA receptor had already been shown to play a critical role in the development of the retinotectal pathway in frogs and other vertebrates.[17] This role was elegantly demonstrated in the so-called three eyed frog in which an extra eye is grafted into the tadpole so that the tectum becomes dually innervated by the optic tracts of both the normal and the grafted eyes. This "competition" between the normal and extra eyes, artificially generated, produces alternating stripes of retinotectal input so that the axons from the two eyes remain segregated. Blocking the NMDA receptor by application of NMDA antagonists completely blocks this segregation.[18] The NMDA receptor was also shown to be important in maintaining the normal topography of the retinotectal pathway in this species, since retinal ganglion cells were found to project to incorrect locations after treatment with NMDA antagonists.[19] The NMDA receptor together with impulse activity was subsequently shown to also contribute to the regrowth of topographically ordered connections, which occurs during regeneration of the optic tract after nerve crush,[20] and to the correct registration that occurs between fibers of isthmotectal axons, which relay visual input from the eye to the ipsilateral tectum, and those of the contralateral retina that project directly to the tectum.[21]

There was thus substantial evidence that the NMDA receptor was important for the development of the proper organization of the retinotectal pathway in cold-blooded vertebrates in 1992, but it was unclear whether this receptor was also important in the development of this pathway in mammals, where the refinement is both more extensive and occurs before eye opening. Simon et al.,[22] were the first to test this hypothesis by applying the NMDA receptor antagonists MK-801 and D-2-amino-5-phosphonopentanoic acid (AP5) to the surface of the SC using an Elvax implant. They reported that this treatment produced an increase in the number of aberrantly directed contralateral retinocollicular fibers within the treated superior colliculus. It is worth examining this paper in detail both because of the extraordinary influence it has had upon our view of retinocollicular development, and also because its conclusions have recently been placed in some doubt. A reread of this germinal paper shows that it was conducted with appropriate controls and careful analysis. Two different antagonists were used (MK-801, AP5), one of these (AP5) at two different drug concentrations (10 m*M* or 100 m*M*). Both active and inactive stereoisomers of AP5 were tested. Nonspecific drug effects (e.g., an MK-801 effect on nicotinic receptors) were examined. A relatively sensitive anterograde tracer (DiI) was used to trace retinocollicular axons, again at several concentrations. Two age groups were studied (P12 and P19). And both the size of the injection site in the retina, and the position of the Elvax implants in SC were examined postmortem. Finally, the extent of branching of aberrant fibers was studied quantitatively as well as qualitatively.

The results seem straightforward. An increase in the number of putative aberrantly positioned fibers was found in a majority of animals at both ages P12 and P19. Fibers were found not only in the topographically incorrect regions of the surface map, but also to arborize within incorrect laminae (e.g., stratum opticum). On the other hand, the effects reported are remarkably modest when viewed in light of the total rearrangement of fibers that occurs normally in this structure. Aberrant arbors were found in only about two-thirds of their cases (18 of 26), and then they ranged from only 1 to a maximum of 4 arbors per animal. In addition, a dense zone of topographically correct arborizations was reported to develop normally in all treated animals, regardless of the dose of AP5, which was varied by 10-fold. The authors nevertheless concluded that the NMDA receptor is probably critical in establishing the correct topographic map in the contralateral retinocollicular pathway and that "chronic treatment with NMDA receptor antagonists interferes with the preferential elimination of entire retinal axon arborizations positioned in topographically inappropriate regions of the SC." [22]

This paper is widely cited as demonstrating that the NMDA receptor mediates refinement of the retinocollicular pathway in mammals, despite the fact that only a very small number of aberrant fibers was found after antagonist treatment while a very large portion of the pathway remained in correctly targeted regions of SC. In fact, the results could also be interpreted to demonstrate that NMDA antagonists do not significantly block refinement of the pathway, and that the bulk of the pathway is correctly targeted in the absence of NMDA receptor function. Subsequent to the publication of this paper, an effect of NMDA antagonists on development of the chick ipsilateral retinocollicular pathway was shown by another group.[23,24] In this study, the authors examined the number of retinal ganglion cells, labeled retrogradely, which were located in topographically incorrect positions in the retinas of chicks treated chronically with the NMDA receptor antagonist MK-801. An increase in the number of incorrectly positioned neurons was found, but again the effect was modest. Another report showing that NMDA receptor blockade alters refinement of subcortical retinal projections in mammals used the ferret lateral geniculate nucleus (LGN) as a model and found that the on and off sublaminae of this nucleus failed to segregate normally after treatment with NMDA antagonists.[25] By contrast, the laminae receiving ipsilateral and contralateral retinal afferents segregate normally after NMDA receptor antagonist treatment.[25]

Recently, the importance of the NMDA receptor in refinement of the ipsilateral and contralateral pathways has been brought into question. Colonnese and Constantine-Paton[26] have shown that application of AP5 beginning at birth, using Elvax implants, produced no alteration whatsoever in the refinement of the ipsilateral retinocollicular pathway in the rat. By P14, the pathway had retracted normally from both the caudal SC and from the dorsal SGL, leaving only the characteristic patches of IRP label within the rostral SC, just as in normal rats. In addition, they found no alteration in the ventral boundary of the contralateral pathway, an effect different than that reported by Simon et al.,[22] in which fibers were observed to arborize into the domain of the ipsilateral pathway. Thus, chronic exposure to a competitive antagonist of the NMDA receptor in this more recent study apparently failed to alter withdrawal of inappropriately targeted axons in the ipsilateral and probably in the contralateral retinocollicular pathways.

NMDA blockade did affect sprouting of the ipsilateral pathway after neonatal lesion of the retina. It is well-established that small lesions of the retina shortly after birth produce a sprouting response in the remaining pathway so that the *scotoma* produced by destroying fibers projecting to a limited area of the contralateral SC is filled by fibers from the ipsilateral retina which sprout into this scotoma even though that projection is topographically incorrect (10). This sprouting was increased after treatment with the NMDA receptor antagonist, AP5, although the effect was modest unless a lesion of the visual cortex was also made in these animals.[26] In summary, evidence that the NMDA receptor is essential to refinement of the retinocollicular pathways is at best controversial. What has been demonstrated is that blocking the receptor can result in failure of a small number of fibers to retract from topographically inappropriate regions of the rat SC, but not that it significantly regulates this process.

9.5 CHANGES IN THE CONTRIBUTION OF *N*-METHYL-D-ASPARTATE RECEPTORS TO VISUAL RESPONSES IN SUPERIOR COLLICULUS DURING DEVELOPMENT

In adult rats and cats, NMDA receptors contribute to the visual responses of SC neurons, together with AMPA receptors.[27–30] In rats, during the third week of life (i.e., just after eye opening), there is an increased contribution of NMDA receptors to visual responses which peak at P21 and then rapidly declines to adult levels by P25.[29,30] This is suggestive of a role for NMDA receptors in the development of the visual response profile and refinement of visual responses. Interestingly, dark-rearing prolongs the elevated NMDA-receptor-mediated contribution into adulthood, where it can be maintained at elevated levels if animals are kept in the dark without visual experience[29,30] (Figure 9.2). Restoration of normal visual experience by exposure of rats to normal lighting conditions results in a reversal of these elevated levels of NMDA-receptor involvement in visual responses. These results suggest that the role of NMDA receptors (apart from normal involvement in synaptic transmission) is to enable experience-dependent plasticity after eye opening, and it may be that this is related to the specific expression of certain subunits (probably NR2B) during development and during extended visual deprivation.[31] Interestingly, the time course of increased NMDA receptor contribution to SC visual responses closely matches the time course of the maturation and refinement of the projection from the visual cortex to the superior colliculus.[32,33] There is evidence that much of the cortical input to the SC is functionally mediated via NMDA receptors.[28] Hence the peak involvement of NMDA receptors in visual responses at P20-21 could reflect an over-exuberant cortico-collicular projection such that an unusually high number of SC neurons are innervated by the cortex. The later reduction in the number of AP5-sensitive responses would then correspond with a retraction of the cortical projection.

The effect of prolonged NMDA receptor blockade during development upon visual responses of neurons has also been investigated in the superior colliculus of hamsters, where NMDA receptor blockade in both normal animals and those with retinal lesions, produces enlarged single unit receptive fields.[34] Further, in the rat, molecular changes in the NMDA receptor subunits have been found to alter synaptic transmission.[35] Thus, changes in the composition of the NMDA receptor do contribute to functional maturation of the retinotectal pathway, but these occur largely after anatomical refinement of the pathway is complete.

9.6 SEROTONIN IS ALSO IMPORTANT FOR RETINAL FIBER SEGREGATION IN THE DEVELOPING SUPERIOR COLLICULUS

Evidence that the aminergic neurotransmitter serotonin (5-HT) is important for refinement of the retinocollicular pathway was first shown in hamster superior colliculus after injections of the neurotoxin, 5,7-dihydroxytryptamine (5,7-DHT). Subcutaneous injection of 5,7-DHT led to an increase in the amount of 5-HT, as

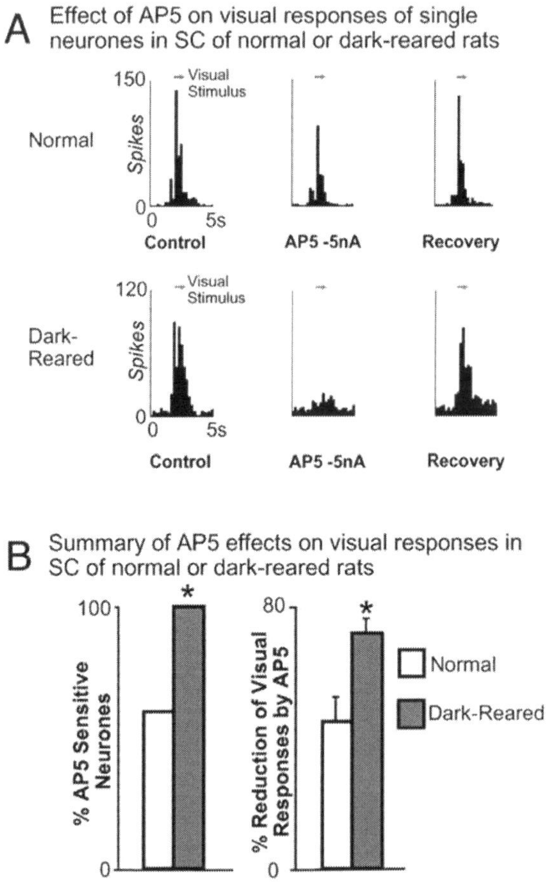

FIGURE 9.2 Effect of dark-rearing on the involvement of NMDA receptors in visual responses in the SC. The NMDA-receptor antagonist AP5 was used as a probe for functional involvement of NMDA receptors in visual responses. (A) Peristimulus time histograms that show counts of action potentials of single SC neurons in response to visual stimulation (horizontal arrow). Records are taken from representative neurons where the NMDA antagonist AP5 had an effect on visual responses, in either a normally reared subject or a dark-reared subject. Records show responses before (control), during and after (recovery) local application of AP5 via iontophoresis. Note that the antagonist had a greater effect on visual responses recorded in the dark-reared subject. (B) Histograms showing the proportion of neurons where AP5 had an effect on visual responses, and the effect of this antagonist in terms of reducing visual responses on those neurons where it had an effect. Note that in dark-reared subjects, visual responses of all neurons were reduced by AP5, whereas only 59% of neurons were affected in normally reared subjects. Furthermore, in those neurons where there was an effect of AP5, the effect was much greater in dark-reared subjects. (From Binns, K.E. and Salt, T.E., Developmental changes in NMDA receptor-mediated visual activity in the rat superior colliculus, and the effect of dark rearing, *Exp. Brain Res.*, 120, 335–344, 1998. With permission.)

measured by HPLC,[36] and to increase both the density of 5-HT containing fibers and the number of synaptic contacts that these fibers make in the superficial layers of the hamster SC.[36,37] This increase in 5-HT and 5-HT containing fibers has a profound effect upon the development of the uncrossed retinocollicular pathway. A single injection of 5,7,-DHT at birth alters the pathway permanently such that there is a significant increase in the distribution of retinal fibers in the mediolateral and rostrocaudal axes, an extension of uncrossed fibers into the more superficial regions of the SGL, and a loss of the characteristic uncrossed fiber patches seen within the rostral SC.[36,38] A similar disruption in the pathway was also found after exogenous application of 5-HT to the developing neonatal hamster SC by use of Elvax implants. There is no corresponding alteration in the distribution of the corticocollicular pathway after this treatment.[39]

The effect of 5-HT appears to be due specifically to an expansion of the terminal arbors of the retinocollicular pathway and not due to alterations in the retina. Thus, 5,7-DHT treated hamsters do not have increased numbers of retinal ganglion cells projecting to SC, nor is the distribution or size of these RGCs altered by this treatment.[40] The number of optic nerve axons is also unaltered by 5-7-DHT injections.[40] Rather the effect appears to be due to a significant expansion of the terminal arbor fields of individual retinocollicular axons. Thus the length, area, and volume of anterogradely labeled retinocollicular axon arbors increases after 5,7-DHT injections at birth while the number of branch points and numbers of boutons per axon do not.[41] In other words, each axon arbor covers a greater territory after exposure to increased levels of 5-HT, while the number of synapses per arbor remains constant.

These authors consider that the effect is mediated by the 5-HT 1B receptor and that it is a presynaptic process since these receptors are thought to be concentrated on retinocollicular axon terminals. Evidence for this includes: 5-HT 1B ligand binding is reduced significantly in the SC opposite unilateral enucleation while that of 5-HT 1A receptors is not[42,43]; EPSP amplitude is reduced by 5-HT even after blockade of 5HT 1A receptors; and visually driven activity in SC neurons is also reduced after iontophoretic application of 5-HT[38,42,43] or after stimulation of the dorsal Raphe nucleus, which increases 5-HT levels in the superficial layers of SC.[43] The disruption in fiber refinement is likely due to inhibition of retinocollicular synaptic transmission, and it is this inhibition that leads to an expansion in axon arbors.

More recently, Gaspar and associates have confirmed that 5-HT contributes to the process of retinocollicular pathway refinement in the rodent SC and have demonstrated some of the mechanisms by which 5-HT helps regulate this process. Initial results from this group demonstrating a role for 5-HT came from experiments using a transgenic knockout mouse that lacks monoamine oxidase A (MAOA), the degradative enzyme of 5-HT.[44] These MAOA knockout mice have abnormally high levels of 5-HT during both pre-and postnatal development, and they thus provide a convenient model for studying how 5-HT can influence pathway refinement. Upton et al.,[45] showed that the ipsilateral retinocollicular pathway fails to retract into distinct rostromedial patches, but remains diffusely spread through the SC as late as six months after birth in this mutant.[44] The ipsilateral and contralateral retinal afferents to the dorsal lateral geniculate nucleus also fail to segregate in these mutants, presumably because of the abnormally high levels of 5-HT in both structures.[45]

To further test how 5-HT levels alter pathway refinement, this group recently examined the refinement of the ipsilateral retinocollicular pathway in knockout mice in which either the 5-HT1B receptor, or 5-HTT transporter genes, or both, were disrupted.[46] The 5-HT1B receptor regulates glutamate release in adult hamsters,[42] and thus disruption of this receptor during development should alter retinocollicular synaptic transmission. Consistent with this argument, mice lacking the 5-HT1B receptor also had a more diffuse ipsilateral retinocollicular projection than normal mice when examined between 6 weeks and 6 months after birth. The effect occurred in both the rostrocaudal and dorsoventral planes of SC, with a continuous band of fibers found within the rostral SC and with a greater spread of ipsilateral fibers from the OL into the ventral SGL in these mutants.[46] These authors also reported that the ipsilateral retinocollicular projection was disrupted in mutants with deletion of the 5-HTT transporter gene, suggesting that altering either the 5-HT receptor or the 5-HT transporter can alter refinement of this glutamatergic pathway.

Gaspar and colleagues have also cross-mated the 5-HT1B knockout mice with MAOA knockout mice, reasoning that knockout of the 5-HT1B receptor should render the increase in 5-HT in the MAOA knockout ineffective. Salican et al.,[47] had previously shown that the defect in retinogeniculate fiber segregation in the LGN produced by the MAOA knockout was corrected in this double knockout where the ipsilateral and contralateral retinal fibers were found to segregate normally. Some restoration was also found in the ipsilateral retinocollicular pathway, although in this case refinement was not completely restored in the most rostral SC.[46] A similar partial recovery was found in 5-HT1B/5HTT double knockouts, suggesting that down-regulation of 5-HT also alters the refinement of this pathway.[46]

In summary, this evidence shows that: (1) the 5-HT1B receptor helps mediate refinement of the ipsilateral retinocollicular pathway. (2) either increased or decreased 5-HT levels or a lack of the 5-HT transporter can reduce segregation of the ipsilateral and contralateral retinal axons in the superior colliculus. Apparently, both an excess or a lack of 5-HT can alter fiber refinement in SC.

9.7 NITRIC OXIDE PARTIALLY MEDIATES RETRACTION OF THE RETINOCOLLICULAR PATHWAY

There is emerging evidence that other receptors, channels, and signaling molecules are also involved in the refinement of the mammalian retinocollicular pathway. Nitric oxide is one of these. A number of years ago it was suggested that nitric oxide could serve as a retrograde messenger by which a postsynaptic neuron could "inform" a presynaptic terminal of the effect that terminal had produced in that neuron.[48] Nitric oxide could thus contribute to pathway refinement by modifying synaptic efficacy, and thereby determine which synapses survive and which die. Synapses, which are correctly targeted, would be stabilized and made permanent, while those which are incorrectly targeted would be weakened and thereby eliminated. The original hypothesis proposed that nitric oxide could serve both to strengthen or weaken synapses, depending upon the state of activity in the presynaptic terminal. Presynaptic terminals whose activity was highly correlated with that of the postsynaptic neuron would be strengthened while those in which activity was not correlated would be weakened.[48,49]

Nitric oxide (NO) seems an ideal candidate to mediate the strengthening and/or weakening of synapses during development. As a free radical gas it is readily diffusible and can therefore travel in a retrograde as well as in an anterograde direction. Because it diffuses readily, it can influence more than one synapse as it has been shown to spread well beyond the site of release. Finally, it is known that the Ca^{++} influx that occurs through the NMDA receptor, as well as through other channels, contributes to the release of NO in some systems under some circumstances. Consistent with this hypothesis, our laboratory has recently shown that NO is involved in refinement of the ipsilateral retinocollicular pathway.

Several lines of evidence point to a role for NO in this process: (1) NOS, the NO synthetic enzyme, is expressed maximally in neurons within the superficial layers of the rodent SC between the ages of P4-P21,[50,51] a time during which the ipsilateral and contralateral retinocollicular pathways are undergoing refinement. This expression is robust, as demonstrated by antibodies directed against the enzyme and by the histochemical marker, NADPH diaphorase. Labeling with both of these markers is particularly heavy within the SGL and is contained in a variety of cell types that are thought to receive input directly from the retinocollicular pathway; (2) disruption of both the neuronal and endothelial isoforms of the NOS gene in double knockout mice produces a significant disruption in refinement of the ipsilateral retinocollicular pathway such that the pathway is more exuberant across both the rostral-caudal and mediolateral SC at later ages than in control mice. This effect can be seen as early as P9 where ipsilateral retinocollicular fibers remain distributed throughout much of the SC in e,nNOS double-knockout mice. The effect is also present between the ages of P15-21 where a conspicuous band of ipsilateral retinocollicular fibers remains spread across the entire mediolateral extent of the rostral SC and with substantial fiber labeling within portions of the lateral SC in more caudal sections (Figure 9.3). The disruption of retraction in this pathway is also seen in the dorsal ventral plane because ipsilateral retinocollicular fibers extend into the dorsal SGL, a territory normally reserved for the contralateral pathway by the age of P15. Although the disruption of pathway refinement in e,nNOS double knockout mice is statistically significant at earlier ages, the pathway does eventually retract and forms a pattern similar to that of normal rodents by P42.[4,5]

These results were the first to describe the effect of NO on development of the retinocollicular pathway in the mammalian SC, the time-course of the effect, and its dependence upon both the endothelial and neuronal isoforms of the enzyme.[4,5,50] They have been confirmed in rats by Vercelli et al.,[52] who have shown in both the rat SC and LGN that inhibitors of NOS alter segregation of the ipsilateral and contralateral inputs. Thus, intraperitoneal injections of the NOS inhibitor N-w-nitro-L-arginine (NO-Arg) for 4-6 weeks beginning at birth results in a substantial increase in the density of ipsilateral fibers within the SC, an increase in the length of these fibers, and an increase in bouton density.[52,53] A similar effect was also shown in the dLGN where there was less obvious segregation of the tracts from the two eyes after NOS inhibition.

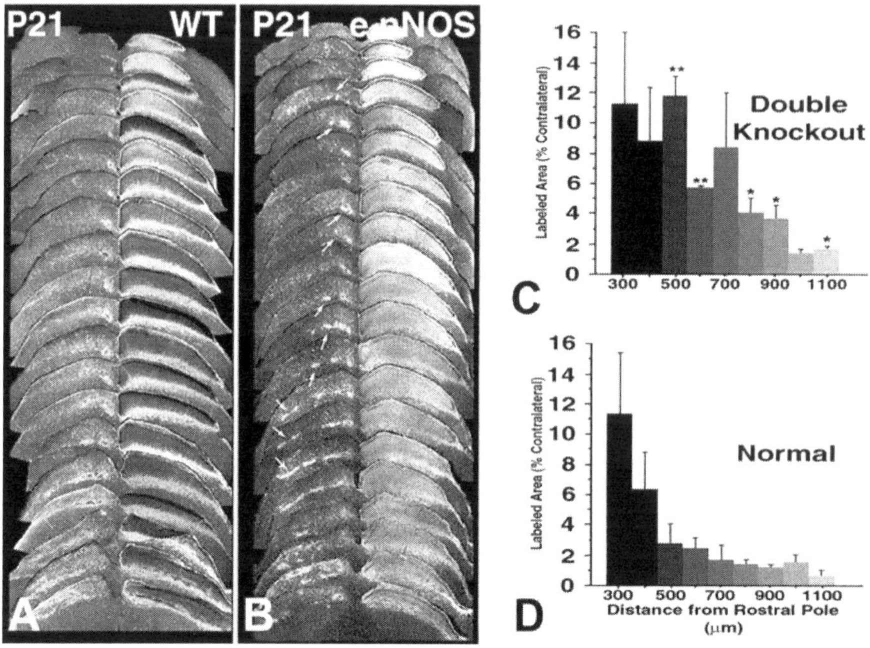

FIGURE 9.3 Distribution of the retinocollicular pathway in C57/BL6 wild type (WT) and e,nNOS double knockout mice at age P21. (A, B) The ipsilateral retinocollicular pathway is more extensive in both the rostro-lateral and caudal SC (arrows) in the double knockout mouse compared to the WT mouse. Sections of the superior colliculus (SC) labeled and stacked as in Figure 9.1. (C, D) Histograms illustrating quantitatively the amount of label in the ipsilateral SC as a percent of the amount measured in the contralateral SC of the same section in WT and double knockout mice. Note that there was significantly greater label in the double knockout mice at five of the intervals measured. ** = ($p < .005$), * = ($p < .05$). (From Mize, R.R. and Lo, F.-S., Nitric oxide, impulse activity, and neurotrophins in visual system development, *Brain Res.*, 886, 15–32, 2000. With permission.)

NO has also been shown to mediate retraction of the chick ipsilateral retinotectal pathway, a pathway that is normally eliminated during development but is partially spared by systemic application of NOS inhibitors during a critical period of development.[23,54,55] This effect is apparently mediated through the NMDA receptor because application of NMDA antagonists also suppresses the elimination of the pathway.[24] NO has also been shown to mediate refinement of the *on* and *off* sublaminae in the ferret lateral geniculate nucleus (LGN) where NOS inhibitors block segregation of the fibers that form these sublaminae,[56–58] an effect that is also mediated by NMDA receptors[25] and impulse activity.[57]

Although nitric oxide has been shown to be a contributing factor in the elimination of incorrectly targeted axons in several brain areas, it appears to play no role in some other regions in which pathway refinement is known to occur. Thus, inhibition of NOS in the three eyed frog does not alter the formation of alternating stripes from the normal and transplanted eyes[59] even though application of NO or

NO analogs *in vitro* produces collapse of retinal axons in this species.[60] The application of NOS inhibitors also fails to disrupt the segregation of ocular dominance columns in visual cortex, visualized anatomically,[61] or the plasticity that occurs in ocular dominance after monocular deprivation.[62,63]

9.8 VOLTAGE GATED CALCIUM CHANNELS ALSO MEDIATE PATHWAY REFINEMENT

We have recently shown that voltage gated calcium channels are also involved in mediating refinement of the retinal pathway.[6,64] As is the case with nitric oxide synthase, the L-type calcium channel is avidly expressed in retinorecipient neurons during the period of ipsilateral retinocollicular refinement. An antibody to the alpha 1C subunit of the Ca^{++} channel (an L-type component) densely labels a band of neurons within the superficial gray layer of the mouse SC from ages P4 to P15,[65] the time of maximal refinement of the pathway. This labeling is postsynaptic, and occurs in the somatic cytoplasm and proximal dendrites of cells with a variety of morphologies, including stellate, bipolar, and vertical fusiform neurons.[65] A few horizontal cells are also labeled, suggesting that both excitatory and inhibitory neurons contain the L-type Ca^{++} channel because horizontal neurons are known to be GABAergic.[66] Both types of neurons are thought to receive direct retinal input.[67] Thus, the L-type Ca^{++} channel is expressed in the appropriate neurons at the appropriate ages to influence pathway refinement.

Secondly, down-regulation of L-type calcium channel function by deletion of the β3 subunit of the Ca^{++} channel in gene knockout mice significantly delays refinement of the ipsilateral retinocollicular pathway.[6] The pathway was much more extensively distributed in the SC of these β3 subunit knockout mice compared to wild-type control mice at age P15 (Figure 9.4). Like the pattern seen in e,nNOS knockout mice, the pattern in the β3 subunit knockouts included a much more extensive distribution of fibers within both the rostral and caudal SC. This expanded distribution included a continuous band of label across the entire mediolateral extent of SC within the rostral SC, and a much more robust distribution laterally and dorsally within the more caudal SC. There is therefore a delay in retraction of these fibers in these mutants in both the rostral caudal and dorsal-ventral dimensions of SC.[6] These differences were statistically significant when labeling area was measured using computer image analysis. A similar delay in retraction of the contralateral retinogeniculate pathway was also found in these animals (Figure 9.4).[6]

Which downstream signaling cascades and calcium second messengers are involved in Ca^{++} channel dependent pathway refinement is largely unknown. One possibility is the activation of the second messenger inositol-1,4,5-trisphosphate (IP3). IP3 in the cytoplasm releases Ca^{++} from internal Ca^{++} stores, acting through the IP3 receptor. While the role of the IP3 pathway in pathway refinement has not been tested, it has been shown to be one signaling pathway contributing to synaptic potentiation and depression. LTP is enhanced in IP3R1 knockout mice while LTD is blocked in these mutants.[68] Whether IP3 is involved in pathway refinement remains to be tested.

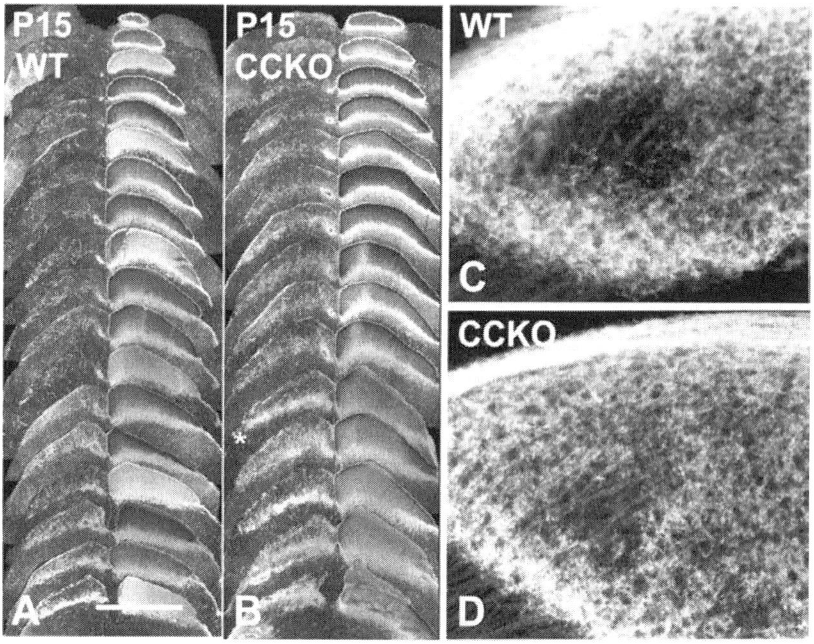

FIGURE 9.4 Distribution of the retinal pathway at age P15 in wild type (WT) and in knockout mice in which the ß3 subunit of the voltage gated calcium channel was deleted (calcium channel knockouts, CCKO). (A, B) The ipsilateral retinocollicular pathway is more extensive in both the rostro-lateral and caudal SC in the knockout mouse compared to the WT mouse. Sections of the superior colliculus (SC) labeled and stacked as in Figure 9.1. (C,D) Illustration of fiber labeling of the contralateral retinogeniculate pathway within the lateral geniculate nucleus (LGN) of WT and CCKO mice. Note that the contralateral fibers remain distributed within the ipsilateral pathway domain of the LGN of the CCKO compared to the WT mouse. * indicates missing section. Scale bar = 1 mm in A and B; 100 μM in C and D. (Modified from Cork, R.J., Namkung, Y., Shin, H.–S., and Mize, R.R., Development of the visual pathway is disrupted in mice with a targeted disruption of the calcium channel β3-subunit gene, *J. Comp. Neurol.*, 440, 177–191, 2001. With permission.)

9.9 ARE NEUROTROPHINS ESSENTIAL FOR PATHWAY STABILIZATION?

Although Gally et al.,[48] originally proposed that nitric oxide could serve as a retrograde message to both stabilize or eliminate synapses, virtually all experiments testing this hypothesis suggest that NO contributes to pathway retraction and synapse elimination but not to pathway or synapse stabilization. Other molecules, notably the neurotrophins, have been implicated in pathway stabilization. The neurotrophins include a variety of molecules, such as nerve growth factor (NGF), brain derived neurotrophic factor (BDNF), and neurotrophins NT-3 and NT-4/5. Each of these is associated with specific tyrosine kinase receptors, the so-called Trk receptors.[69,70]

The best evidence for their role in pathway stabilization has been found in the formation of ocular dominance columns in visual cortex.

Ocular dominance columns result from the segregation of geniculocortical afferents from the LGN within layer IV of visual cortex. Although NGF was the neurotrophin originally thought to be involved in the formation of ocular dominance columns in visual cortex,[71] more recent evidence suggests that BDNF may be the critical neurotrophic factor involved in the process of segregating geniculocortical inputs from the contralateral and ipsilateral eyes. Evidence for this includes the following. (1) both BDNF and its TrkB receptor are expressed during the critical period of ocular dominance formation,[72,73] thus demonstrating that the molecules are present when required; (2) levels of BDNF and TrkB mRNA and protein can be altered by visual experience. Both are decreased by monocular deprivation (MD) or dark rearing[73–78] and by blocking impulse activity.[74,77] (3) application of BDNF disrupts segregation of ocular dominance columns in kitten visual cortex[79] as do trkB receptor bodies that bind trkB ligands, rendering them inactive.[80] (4) BDNF also blocks the shift in ocular dominance (OD) that normally occurs after monocular deprivation (MD).[81] (5) Although over-expression of BDNF in transgenic mice fails to block the shift in OD in response to MD, it does alter the critical period for OD such that plasticity can occur at an earlier age.[82] (6) Application of exogenous BDNF to the visual cortex also desegregates OD columns in both normal kittens and kittens with one eye closed, which shows that high concentrations of BDNF can promote exuberant growth of fibers from both the normal and the deprived eye.[83] There is thus substantial evidence that BDNF promotes outgrowth of LGN axons within mammalian visual cortex, although its exact action in segregating fibers from the two eyes is not yet well-understood.

The effect of BDNF on retinocollicular refinement has been less widely studied, but evidence to date suggests that it may also serve to promote axon outgrowth and to stabilize this pathway.[84] In the hamster, BDNF mRNA has been localized within the SC beginning at about the time that retinal ganglion cell axons reach the SC, and BDNF protein reaches a maximal concentration in SC between P4-P15. It then declines about two-thirds to adult levels by P18.[84,85] TrkB protein, as measured by antibody immunocytochemistry, is also present within both cell bodies and putative axonal boutons within the mouse SC during this period of retinocollicular pathway refinement.[86] Our results show that it is especially heavily concentrated within neurons within the SGL during the first two weeks after birth, although it is present also in adults. BDNF has been shown directly to alter axon arbor outgrowth in vertebrates. Thus, infusion of BDNF onto chick retinotectal co-cultures promotes the growth of retinal ganglion cell axon arbors[87,88]; and BDNF also promotes branching of RGC axons in the frog optic tectum, an effect that can be blocked by BDNF antibodies.[89] In addition, over-expression of BDNF using an adenoviral vector prevents retraction of the rat ipsilateral retinocollicular pathway in regions where the vector is expressed.[90] This result is intriguing because the ipsilateral axons are retained in ectopic sites where the vector is expressed, regardless of whether the sites are in topographically correct or incorrect locations.[90] Thus, BDNF can promote axon growth into functionally inappropriate regions of SC. How BDNF can then contribute to correct targeting in normal animals is unclear, although it could do so by being released only when impulse activity is appropriate to establish a correct connection.

BDNF may also serve as a target-derived factor that can be retrogradely transported back to the retina in order to regulate retinal ganglion cell death and survival. BDNF is expressed in the hamster retina, but with a different time course than that seen in the SC. Protein levels of BDNF in retina remain low during the first postnatal week of life, a time in which BDNF levels are maximal in SC.[91] BDNF then increases 4- to 5-fold in retina to achieve adult levels by P18. If BDNF is injected exogenously into the SC at P4, there is a significant reduction in the death of retinal ganglion cells,[92] which suggests that BDNF could signal to the retina which ganglion cells have established correct connections and should therefore survive. Decreases in BDNF have the opposite effect. Thus, reductions in BDNF and its associated TrkB receptor ligands in SC produced by injection of TrkB fusion proteins cause an increase in the rate of retinal ganglion cell death compared to normal animals. This increase in retinal ganglion cell death also occurs in mice in which the full length trkB receptor has been mutated.[78] Thus, the mechanism of action of BDNF must work through a signaling cascade that is initiated by activation of the trkB receptor. BDNF may also regulate cell number in the SC itself during development. Intraocular injections of physiological concentrations of BDNF or NT-4/5 significantly decrease the numbers of pyknotic cells in the developing superior colliculus (SC); and also increase total cell number in the SC contralateral to the treated eye.[93] Thus, both BDNF and NT-4/5 may serve as anterograde survival factors for postsynaptic cells in the SC as well as retrograde factors to promote cell survival in the retina.

Recently, Ernst et al.,[87] have shown how BDNF and NO may work cooperatively in refinement of the developing chick retino-collicular pathway. In retinotectal co-cultures, application of NO donors produced growth cone collapse and axon retraction. Application of BDNF one hour before application of NO donors protected these axons from NO mediated collapse. If both BDNF and NO donors were applied, the growth cones were *stabilized*, so that neither axon outgrowth nor retraction occurred. Axons in contact with BDNF coated latex beads also failed to collapse upon application of NO donors,[87] suggesting that the action of BDNF is mediated locally, probably on receptors located on the membranes of the growth cones themselves. This growth cone *stabilization* in the presence of both BDNF and NO donors was accompanied by a change in actin filaments to a cytochalasin D-resistant form, suggesting that NO and/or BDNF may alter the ADP-ribosylation of actin, resulting in a more structurally stable axon. This data can be interpreted to demonstrate that BDNF promotes outgrowth, NO promotes retraction, and that the two acting together promote stabilization by altering the actin composition of the axon, possibly by reducing NO-induced depolymerization. Most recently, these authors have found that the mechanism of action of BDNF appears to be through a signaling cascade involving activation of protein kinase since treatment with PKA inhibitors blocked the effects of BDNF on growth cone mobilization.[94] PKA was also required in order to protect actin filaments from depolymerization by cytochalasin D. Together these results provide substantial support for the hypothesis that NO and BDNF act cooperatively in the refinement of the chick retinotectal pathway.

The source(s) of BDNF and the location(s) of the TrkB receptor in SC are not yet firmly established. Since BDNF is present in high concentrations in SC before it is found in retina,[84] it is likely that it is released from postsynaptic neurons in SC, at least during the first week of life. It presumably acts upon TrkB or p75 receptors, but whether these are pre or postsynaptic or both is not yet clear. TrkB immunocytochemistry shows that considerable receptor protein is located postsynaptically, and Maffei and colleagues[95] have recently shown that BDNF travels in an anterograde direction in the retinocollicular pathway within the first week after birth and that the transported BDNF can rapidly alter target neurons in both the SC and LGN. Inactivation of BDNF increased the rate of programmed neuronal cell death in SC, while transport of BDNF in retinal fibers after injections of BDNF into the eye prevented degeneration of LGN neurons. Thus, there is accumulating evidence that BDNF can travel either anterogradely or retrogradely in retinal ganglion cells, can be released from these cells, and can act as a survival factor for postsynaptic neurons in both subcortical target structures and also within the retina itself. Exactly how this contributes to pathway refinement remains to be worked out.

9.10 IS PATHWAY REFINEMENT A CONSEQUENCE OF SYNAPTIC POTENTIATION AND DEPRESSION?

A number of reviews have proposed that changes in the efficacy of synaptic transmission, specifically long term potentiation (LTP) and long term depression (LTD), could be the synaptic mechanisms that mediate pathway refinement during development.[96–100] The idea is that LTP produces strengthening of synapses and consequent stabilization of the parent fibers of these synapses; and that LTD leads to weakening of synapses that produces retraction and elimination of the fibers establishing these weakened synapses.[96,101] Despite the appeal of this hypothesis, there is to date only limited evidence to support it.[102] There are a number of reports that LTP and LTD can be induced in the developing brain, but whether they contribute to the process of pathway refinement remains controversial. LTP has been shown at synapses during the critical periods of barrel field and ocular dominance development.[103,104] This developmental form of LTP is homosynaptic and NMDA dependent, but does not appear to be affected by NO.[103–105] This developmental LTP can be regulated by BDNF. Thus, application of BDNF in neonatal slices of rat visual cortex increases the magnitude of LTP as well as enhancing field potentials and EPSCs, actions that can be blocked by TrkB-IgG fusion proteins or BDNF scavengers.[106,107] This effect of BDNF is apparently mediated through TrkB receptors because blocking them impairs LTP in the visual cortex.[108]

Recent studies in our laboratory suggest that LTP and LTD may also be involved in pathway refinement in the retinocollicular pathway of rodents. We have demonstrated that both LTP and LTD can be induced in an isolated brainstem preparation that includes the entire superior colliculus, as well as the tegmentum, and thalamus, with much of the optic tract intact.[64,109–111] LTP is induced only by high frequency stimulation at low stimulus intensities, while LTD is induced by low or high frequency tetanus, but only at higher stimulus intensities.[111] Perinatal LTP is both

NMDA and L-type Ca^{++} channel dependent because it is completely blocked by the NDMA antagonist APV and also by the Ca^{++} channel inhibitor nitrendipine. Application of both produces a slight depression in response. By contrast, perinatal LTD in SC does not depend upon the NMDA receptor, although it is partially mediated by L-type calcium channels[110,111] (Figure 9.5); LTD may also be mediated by NO because LTD magnitude is reduced in a subset of e,nNOS double knockout mice and is greatly enhanced in some animals when SNP, an NO donor, is applied to the SC.[64] However, this result must be replicated in a larger sample of knockout mice.

Both LTP and LTD are developmentally regulated in that there is a peak in magnitude during early postnatal life with a decline at older ages. The time course of expression differs for the two. For LTD, depression is maximal between 3-9 days after birth, and smaller at later ages (P10-P25). By contrast, the magnitude of LTP is maximal between P7-13, and less at both earlier and later ages. The time course of expression and NMDA dependence of perinatal LTP in the retinocollicular pathway are consistent with its role in ipsilateral retinocollicular pathway stabilization. The time course of expression and possible NO dependence of LTD further suggest that it could contribute to pathway retraction. However, these data are only correlative, and studies showing a direct relationship between these synaptic phenomena and pathway refinement are required.

That impulse activity in the retinocollicular pathway contributes to refinement is more generally established. Spontaneous electrical activity in the retina is known to occur as waves of correlated impulse activity that sweep across the retina prior to eye opening.[112–114] This activity is thought to lead to correlated firing in neighboring retinal ganglion cells, which should in turn lead to synaptic strengthening in neighboring fibers that fire together and synaptic weakening in fibers from other retinal sites in which the activity is uncorrelated.[113,115] Impulse activity has been shown to be important in the formation of OD in visual cortex[116] and in the segregation of retinal axons into eye-specific layers in the LGN.[117] The pattern of activity is important, since artificially induced synchronous activity in the two optic nerves blocks formation of OD columns while asynchronous activity sharpens them.[118]

FIGURE 9.5 (See facing page.) Long-term potentiation (LTP) and long-term depression (LTD) in the SC of the perinatal rat. (A) A medium frequency (50 Hz) tetanus of low intensity produces an immediate short-term potentiation followed by a long term potentiation of the field potential (FP) that lasts for at least 90 minutes (black circles). This potentiation is completely blocked by application of the NMDA antagonist D-APV. (B) A medium frequency (50 Hz) tetanus of high intensity produces an immediate short-term depression followed by a long term depression of the field potential (FP) that lasts for at least 90 minutes (black circles). The L-type Ca^{++} channel blocker nitrendipine partially reduces this LTD (black triangles). Addition of D-APV has no further effect (open circles), showing that the LTD is partially dependent upon the L-type Ca^{++} channel but not upon the NMDA receptor. Insets are examples of FP before (1) and after (2) tetanus in control and drug conditions. (From Lo, F.-S and Mize, R.R., Properties of long-term depression (LTD) and potentiation (LTP) of retinocollicular synaptic transmission in developing rat superior colliculus, *Eur. J. Neurosci.*, 15, 1421–1433, 2002. With permission.)

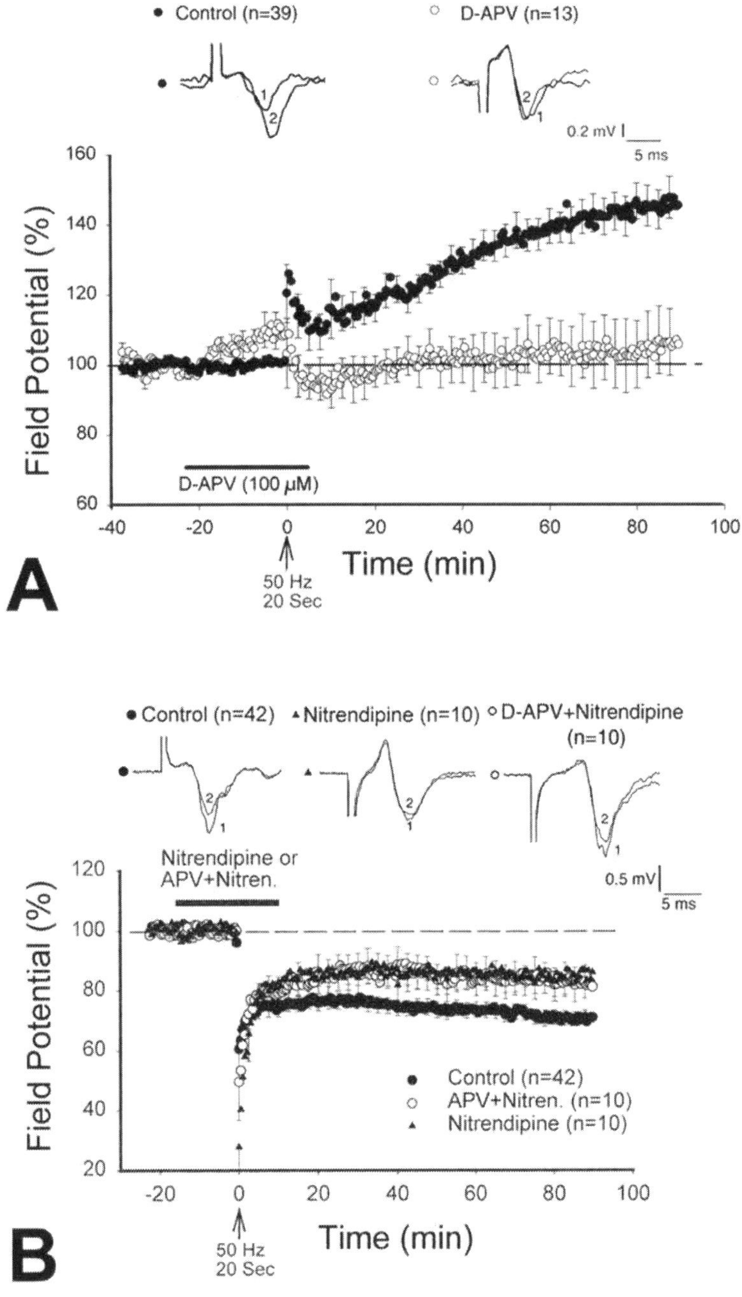

Control (n=39)
D-APV (n=13)
0.2 mV
5 ms
Field Potential (%)
D-APV (100 μM)
Time (min)
50 Hz
20 Sec
A
Control (n=42)
Nitrendipine (n=10)
D-APV+Nitrendipine (n=10)
Nitrendipine or APV+Nitren.
0.5 mV
5 ms
Field Potential (%)
Control (n=42)
APV+Nitren. (n=10)
Nitrendipine (n=10)
Time (min)
50 Hz
20 Sec
B

FIGURE 9.5

Activity patterns are also important in refinement of the retinocollicular pathway in rodents, and in the retinotectal pathway in vertebrates where either monocular deprivation or block of impulse activity with TTX can delay the refinement of the retinocollicular pathways.[11,20,21,119–121]

Recently, Zhang et al.,[122] have demonstrated directly that the temporal order of activation of synapses in convergent pathways is a critically important determinant of whether postsynaptic activity will be potentiated or depressed. Thus, repetitive suprathreshold stimulation of one retinotectal projection that produces spiking of a postsynaptic neuron in frog tectum potentiates that synapse. Pairing this stimulus with a weak stimulus from a second convergent input to the same eye potentiates the postsynaptic response if the weak stimulus occurs before, but depresses the response if the weak stimulus occurs after the suprathreshold stimulus.[122] Thus, both the temporal sequencing of impulse activity in the convergent inputs and the level of spiking in the postsynaptic neuron determine whether convergent synapses will be potentiated or depressed.

9.11 CONCLUSIONS AND SUMMARY

In this chapter we have reviewed evidence that a variety of molecules are involved in the refinement of the retinal pathway to the superior colliculus that occurs during early postnatal life. We have argued that none of these is the *magic bullet*; that there is no single receptor, channel, or signaling molecule that alone sculpts this pathway, that there is no single antagonist or genetic deletion that can completely block the significant rearrangement and elimination of axons that occurs in the ipsilateral and contralateral retinocollicular pathways during development. Blockade of the NMDA receptor using antagonists leads to a small increase in incorrectly targeted axons in the contralateral pathway,[22] but largely preserves correctly targeted axons within the rostral SC. Further, NMDA antagonists marginally disrupt retraction of the ipsilateral retinocollicular pathway or fail to disrupt it altogether.[23,26] Down-regulation of nitric oxide production via NOS inhibitors or with NOS knockout mice significantly delays but does not prevent retraction of the ipsilateral retinocollicular pathway.[4,5,52] A similar partial and transient delay in retraction of this pathway is seen after manipulation of the L-type Ca^{++} channel.[6] Each of these manipulations only partially disrupts refinement, and none apparently acts alone in producing the final topographic precision that occurs in this pathway.

Manipulation of the neurotransmitter 5-HT and its associated 5-HT1B receptor and 5HT transporter has a more profound effect,[43] but even in this case some refinement occurs in both the ipsilateral and contralateral pathways. Spontaneous electrical activity in presynaptic afferents also plays an important role, as does patterned visual input, but here again some refinement can occur in the absence of synaptic transmission, as shown by blocking it with TTX and other substances. The picture that thus emerges is that a variety of receptors, channels, and signaling molecules are involved in this process, that they operate through a variety of signal transduction pathways, and that alternative mechanisms can be recruited that can contribute to pathway refinement in the absence of one or several of these molecules and pathways.

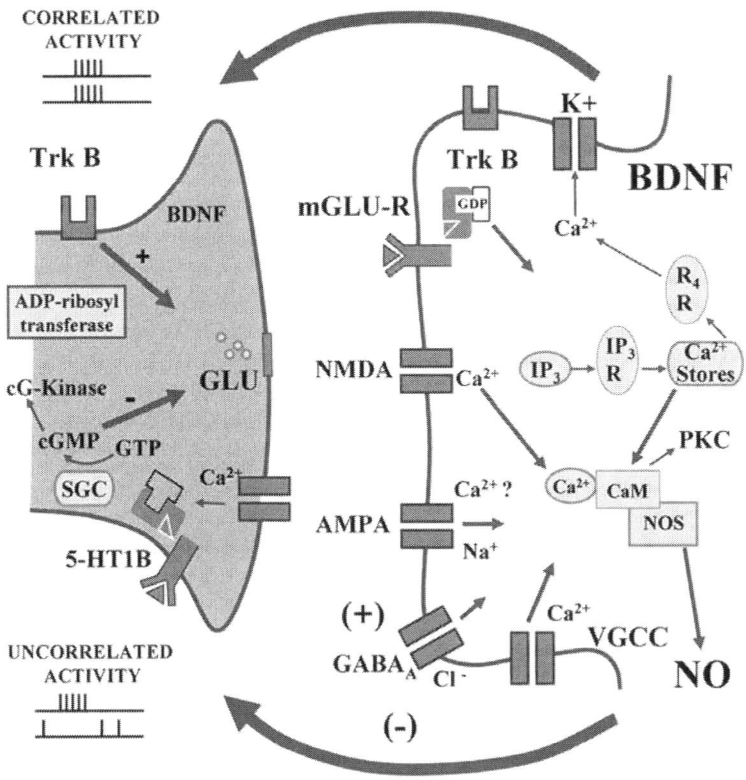

FIGURE 9.6 Diagram showing the receptors, channels, and signaling pathways likely to be involved in refinement of the retinocollicular pathway during development. The presynaptic terminal releases glutamate, and regulation of this release and other events that occur during stabilization or elimination of the developing synapse probably involves the 5-HT 1B and TrkB receptors as well as other unidentified receptors/channels. Membrane components involved in the postsynaptic cell include the ionotropic AMPA and NMDA receptors, the L-type voltage gated Ca^{++} channel, and possibly the mGlu receptors. GABA receptors may or may not be involved in this process. Nitric oxide (NO) is also contained in SC neurons and its release probably occurs through a calcium-calmodulin dependent activation of NOS. Postsynaptic TrkB receptors may mediate release of BDNF, while the illustrated IP3 pathway may contribute to release of intracellular Ca^{++} stores, which may also contribute to production of NO and/or other events responsible for strengthening or weakening the developing synapse. We propose that BDNF is released when impulse activity is correlated and that this coincident activity would stabilize the synapse. NO would be released when impulse activity is uncorrelated that would lead to synapse elimination.

Figure 9.6 summarizes the candidates most likely to be involved in refinement of the ipsilateral retinocollicular pathway. Glutamate is well-established as the excitatory neurotransmitter of the retinocollicular pathway, as illustrated in the presynaptic terminal in Figure 9.6. The 5HT-1B receptor is present on this terminal where it helps regulate release of the transmitter. We suggest that nitric oxide also influences the presynaptic terminal following release from the postsynaptic neuron. We do not yet know by which signaling cascade(s) this action is mediated postsynaptically. Likely candidates include soluble guanylate cyclase or ADP-ribosyl transferase (Figure 9.6). Various lines of evidence suggest that the TrkB receptor is located both pre-and postsynaptically. On the presynaptic membrane, this receptor could contribute to the retrograde transport of BDNF in retinal axons, or it could contribute to the regulation of glutamate release.

On the postsynaptic side, there is solid physiological and anatomical evidence for the presence of both ionotropic and metabotropic glutamate receptors. The NMDA receptor is clearly present on some neurons that are postsynaptic to retinal terminals, and it plays a role, although it is not essential, for refinement of the retinocollicular pathway. The AMPA receptor is also present, but we do not yet know whether it contributes to fiber stabilization or retraction. It should be noted that trafficking of this receptor has been shown to be an important mechanism for mediating synaptic plasticity in other brain regions.[123] Metabotropic glutamate receptors (mGluRs), which also have been shown to be important in synaptic plasticity, are present at the developing retinocollicular synapse, but whether they influence pathway refinement or synaptic stabilization has not yet been established. GABA receptors have been demonstrated to be functionally active at or shortly after birth, but the extent to which GABA and the inhibitory circuitry related to it contribute to pathway refinement is unknown.

Both nitric oxide and the L-type voltage gated Ca^{++} channel also participate in pathway refinement as well as synaptic transmission in the developing retinocollicular pathway. We have shown that L-type Ca^{++} channels are located postsynaptically at the appropriate ages, and that down-regulating this channel can disrupt refinement of the ipsilateral retinocollicular pathway. NO is also present in postsynaptic neurons within the retinorecipient region of the developing SC, and its maximal expression occurs at the appropriate ages as well. Altering NO levels in knockout animals or by the use of NOS inhibitors has also been shown to delay pathway refinement. The downstream signaling cascades involved are not yet established in the SC, but Ca^{++} influx through the NMDA receptor could be one source of the Ca^{++} calmodulin dependent production of NO. Other cascades that could be involved in this release include the IP3 pathway that would lead to the release of intracellular Ca^{++} from intracellular stores, or Ca^{++} influx through other VGCCs or even the AMPA receptor (Figure 9.6). Whatever the second messengers involved, postsynaptic release of NO is suggested in our model to result in pathway elimination, presumably by modifying synaptic transmission in the presynaptic axon and/or by altering the mechanical stability of the axon.

In our model we suggest that BDNF could be the molecule that provides a positive signal for strengthening synaptic connections and stabilizing appropriately targeted axons (Figure 9.6). We show that the TrkB receptor is present postsynaptically as well as presynaptically as shown by antibody immunocytochemistry, but we do not know the effect that the TrkB receptor has on the postsynaptic membrane. The downstream signaling pathways involved are only beginning to be explored, and it is likely that other receptors will also be identified that are important for this process. It is, in sum, remarkable how much is yet to be learned about this well-studied but highly complex process of pathway refinement.

ACKNOWLEDGMENTS

We wish to thank our colleagues who performed many of the experiments in our laboratories that are cited in this review. They include Drs. John Cork, Fu Sun Lo, and Hope Wu in New Orleans and Drs. Esther Binns and Jon Turner in London. Our research in New Orleans was supported by USPHS grants NS-36000 from the National Institute of Neurological Disorders and Stroke and EY-02973 from the National Eye Institute and in London by the Wellcome Trust.

REFERENCES

1. O'Leary, D.D., Fawcett, J.W., and Cowan, W.M., Topographic targeting errors in the retinocollicular projection and their elimination by selective ganglion cell death, *J. Neurosci.,* 6, 3692–705, 1986.
2. Simon, D.K. and O'Leary, D.D., Relationship of retinotopic ordering of axons in the optic pathway to the formation of visual maps in central targets, *J. Comp. Neurol.,* 307, 393–404, 1991.
3. Land, P.W. and Lund, R.D., Development of the rat's uncrossed retinotectal pathway and its relation to plasticity studies, *Science.,* 205, 698–700, 1979.
4. Wu, H.H., Cork, R.J., Shuman, D., Huang, P.L., and Mize R.R., Refinement of the ipsilateral retinocollicular projection is delayed in double endothelial and neuronal nitric oxide synthase gene knockout mice, *Dev. Brain Res.,* 120, 105–111, 2000a.
5. Wu, H.H., Cork, R.J., and Mize R.R., Development of the ipsilateral retinocollicular pathway in normal and NOS gene knockout mice, *J. Comp. Neurol.,* 426, 651–665, 2000b.
6. Cork, R.J., Namkung, Y., Shin, H.–S., and Mize, R.R., Development of the visual pathway is disrupted in mice with a targeted disruption of the calcium channel β3-subunit gene, *J. Comp. Neurol.,* 440, 177–191, 2001.
7. Lund, R.D., Land, P.W., and Boles, J., Normal and abnormal uncrossed retinotectal pathways in rats: an HRP study in adults, *J. Comp. Neurol.,* 189, 711–720, 1980.
8. Edwards, M.A., Schneider, G.E., and Caviness, V.S., Jr., Development of the crossed retinocollicular projection in the mouse, *J. Comp Neurol.,* 248, 410–421, 1986.
9. Godement, P., Salaun, J., and Imbert, M., Prenatal and postnatal development of retinogeniculate and retinocollicular projections in the mouse, *J. Comp. Neurol.,* 230, 552–575, 1984.
10. Lund, R.D. and Lund, J.S., Plasticity in the developing visual system: the effects of retinal lesions made in young rats, *J. Comp. Neurol.,* 169, 133–154, 1976.
11. Thompson, I. and Holt, C., Effects of intraocular tetrodotoxin on the development of the retinocollicular pathway in the Syrian Hamster, *J. Comp. Neurol.,* 282, 371–388, 1989.
12. Simon, D.K. and O'Leary, D.D.M., Development of topographic order in the mammalian retinocollicular projection, *J. Neurosci.,* 12, 1212–1232, 1992.
13. Masu, M., Nakajima, Y., Moriyoshi, K., Ishii, T., Akazawa, C., and Nakanashi, S., Molecular characterization of NMDA and metabotropic glutamate receptors, *Ann. N. Y. Acad. Sci.,* 707, 153–164, 1993.
14. Nakanishi, S., Masu, M., Bessho, Y., Nakajima, Y., Hayashi, Y., and Shigemoto, R., Molecular diversity of glutamate receptors and their physiological functions, *EXS.,* 71, 71–80, 1994.

15. Cull-Candy, S., Brickley, S., and Farrant, M., NMDA receptor subunits, diversity, development and disease, *Curr. Opin. Neurobiol.*, 11, 327–335, 2001.
16. Scheetz, A.J. and Constantine-Paton, M., Modulation of NMDA receptor function: implications for vertebrate neural development, *FASEB J.*, 8,745–752, 1994.
17. Hofer, M., Prusky, G.T., and Constantine-Paton, M., Regulation of NMDA receptor mRNA during visual map formation and after receptor blockade, *J. Neurochem.*, 62, 2300–2307, 1994.
18. Cline, H.T., Debski, E.A., and Constantine-Paton, M., The role of the NMDA receptor in the development of the frog visual system, *Adv. Exp. Med. Biol.*, 268, 197–203, 1990.
19. Cline, H.T. and Constantine-Paton, M., NMDA receptor antagonists disrupt the retinotectal topographic map, *Neuron*, 3, 413–426, 1989.
20. Schmidt, J.T. and Buzzard, M., Activity-driven sharpening of the regenerating retinotectal projection: effects of blocking or synchronizing activity on the morphology of individual regenerating arbors, *J. Neurobiol.*, 21, 900–917, 1990.
21. Scherer, W.J. and Udin, S.B., N-methyl-D-aspartate antagonists prevent interaction of binocular maps in Xenopus tectum, *J. Neurosci.*, 9, 3837–3843, 1989.
22. Simon, D.K., Prusky, G.T., O'Leary, D.D.M., and Constantine-Paton, M., N-methyl-d-aspartate receptor antagonists disrupt the formation of a mammalian neural map, *Proc. Natl. Acad. Sci. USA 89*, 10593–10597, 1992.
23. Ernst, A.F., Jurney, W.M., and McLoon, S.C., Mechanisms involved in development of retinotectal connections: roles of Eph receptor tyrosine kinases, NMDA receptor and nitric oxide, *Prog. Brain Res.*, 118, 115–132, 1998.
24. Ernst, A.F., Wu, H.H., El-Fakahany, E.E., and McLoon, S.C., NMDA receptor mediated refinement of a transient retinotectal projection requires nitric oxide, *J. Neurosci.*, 19, 229–235, 1999.
25. Hahm, J.-O., Langdon, R.B., and Sur, M. Disruption of retinogeniculate afferent segregation by antagonists to NMDA receptors. *Nature*, 351, 568–570, 1991.
26. Colonnese, M.T. and Constantine-Paton, M., Chronic NMDA receptor blockade from birth increases the sprouting capacity of ipsilateral retinocollicular axons without disrupting their early segregation, *J. Neurosci.*, 21, 1557–1568, 2001.
27. Binns, K.E. and Salt, T.E., Excitatory amino acid receptors participate in synaptic transmission of visual responses in the superficial layers of the cat superior colliculus, *Eur. J. Neurosci.* 6, 161–169, 1994.
28. Binns, K.E. and Salt, T.E., Corticofugal influences on visual responses in cat superior colliculus: the role of NMDA receptors, *Vis. Neurosci.* 13, 683–694, 1996.
29. Binns, K.E. and Salt, T.E. Developmental changes in NMDA receptor-mediated visual activity in the rat superior colliculus, and the effect of dark rearing, *Exp. Brain Res.* 120, 335–344, 1998a.
30. Binns, K.E. and Salt, T.E., Experience-dependent changes in the importance of N-methyl-D-aspartate (NMDA) receptors for visual transmission in superior colliculus, *Dev. Brain Res.* 110, 241–248, 1998b.
31. Binns, K.E., Turner, J.P., and Salt, T.E., Visual experience alters the molecular profile of NMDA-receptor-mediated sensory transmission, *Eur. J. Neurosci.*, 11, 1101–1104, 1999.
32. Thong, I.G. and Dreher, B., The development of the corticotectal pathway in the albino rat, *Brain Res.*, 390, 227–238, 1986.
33. Kasper, E.M., Lubke, J., Larkman, A.U., and Blakemore C., Pyramidal neurons in layer 5 of the rat visual cortex. III. Differential maturation of axon targeting, dendritic morphology, and electrophysiological properties, *J. Comp. Neurol.*, 339, 495–518, 1994.

34. Huang, L. and Pallas, S.L., NMDA antagonists in the superior colliculus prevent developmental plasticity but not visual transmission or map compression, *J. Neurophysiol.*, 86, 1179–1194, 2001.
35. Shi, J., Aamodt, S.M., and Constantine-Paton, M. Temporal correlations between functional and molecular changes in NMDA receptors and GABA neurotransmission in the superior colliculus, *J. Neurosci.,* 17, 6264–6276, 1997
36. Rhoades, R.W., Bennett-Clarke, C.A., Lane, R.D., Leslie, M.J., and Mooney, R.D,. Increased serotoninergic innervation of the hamster's superior colliculus alters retinotectal projections, *J. Comp. Neurol.,* 334, 397–409, 1993.
37. Arce, E.A., Rhoades, R.W., and Mooney, R.D., Neonatal administration of the neurotoxin 5, 7-dihydroxytryptamine results in synaptic reorganization in the superficial gray layer of the hamster's superior colliculus, *Vis. Neurosci.,* 12, 779–783, 1995.
38. Mooney, R.D., Crnko-Hoppenjans, T.A., Ke, M., Bennett-Clarke, C.A., Lane, R.D., Chiaia, N.L., and Rhoades, R.W., Augmentation of serotonin in the developing superior colliculus alters the normal development of the uncrossed retinotectal projection, *J. Comp. Neurol.,* 393, 84–92, 1998.
39. Crnko-Hoppenjans, T.A., Yaggie, J., Mooney, R.D., and Rhoades, R.W., Organization of the corticotectal projection in hamsters with neonatally elevated levels of serotonin in the superior colliculus, *Neurosci. Lett.*, 276, 111–114, 1999.
40. Crnko, T.A., Mooney, R.D., Crissman, R.S., Zheng, L., and Rhoades, R.W., Increased serotonin in the developing superior colliculus does not alter the number or distribution of retinotectal ganglion cells, *J. Comp. Neurol.*, 364, 414–424, 1996.
41. Crnko-Hoppenjans, T.A., Mooney, R.D., and Rhoades, R.W., Neonatally elevated serotonin levels alter terminal arbors of individual retinal ganglion cells in superior colliculus of hamsters, *J. Comp. Neurol.,* 432, 528–536, 2001.
42. Mooney, R.D., Shi, M.Y., and Rhoades, R.W., Modulation of retinotectal transmission by presynaptic 5-HT1B receptors in the superior colliculus of the adult hamster, *J. Neurophysiol.,* 72, 3–13, 1994.
43. Mooney, R.D., Huang, X., Shi, M.Y., Bennett-Clarke, C.A., and Rhoades R.W., Serotonin modulates retinotectal and corticotectal convergence in the superior colliculus, *Prog Brain Res.,* 112, 57–69, 1996.
44. Cases, O., Vitalis, T., Seif, I., De Maeyer, E., Sotelo, C., and Gaspar, P., Lack of barrels in the somatosensory cortex of monoamine oxidase A-deficient mice: role of a serotonin excess during the critical period, *Neuron*, 16, 297–307, 1996.
45. Upton, A.L., Salichon, N., Lebrand, C., Ravary, A, Blakely, R., Seif, I., and Gaspar, P., Excess of serotonin (5-HT) alters the segregation of ispilateral and contralateral retinal projections in monoamine oxidase A knock-out mice: possible role of 5-HT uptake in retinal ganglion cells during development, *J. Neurosci.,* 19, 7007–7024, 1999.
46. Upton, A.L., Ravary, A., Salichon, N., Moessner, R., Lesch, K.P., Hen, R., Seif, I., and Gaspar, P., Lack of 5-HT(1B) receptor and of serotonin transporter have different effects on the segregation of retinal axons in the lateral geniculate nucleus compared to the superior colliculus, *Neuroscience,* 111, 597–610, 2002.
47. Salichon, N., Gaspar, P., Upton, A.L., Picaud, S., Hanoun, N., Hamon, M., DeMaeyer, E., Murphy, D.L., Mossner, R., Lesch, K.P., Hen, R., and Seif, I., Excessive activation of serotonin (5-HT) 1B receptors disrupts the formation of sensory maps in monoamine oxidase A and 5-HT transporter knock-out mice, *J. Neurosci.*, 21, 884–896, 2001.
48. Gally, J.A., Montague, P.R., Reeke, G.N., Jr., and Edelman, G.M., The NO hypothesis: possible effects of a short-lived, rapidly diffusible signal in the development and function of the nervous system, *Proc. Natl. Acad. Sci. U S A.,* 87, 3547–3551, 1990.

49. Montague, P.R., Gally, J.A., and Edelman, G.M., Spatial signaling in the development and function of neural connections, *Cereb. Cortex*, 1, 199–220, 1991.
50. Mize, R.R., Wu, H.H., Cork, R.J., and Scheiner, C.A., The role of nitric oxide in development of the patch-cluster system and retinocollicular pathways in the rodent superior colliculus, *Prog. Brain Res.,* 118, 133–154, 1998.
51. Scheiner, C.A., Cork, R.J., and Mize, R.R., Failure to disrupt development of cholinergic patches in the superior colliculus in nitric oxide synthase deficient mice, *Devel. Brain Res.,* 118, 217–220, 1999.
52. Vercelli, A., Garbossa, D., Biasiol, S., Repici, M., and Jhaveri, S., NOS inhibition during postnatal development leads to increased ipsilateral retinocollicular and retinogeniculate projections in rats, *Eur. J. Neurosci.,* 12, 473–490, 2000.
53. Vercelli, A., Garbossa, D., Repici, M., Biasiol, S., and Jhaveri, S., Role of nitric oxide in the development of retinal projections, *Ital. J. Anat. Embryol.,* 106, 489–498, 2001.
54. Wu, H.H., Williams, C.V., and McLoon, S.C., Involvement of nitric oxide in the elimination of a transient retinotectal projection in development, *Science,* 265, 1593–1596, 1994.
55. Wu, H.H., Waid, D.K., and McLoon, S.C., Nitric oxide and the developmental remodeling of retinal connections in the brain, *Prog. Brain Res.,* 108, 273–286, 1996.
56. Cramer, K.S. and Sur, M., The role of NMDA receptors and nitric oxide in retinogeniculate development, *Prog. Brain Res.,* 108, 235–244, 1996.
57. Cramer, K.S. and Sur, M., Blockade of afferent impulse activity disrupts on/off sublamination in the ferret lateral geniculate nucleus, *Dev. Brain Res.,* 98, 287–290, 1997.
58. Cramer, K.S. and Sur, M., The neuronal form of nitric oxide synthase is required for pattern formation by retinal afferents in the ferret lateral geniculate nucleus, *Dev. Brain Res.,* 116, 79–86, 1999.
59. Renteria, R.C. and Constantine-Paton, M., Nitric oxide in the retinotectal system: a signal but not a retrograde messenger during map refinement and segregation, *J. Neurosci.,* 19, 7066–7076, 1999.
60. Renteria, R.C. and Constantine-Paton, M., Exogenous nitric oxide causes collapse of retinal ganglion cell axonal growth cones *in vitro*, *J. Neurobiol.,* 29, 415–428, 1996.
61. Finney, E.M., and Shatz, C.J., Establishment of patterned thalamocortical connections does not require nitric oxide synthase, *J. Neurosci.*, 18, 8826–8838, 1998.
62. Ruthazer, E.S., Gillespie, D.C., Dawson, T.M., Snyder, S.H., and Stryker, M.P., Inhibition of nitric oxide synthase does not prevent ocular dominance plasticity in kitten visual cortex, *J. Physiol. (Lond.)*, 494, 519–527, 1996.
63. Reid, S.N.M., Daw, N.W., Czepita, D., Flavin, H.J., and Sessa, W.C., Inhibition of nitric oxide synthase does not alter ocular dominance shifts in kitten visual cortex, *J. Physiol. (Lond),* 494, 511–517, 1996.
64. Mize, R.R. and Lo, F.-S., Nitric oxide, impulse activity, and neurotrophins in visual system development, *Brain Res.,* 886, 15–32, 2000.
65. Mize, R.R., Graham, S.K., and Cork, R.J., Expression of the L-type calcium channel in the developing mouse visual system by use of immunocytochemistry, *Dev. Brain Res.*, 136, 185–195, 2002.
66. Mize, R.R., Immunocytochemical localization of gamma-aminobutyric acid (GABA) in the cat superior colliculus, *J. Comp. Neurol.,* 276, 133–155, 1988.
67. Mize, R.R., Spencer, R.F., and Sterling, P., Two types of GABA-accumulating neuron in the superficial gray layer of the cat superior colliculus, *J. Comp. Neurol.,* 206, 180–192, 1982.

68. Nishiyama, M., Hong, K., Mikoshiba, K., Poo, M.-M., and Kato, K., Calcium stores regulate the polarity and input specificity of synaptic modification, *Nature Lett.,* 408, 584–588, 2000.
69. McAllister, A.K., Katz, L.C., and Lo, D.C., Neurotrophins and synaptic plasticity, *Annu. Rev. Neurosci.,* 22, 295-318, 1999.
70. Thoenen H. Neurotrophins and neuronal plasticity, *Science,* 270, 593–598, 1995.
71. Maffei, L., Berardi, N., Domenici, L., Parisi, V., and Pizzorusso, T., Nerve growth factor (NGF) prevents the shift in ocular dominance distribution of visual cortical neurons in monocularly deprived rats, *J. Neurosci.,* 12, 4651–4662, 1992.
72. Cabelli, R.J., Allendoerfer, K.L., Radeke, M.J., Welcher, A.A., Feinstein, S.C., and Shatz, C.J., Changing patterns of expression and subcellular localization of TrkB in the developing visual system, *J. Neurosci.*, 16, 7965–7980, 1996.
73. Rossi, F.M., Bozzi, Y., Pizzorusso, T., and Maffei, L., Monocular deprivation decreases brain-derived neurotrophic factor immunoreactivity in the rat visual cortex, Neuroscience, 90, 363–368, 1999.
74. Bozzi, Y., Pizzorusso, T., Cremisi, F., Rossi, F.M., Barsacchi, G., and Maffei, L., Monocular deprivation decreases the expression of messenger RNA for brain-derived neurotrophic factor in the rat visual cortex, *Neuroscience,* 69, 1133–1144, 1995.
75. Schoups, A.A., Elliott, R.C., Friedman, W.J., and Black, I.B., NGF and BDNF are differentially modulated by visual experience in the developing geniculocortical pathway. *Dev. Brain Res.,* 86,326–334, 1995.
76. Lein, E.S., Hohn, A., and Shatz, C.J., Dynamic regulation of BDNF and NT-3 expression during visual system development, *J. Comp. Neurol.,* 420, 1–18, 2000.
77. Lein, E.S. and Shatz, C.J., Rapid regulation of brain-derived neurotrophic factor mRNA within eye-specific circuits during ocular dominance column formation, *J. Neurosci.,* 20, 1470–1483, 2000.
78. Pollock, G.S., Vernon, E., Forbes, M.E., Yan, Q., Ma, Y.T., Hsieh, T., Robichon, R., Frost, D.O., and Johnson, J.E., Effects of early visual experience and diurnal rhythms on BDNF mRNA and protein levels in the visual system, hippocampus, and cerebellum, *J. Neurosci.,* 21, 3923–3931, 2001.
79. Cabelli, R.J., Hohn, A., and Shatz, C.J., Inhibition of ocular dominance column formation by infusion of NT-4/5 or BDNF, *Science*, 267, 1162–1166, 1995.
80. Cabelli, R.J., Shelton, D.L., Segal, R.A., and Shatz, C.J., Blockade of endogenous ligands of trkB inhibits formation of ocular dominance columns, *Neuron*, 19, 63–76, 1997.
81. Galuske, R.A., Kim, D.S., Castren, E., Thoenen, H., and Singer, W., Brain-derived neurotrophic factor reversed experience-dependent synaptic modifications in kitten visual cortex, *Eur. J. Neurosci.,* 8, 1554–1559, 1996.
82. Hanover, J.L., Huang, Z.J., Tonegawa, S., and Stryker, M.P., Brain-derived neurotrophic factor overexpression induces precocious critical period in mouse visual cortex, *J. Neurosci.*, 19, RC40, 1999.
83. Hata, Y., Ohshima, M., Ichisaka, S., Wakita, M., Fukuda, M., and Tsumoto, T., Brain-derived neurotrophic factor expands ocular dominance columns in visual cortex in monocularly deprived and nondeprived kittens but does not in adult cats, *J. Neurosci.,* 20, RC57, 2000.
84. Frost, D.O., BDNF/trkB signaling in the developmental sculpting of visual connections, *Prog. Brain Res.,* 134, 35–49, 2001.
85. Frost, D.O., Ma Y.T., Hsiech, T., Forbest, M.E., and Johnson, J.E., Developmental changes in BDNF protein levels in the hamster retina and superior colliculus, *J. Neurobiol.* 49, 173–187, 2001.

86. Mize, R.R., Distribution of TrkB receptor in the mouse visual system using antibody immunocytochemistry, *Neurosci. Abs.*, 27, 252. 12, 2001.
87. Ernst, A.F., Gallo, G., Letourneau, P.C., and McLoon, S.C., Stabilization of growing retinal axons by the combined signaling of nitric oxide and brain-derived neurotrophic factor, *J. Neurosci.*, 20, 1458–1469, 2000.
88. Inoue, A. and Sanes, J.R., Lamina-specific connectivity in the brain: regulation by N-cadherin, neurotrophins and glycoconjugates, *Science*, 276, 1428–1431, 1997.
89. Cohen-Corey, S. and Fraser, S.E., Effects of brain-derived neurotrophic factor on optic axon branching and remodeling *in vivo, Nature*, 378, 192–196, 1995.
90. Isenmann, S., Cellerino, A., Gravel, C., and Bahr, M., Excess target-derived brain-derived neurotrophic factor preserves the transient uncrossed retinal projection to the superior colliculus, *Mol. Cell. Neurosci.,* 14, 52–65, 1999.
91. Ma, Y.-T., Hsieh, T., Forbes, M.E., Johnson, J.E., and Frost, D.O., BDNF protein levels in developing hamster visual system, *Soc. Neurosci. Abstr.,* 23, 329, 1998.
92. Ma, Y.-T., Hsieh, T., Forbes, M.E., Johnson, J.E., and Frost, D.O., BDNF injected into the superior colliculus reduces developmental retinal ganglion cell death, *J. Neurosci.* 18, 2097–2107, 1998.
93. Spalding, K.L., Tan, M.M., Hendry, I.A., and Harvey, A.R., Anterograde transport actions of BDNF and NT-4/5 in the developing rat visual system, *Mol. Cell. Neurosci.,* 19, 485–500, 2002.
94. Gallo, G., Ernst, A.F., McLoon S.C., and Letourneau, P.C., Transient PKA activity is required for initiation but not maintenance of BDNF-mediated protection from nitric oxide-induced growth-cone collapse, *J. Neurosci.*, 22, 5016–5023, 2002.
95. Caleo, M., Menna, E., Chierzi, S., Cenni, M.C., and Maffei, L., Brain-derived neurotrophic factor is an anterograde survival factor in the rat visual system, *Curr. Biol.*, 10, 1155–1161, 2000.
96. Bear, M.F., A synaptic basis for memory storage in the cerebral cortex, *Proc. Natl. Acad. Sci. USA*, 93, 13453–13459, 1996.
97. Bear, M.F., Cooper, L.N., and Ebner, F.F., A physiological basis for a theory of synapse modification, *Science*, 237, 42–48, 1987.
98. Goodman, C.S. and Shatz, C.J., Developmental mechanisms that generate precise patterns of neuronal connectivity, *Cell*, 72, 77–98, 1993.
99. Miller, K.D., Keller, J.B., and Stryker, M.P., Ocular dominance column development, analysis and simulation, *Science*, 245, 605–615, 1989.
100. Singer, W., Development and plasticity of cortical processing architectures, *Science*, 270, 758–764, 1995.
101. Katz, L.C. and Shatz, C. J., Synaptic activity and the construction of cortical circuits, *Science,* 274, 1133–1138, 1996.
102. Constantine-Paton, M. and Cline, H.T., LTP and activity-dependent synaptogenesis: the more alike they are, the more different they become, *Curr. Opin. Neurobiol.*, 8, 139–148, 1998.
103. Crair, M.C. and Malenka, R.C., A critical period for long-term potentiation at thalamocortical synapses, *Nature*, 375, 325–328, 1995.
104. Kirkwood, A., Lee, H.K., and Bear, M.F., Co-regulation of long-term potentiation and experience-dependent synaptic plasticity in visual cortex by age and experience, *Nature,* 375, 328–331, 1995.
105. Dudek, S.M. and Friedlander, M.J., Developmental down-regulation of LTD in cortical layer IV and its independence of modulation by inhibition, *Neuron*, 16, 1097–1106, 1996.

106. Akaneya, Y., Tsumoto, T., Kinoshita, T., and Hatanaka, H., Brain-derived neurotrophic factor enhances long-term potentiation in rat visual cortex, *J. Neurosci.,* 17, 6707–6716, 1997.
107. Huber, K.M., Sawtell, N.B., and Bear, M.F., Brain-derived neurotrophic factor alters the synaptic modification threshold in visual cortex, *Neuropharmacology,* 37, 571–579, 1998.
108. Sermasi, E., Margotti, E., Cattaneo, A., and Domenici, L., Trk B signaling controls LTP but not LTD expression in the developing rat visual cortex, *Eur. J. Neurosci.,* 12, 1411–1419, 2000.
109. Lo, F.-S. and Mize, R.R., Retinal input induces three firing patterns in neurons of the superficial colliculus of neonatal rats, *J. Neurophysiol.,* 81, 954–958, 1999.
110. Lo, F.-S and Mize, R.R., Synaptic regulation of L-type calcium channel activity and long term depression during refinement of the retinocollicular pathway in developing rodent superior colliculus, *J. Neurosci.,* 20, RC58, 1–6, 2000.
111. Lo, F.-S and Mize, R.R., Properties of long-term depression (LTD) and potentiation (LTP) of retinocollicular synaptic transmission in developing rat superior colliculus, *Eur. J. Neurosci.,* 15, 1421–1433, 2002.
112. Galli, L. and Maffei, L., Spontaneous impulse activity of rat retinal ganglion cells in prenatal life, *Science*, 242, 90–91, 1998.
113. Meister, M., Wong, R.O., Baylor, D.A., and Shatz, C.J., Synchronous bursts of action potentials in ganglion cells of the developing mammalian retina, *Science*, 252, 939–943, 1991.
114. Shatz, C.J., Impulse activity and the patterning of connections during CNS development, *Neuron*, 5, 745–756, 1990.
115. Blais, B.S., Shouval, H.Z., and Cooper, L.N., The role of presynaptic activity in monocular deprivation: comparison of homosynaptic and heterosynaptic mechanisms, *Proc. Natl. Acad. Sci. USA*, 96, 1083–1087, 1999.
116. Stryker, M.P. and Harris, W.A., Binocular impulse blockade prevents the formation of ocular dominance columns in cat visual cortex, *J. Neurosci.,* 6, 2117–2133, 1986.
117. Shatz, C.J. and Stryker, M.P., Prenatal tetrodotoxin infusion blocks segregation of retinogeniculate afferents, *Science*, 242, 87–89, 1988.
118. Stryker, M.P. and Strickland, S.L., Physiological segregation of ocular dominance columns depends on the pattern of afferent electrical activity, *Invest. Ophthalmol. Vis. Sci.,* 25, 278, 1984.
119. Brickley, S.G., Dawes, E.A., Keating, M.J., and Grant, S., Synchronizing retinal activity in both eyes disrupts binocular map development in the optic tectum, *J. Neurosci.,* 18, 1491–1504, 1998.
120. Godement, P., Saluan, J., and Metin, C., Fate of uncrossed retinal projections following early or late prenatal monocular enucleation in the mouse, *J. Comp. Neurol.,* 255, 97–109, 1987.
121. Reh, T.A. and Constantine-Paton, M., Eye-specific segregation requires neural activity in three-eyed *Rana pipiens, J. Neurosci.,* 5, 1132–1143, 1985.
122. Zhang, L.I., Tao, H.W. Holt, C.E., Harris, W.A., and Poo, M., A critical window for cooperation and competition among developing retinotectal synapses, *Nature,* 395, 37–44, 1998.
123. Malinow, R. and Malenka, R.C., AMPA receptor trafficking and synaptic plasticity, *Annu. Rev. Neurosci.,* 25, 103–126, 2002.

10 On the Role of Subcortical Feedback Mechanisms in the Control of Head-Unrestrained Gaze Saccades

Daniel Guitton, André Bergeron, and Woo Young Choi

CONTENTS

0-8493-0097-5/04/$0.00+$1.50

10.1 INTRODUCTION

A central problem in motor control is to understand how the brain coordinates the trajectories of simultaneously moving and coupled body segments, so as to achieve a common goal. "Both the difficulty and fascination of this problem lie in the apparent conflict between two fundamental properties of the motor system: the ability to accomplish high-level goals reliably and repeatedly, versus variability on the level of movement details."[1] We focus here primarily on a striking example of multi-segment motor control, namely the use of coordinated eye and head movements to displace rapidly the visual axis (called a gaze shift). The important constraints on end-point accuracy imposed by the need to foveate a target within the context of potentially highly variable eye and head trajectories[2] exemplify the challenge to the neural gaze controller. Particularly striking is that perturbations of a gaze trajectory are compensated such that movement performance is maintained. While research on the limb and eye/gaze motor systems has evolved largely independently, current theoretical models in both fields have emphasized feedback control to address coordination and accuracy. In this chapter, we consider the feedback controller approach for eye-head gaze saccades made by head-unrestrained subjects.

In discussing gaze feedback it is important to distinguish between the online control of a gaze trajectory versus the calculation, during periods of fixation, of where gaze is relative to the desired goal. For example if, after the brief presentation of a target, a nontarget directed gaze shift occurs, the system can combine initial retinal error and new gaze position to generate in the dark an accurate gaze saccade.[3,4] In a similar experiment in the head-fixed monkey[5] the locus of activity on the motor map of the superior colliculus (SC), initially determined by retinal error, is updated to encode an accurate target-directed eye saccade.[6]

Here we focus on the online control of gaze saccades. To introduce this subject matter, we focus first on the control of eye saccades made head-fixed.

10.2 RECENT STUDIES RELATING TO FEEDBACK CONTROL OF SACCADIC EYE MOVEMENTS MADE HEAD-FIXED

The hypothesis that saccades are controlled by a nonvisual feedback loop was originally proposed by Vossius[7] who, realizing that vision was too slow to accurately control saccades, used eye muscle proprioception in the feedback pathway. Indeed the delays in the visual pathway are of the same order of magnitude as saccade duration, implying that a visually controlled saccade would move past a target before a visual signal could arrive in the brainstem saccade generator in time to turn it off. Contemporary thinking and models using feedback are largely due to the work of Robinson[8] and Zee et al.,[9] who studied the abnormally slow saccades of patients with spinocerebellar degeneration using double target jumps. The latter authors showed that the saccade to the first target could be turned around in mid-flight so as to reach the second target via a new trajectory. They proposed that saccades are controlled by a *local feedback loop*; a brainstem circuit using efference copy that drives the brainstem saccade burst generator with an *eye motor error* signal. In the next section we will review recent evidence for and against feedback control of eye saccades. An extensive review of this subject is in Moschovakis et al.,[10] (see also Reference 11).

10.2.1 Compensations for Natural Trajectory Variability

A given target location relative to the fovea defines the amplitude and direction of a saccade, but once the saccade is triggered we have no voluntary control on either the temporal or spatial properties of its trajectory. Indeed, saccadic eye movements do not have the richness of limb movements and as such may be considered highly stereotyped. Note by comparison that the speed of a gaze saccade, made head-unrestrained, can be voluntarily controlled.[2] However, natural saccades of a given amplitude and direction do have some trajectory variability and the question arises as to how their accuracy is maintained. There are numerous published descriptions of how the saccadic system compensates for natural deviations in ocular speed and position during a saccade (reviewed in References 11 and 12). The latest study by Quaia et al.,[12] investigated this in monkey with a view to reconciling discrepancies in previous studies. They found that the monkey saccadic system compensates, only partially, for deviations in velocity and position; e.g., 61% of the variability in speed is compensated for. They argue that the saccade signal is adjusted dynamically during saccades, but because compensation is not complete they reject the ideal closed loop position feedback controller.

10.2.2 Compensation for Trajectory Perturbations Induced by Brain Stimulation

Stimulating the omnipause neuron (OPN) region interrupts a saccade in-flight, because OPNs inhibit the saccade generator. When stimulation ceases the saccade resumes and the eye movement sequence appears as two distinct saccades. Using this technique Keller et al.,[13] interrupted saccades made in the dark to remembered targets, an approach that eliminates the possibility that visual feedback influences the resumed saccade. They stopped saccades for up to 100 ms, longer than their normal duration. They found that final endpoint accuracy following the resumed saccade was slightly hypermetric relative to unperturbed control saccades; the mean for three monkeys was +0.5°. Interestingly, this compensation is better than that described by Quaia et al.,[12] for natural trajectory variations. Recordings in the SC during this paradigm showed that OPN stimulation interrupted collicular saccade-related burst activity at the site encoding the saccade.[14] After OPN stimulation stopped, activity resumed at the same site, shortly before the saccade resumed. Note that activity did not appear at the collicular site that would normally encode saccades with metrics similar to that of the resumed saccade; i.e., there was no remapping. Saccades can also be interrupted in-flight by stimulating the *fixation* zone of the rostral SC. Munoz et al.,[15] showed for memory-guided saccades, made in the dark, that accuracy after interruption was not statistically different from control. Recording from collicular *burst* neurons showed results similar to Keller and Edelman.[14] However, discharges of SC *buildup* neurons were different in that they were only momentarily interrupted and resumed firing tonically until the resumed saccade was generated (see also Reference 16).

10.2.3 Compensation for Blink-Induced Trajectory Perturbations

The *interruption* paradigms discussed above have the disadvantage that they utilize electrical stimulation of the brainstem saccade circuits, an approach that directly interferes with these circuits in ways that are unknown. For example, OPN stimulation may antidromically activate collicular efferent fibers that project to OPNs[17] and this in turn may unnaturally modify the activity pattern on the SC map. In contrast, blinks present a noninvasive and controllable method for perturbing saccades. Goossens and Van Opstal[18,19] used air puffs to the eye to generate blinks during saccades made by head-fixed monkeys in the dark to previously flashed targets. The blinks induced complex trajectory deviations from the control trajectory: e.g., for 15 to 25° saccades, the trajectory could be deviated orthogonally to the control direction, by as much as 15°, peak velocity was considerably reduced and saccade duration increased two to threefold. Such important perturbations were largely compensated for. For horizontal target eccentricities between 20° and 60°, compensation for amplitude was hypometric and within 5° of target. The compensation for orthogonal deviations from the target's direction depended on direction but the orthogonal final-point errors were uniformly small and on average less than about 3°.

Thus, in spite of trajectory perturbations that were much more profound than the natural variations in trajectory described in Quaia et al.,[12] the compensations — although not explicitly given in Goossens and Van Opstal[18] — seem better than the 61% calculated by the former authors. Although such excellent compensations for large trajectory perturbations are suggestive of feedback control, an alternative explanation illustrates the complexity of the problem and the danger of inferring neural mechanisms based on behavioral observations alone, without adequate knowledge of the underlying neural mechanisms. Indeed, Goossens and Van Opstal[19] argued that blink-related signals themselves modified the firing frequency profiles of SC output cells. They proposed (see also Reference 15) that the SC internal circuitry emits a site-specific constant number of spikes without receiving feedback from the saccade burst generator about the ongoing saccade trajectory.

10.2.4 Compensation for Slow Saccades due to OPN Area Inactivation

Lesions of neurons in the nucleus raphe interpositus (nRIP), where OPNs are located, result in slower saccades.[20] The reason eliminating part of the OPN population slows saccades is unclear. The slow saccades that follow reversible OPN lesions are nevertheless accurate when directed to continuously visible targets.[21] The properties of these abnormal saccades and their accuracy when made in the dark to remembered targets were not tested. The latter authors also found, in this paradigm, that collicular saccade-related burst neurons discharged at lower frequency and for a longer time than normal. Like Goossens and Van Opstal,[19] they proposed that a site in the SC emits a fixed number of spikes appropriate for the amplitude it encodes. However, in contrast to Goossens and Van Opstal[19] they proposed that this is implemented via a positive feedback loop between the central mesencephalic reticular formation (cMRF) and the SC, gated by OPNs.

10.3 EVIDENCE FOR FEEDBACK CONTROL OF SACCADIC GAZE MOVEMENTS MADE HEAD-UNRESTRAINED

We consider now the control of coordinated eye-head orienting movements, called saccadic gaze shifts, that rapidly displace the visual axis through space. In our discussion of this subject we define gaze (G) = eye-in-space = eye-in-head (E/H) + head-in-space (H/S). It has been proposed that the feedback concepts developed for the saccadic eye movement system, head-fixed, can be extended to gaze control when the head is unrestrained,[2,22–28] reviewed in Reference 29. In subsequent sections, we discuss this gaze feedback hypothesis.

10.3.1 Compensation for Natural Trajectory Variations

Gaze shifts of a given vector made by Rhesus monkeys can have considerable variability in their average velocity.[30] Furthermore, in contrast to eye saccades made head-fixed, the trajectory of gaze saccades can be easily controlled voluntarily. Figure 10.1 shows different eye, head, and gaze trajectories that result when a human subject in the light makes three very large gaze shifts of the same amplitude (120°) but with different head velocities.[2] Initially, gaze was directed to a fixation point 60° to the subject's left, and when it was extinguished the subject oriented to the remembered location of a peripheral target at 60° to the right. The traces have been aligned on the onset of the gaze shifts. Trajectories that correspond to the same

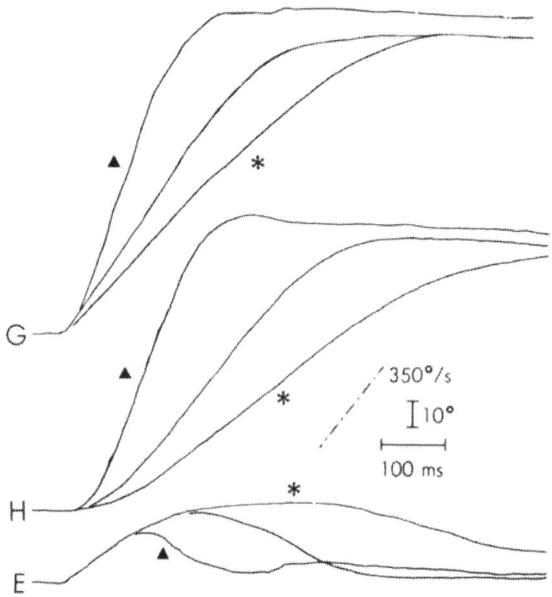

FIGURE 10.1 Large gaze shifts of same amplitude in a human subject instructed to move head at different velocities. Associated eye (E) head (H) and Gaze (G) traces are marked by the same symbol. Initial eye trajectories are the same irrespective of head trajectory, indicating absence of vestibulo-ocular reflex. (Adapted from Guitton, D. et al., On the feedback control of orienting gaze shifts made with eye and head movements, *Prog. Brain Res.*, 142, 55, 2003. With permission.)

movement are identified by the same symbol. Consider first the gaze trace resulting from the fastest head movement, marked by the black triangle. The early part of the gaze shift is driven by an ocular saccade. As the gaze shift progresses, the eye soon begins to slow down and turn around in the orbit, such that for the final one-third of the gaze trajectory the eye is moving in a direction opposite to head motion. The ocular counter-rotation, whose velocity is too low to counter head motion, may be the result of a low gain VOR and/or the summing of oppositely directed signals that drive respectively an eye-saccade and the VOR. The precise mechanisms underlying this are not the subject of this chapter; here we are concerned with how the controller assures gaze accuracy. Consider now the slowest head movement. In this case, the eye saccade reaches a position saturation in the orbit where the eye remains immobile until gaze is in the final one-third of its trajectory, after which the eye begins counter-rolling. The saturation position (at about 35°) is not at the mechanical limits (about 52° in this subject) of eye motion.[2] In all three examples gaze gets on target within a few degrees, irrespective of the very different head and eye trajectories. In particular, for the slow head movements the eye hovers, immobile in the orbit, until gaze approaches the target, at which point the counter rotation begins. Gaze accuracy in these movements suggests the action of a feedback loop. However, there is the possibility that vision may act during the gaze shift itself (for review see Reference 31).

10.3.2 Compensation for Mechanical Perturbations of Head/Gaze Motion

The ability to maintain accuracy despite natural trajectory variations suggests, but does not prove feedback. A more rigorous test is to perturb, in-flight and unexpectedly, gaze trajectories made in the dark and verify that the system compensates for the perturbations such that gaze accuracy remains unaltered relative to control. Such experiments have been done in cat, monkey and human subjects and will be reviewed in this section.

A simple method of perturbing a gaze trajectory is to mechanically interfere with head motion. This occurs naturally in real life, for example in human sporting activities, or as animals bump on obstacles, or carry objects/ food in their mouth as they orient their head to new positions. The effect and compensation for strong perturbations of *monkey* gaze shifts[32–35] is shown in Figure 10.2. In this laboratory task the monkey looked at a central fixation point (FP) during which time a peripheral target (T) was briefly flashed (Figure 10.2A). The extinction of the FP signaled to the head-unrestrained monkey to look, in the dark, at the remembered location of T. The typical pattern of normal eye-head coordination and the resulting gaze shift to a target at 50°, horizontally from FP, is shown in Figure 10.2A. Gaze shift onset and the rapid initial gaze movement were determined by the eye saccade. At the end of the latter, final eye position saturated in the orbit, and the head carried gaze. When gaze in the dark arrived at the intended position, the eye immediately counter-rotated to fully oppose head motion, thereby stabilizing the visual axis in space.

Figure 10.2B shows an interrupted gaze shift, obtained by briefly halting head motion. This was done by using a friction clutch, which immobilized a vertical shaft to which the monkey's head was attached via multiple universal joints. The brake was applied early during the head movement and prevented significant early head motion. Consequently, the initial gaze saccade was due essentially to the eye saccade. The latter reached its terminal position in the orbit, well under the mechanical limits of eye motion, where it remained immobile until head release. At that moment, the eye counter-rotated in the orbit to compensate for ongoing head rotation thereby stabilizing gaze in space, a phenomenon called here a gaze plateau. During the plateau the oculomotor system was out of the saccade mode and into the slow-phase mode; i.e., the VOR was on at unity gain, burst neurons were off and OPNs on.[36–38] In all trials the duration of the gaze plateau was longer than the brake duration because, due to the VOR phase, the onset of corrective eye saccades occurred at variable times after the head-brake release.

The sum of the second eye saccade plus head motion generated a second gaze saccade that compensated for the perturbation and brought final gaze position close to the target zone. This is shown in Figure 10.2C for the same animal. In the control, unperturbed condition, this monkey, in total darkness, overshot and undershot remembered targets at <40° and >40°, respectively. The amplitudes of the second

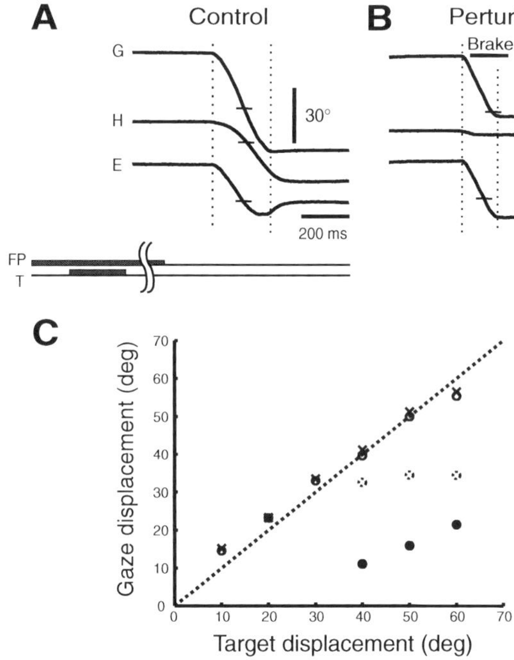

FIGURE 10.2 Examples of control and perturbed gaze shifts in primate, and comparison of accuracies. (A) The trained monkey looks at fixation point (FP) while target (T) flashes for 200 ms in peripheral vision. FP stays on for random time. When FP off, the monkey makes gaze shift (G), in the dark, to the remembered location of the previously flashed target. Typical classic gaze shift consists of a head movement relative to space (H) and an eye movement relative to the head (E). $G = E + H$. Horizontal tic marks on G and H traces show the *zero* central position in front of monkey. Tic mark on E trace shows central eye position in orbit. (B) Example of a gaze trajectory perturbed by momentarily stopping the head movement. Shortly after gaze shift onset, brake stops head for about 150 ms, but eye continues to move and makes a 30° saccade. End of saccade marks beginning of gaze plateau. Gaze remains immobile, in spite of brake release and subsequent head acceleration, because counter-rotation of eye compensates for head motion. During head motion a new eye saccade is triggered that together with the head movement generates a corrective gaze saccade that brings gaze near the remembered target location. (C) Corrective gaze saccades bring gaze onto remembered target location with an accuracy similar to control movements. Cross, control gaze shift; open circle, perturbed gaze shift; dotted circle, first gaze step in perturbed gaze shift; filled circle, corrective gaze shift.

gaze saccades were sufficient to compensate for the gaze perturbations by bringing gaze to positions indistinguishable from control values. Notice that in the perturbed condition, the first gaze saccade undershot the target, even when the target was within the oculomotor range (e.g., at 40°).

Similar experimental results have been reported previously in cat, monkey and humans. These will be briefly reviewed here. Laurutis and Robinson[25] perturbed human head motion very briefly during large gaze saccades of about 180°, made between two targets in a dimly lit room. The perturbation did not halt head motion nor cause a plateau in gaze position; the gaze velocity was, however, slowed by about 180°/s. In the perturbed gaze trajectories, the duration of the overall gaze shift was longer and final gaze position compensated for the perturbation: control gaze shifts were within about 3° of the target compared to 4° for the perturbed trials, a non significant difference.

Guitton and Volle[2] studied gaze accuracy in two human subjects that made active eye-head gaze shifts to targets briefly flashed *in the dark*. In the control condition, gaze undershot targets, located 30 to 50° from the fovea, by about 4° and overshot targets in the 60-80° range by about 2°, on average for both subjects. By comparison, when the head movement was mechanically braked — sometimes halting gaze motion and creating gaze plateaus as described above, sometimes slowing it as in Laurutis and Robinson[25] — subjects undershot targets in the 30 to 50° range on average by only 0.5°, and undershot targets in the 60 to 80° range by 1°. Clearly, for each subject, gaze accuracy *in the dark* was not significantly dependent on whether or not the movements were perturbed.

Tomlinson[28] perturbed, in the monkey, 65° gaze saccades to visible targets, with a torque motor that could briefly either slow down or speed up head movements. Perturbations of the gaze trajectory amounted to about ±10% of the intended amplitude, and were compensated-for within about 1°.

In similar experiments to those in monkey, Matsuo et al.,[39,40] (see also Reference 23) have found, in the cat, that gaze shifts in the dark to the remembered location of previously flashed targets at an eccentricity of 50° showed excellent compensation for gaze trajectory perturbations that halted gaze trajectories in mid-flight for periods ranging from 50 to 400 ms. The gaze shift amplitude was 43.2 ± 2.4° for control and 43.0 ± 2.6° for perturbed gaze shifts.

Without using feedback concepts, it is difficult to account for the excellent compensation of gaze trajectory perturbations, particularly those imposed by mechanically halting gaze motion (Figure 10.2). This is because the corrective eye saccade does not simply start off where the last saccade ended, nor does the head movement that accompanies it have a trajectory similar to the head movement that accompanies the equivalent final portion of the control trajectory. Indeed, the corrective eye saccade begins at locations that depend on the amplitude of the eye counter-rotation, a *noisy* quantity that depends on the head velocity that follows brake release as well as other factors unknown to us. We have hypothesized elsewhere[2,23,32,33,41–43] that the second eye saccade may be driven by brainstem circuits that produce the quick-phases of vestibular nystagmus.

10.3.3 Compensation for Gaze Shifts Perturbed by Electrical Stimulation of Superior Colliculus

Additional support for the hypothesis that gaze saccades in the head-unrestrained cat are controlled on-line by feedback was provided by the experiments of Pélisson et al.,[44] who perturbed goal-directed gaze shifts *in-flight* by briefly stimulating the SC. The movements were not halted. Following stimulation offset, the transient perturbations were compensated for, *on-line*, before gaze shifts terminated.

10.3.4 Neural Discharge Properties in Relation to Gaze Feedback Control: Role of Cat Superior Colliculus

It has been known for some time that the cat's SC encodes *gaze* displacements (e.g., References 27 and 45 to 47). For the cat, we have proposed that a gaze position error (GPE) signal is continuously encoded on the motor map of the SC. Specifically, we have proposed[36,41–43,45,46,48] that: (1) Activity begins in the SC at the site appropriate for encoding the desired gaze saccade vector. (2) The locus of an ensemble of discharging neurons moves on the topographically organized collicular map and instructs brainstem eye and head circuits as to the current GPE (the *moving hill* hypothesis). (3) When the activity invades the rostral SC, collicular *fixation* neurons are engaged, which activate omnipause neurons that in turn suppress gaze saccades. We will now briefly review some key observations that support the moving hill hypothesis in the head-unrestrained *cat*. Related observations in the head unrestrained monkey will be considered in a subsequent section.

We have recorded in the cat from antidromically identified tecto-reticular neurons and nonidentified neurons with identical properties in both the SC fixation zone and motor map. SC fixation neurons (SCFNs; reviewed in References 41 to 43) are located in the rostral pole of the SC. They discharge tonically during steady fixation of a behaviorally significant target, irrespective of whether it is visible or not. Figure 10.3A shows SCFN activity during a large single-step gaze shift made by a cat in a direction contralateral to the recording site. Note that this gaze shift is qualitatively similar to that in humans (Figure 10.1) and monkeys (Figure 10.2). Before the gaze shift, the cat faced an opaque barrier (Figure 10.6A) on which there was no target of significance and the SCFN had a characteristically low tonic discharge, lower than during attentive fixation in the dark. A food target was shown at the edge opposite from the cat's initial gaze position and ambient lighting quickly turned off. During the illustrated contralateral gaze shift in the dark (Figure 10.3A), the fixation cell was silent until near the end of the gaze trajectory. We focus now on the processes that determine the onset of SCFN firing and how and why it subsequently intensifies. Much insight into this phenomenon is provided by studying multiple-step gaze shifts in the dark, a typical example being shown in Figure 10.3B. In this example, the target at about 50° from initial gaze position, was attained using a sequence of 3 gaze saccades separated by short periods (gaze position plateaus) during which gaze was either immobile or moving at velocities much slower than those of saccades. An important feature of multiple-step gaze shifts is that, for a given overall amplitude, they vary considerably in terms of the number of gaze saccades, their amplitude and the duration of the gaze plateaus.

Figure 10.4A shows another multiple-step gaze shift, now plotted as GPE versus time. The associated SCFN discharge is shown below the gaze trace and clearly increases as GPE $\rightarrow 0°$. This is shown quantitatively in Figure 10.4B where each numbered point corresponds to a plateau in Figure 10.4A. An analysis of many multiple-step gaze shifts, while recording from the same cell, produced the group of points in Figure 10.4B through which it was possible to fit a Gaussian function.[42] As expected from Figure 10.4A, after the first spike in activity the firing frequency increased as GPE $\rightarrow 0°$ and the Gaussian peaked at GPE $= 0°$. We showed in Bergeron

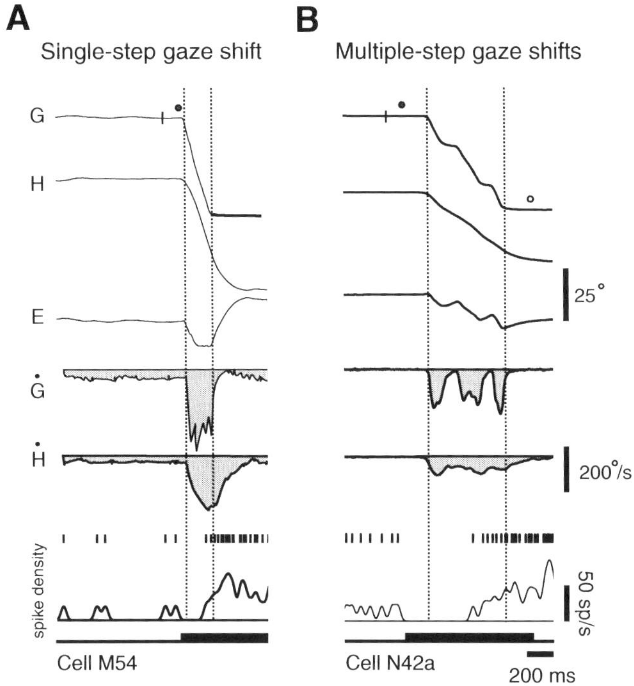

FIGURE 10.3 Discharge of a superior colliculus fixation neuron (SCFN). (A) A 45° single-step gaze shift and associated discharge of an example SCFN, cell M54 recorded in right SC. Associated action potentials shown as tick-marks below eye position trace. Vertical mark on G trace indicates when the target was shown to the cat. Left vertical dotted line marks onset of gaze shift. Right vertical dotted line marks when gaze, in the dark, arrives at the remembered target location. (B) A 45° multiple-step gaze shift and associated discharges of another SCFN, cell N42a in right SC. Dark circle indicates when ambient lighting was extinguished; open circle when it was re-illuminated. Cell pauses during first saccade, ensuing plateau and part of second saccade. Discharge peaks at gaze shift end. Light onset generates burst-discharge. $\dot{G}$, $\dot{H}$: gaze and head velocities. In all traces, down means leftward movement. Line of vertical tic marks shows action potentials. Below this trace is the instantaneous firing frequency in spikes/s. Lowest trace shows when cat in total darkness. (Adapted from Guitton, D. et al., On the feedback control of orienting gaze shifts made with eye and head movements, *Prog. Brain Res.,* 142, 55, 2003. With permission.)

and Guitton[41–42] that the first spike, after the initial pause in activity at the onset of a gaze shift, occurred when the visual axis arrived at a particular distance from the target; i.e., at a particular GPE. The GPE at which, on average, the first spike occurred (for the illustrated cell, open triangle in Figure 10.4B) was called the cell's *preferred* GPE and its value varied from cell to cell. The average for all cells was 13°.

All SCFNs that we recorded (Figure 10.5A) had a frequency of discharge that increased as GPE → 0° (Figure 10.5A), similar to that shown in Figure 10.4B. All Gaussian fits peaked at GPE ≈ 0° (we could not distinguish small offsets from 0°), and the width of the Gaussian curve varied from cell to cell. These results resemble

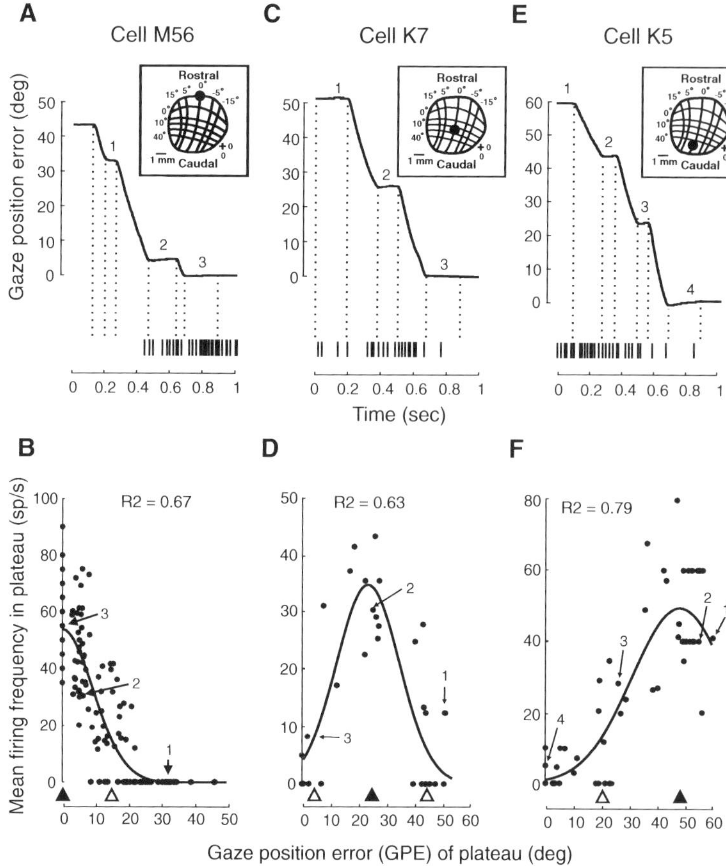

FIGURE 10.4 Firing frequency during gaze plateaus depends on gaze position error (GPE). (A) Example of a multiple-step gaze shift. Associated action potentials of a superior colliculus fixation neuron (cell M56) shown by line of short vertical lines below the gaze trace. Gaze is plotted in relation to time; when GPE = 0°, gaze is on target. Inset shows where cell was recorded on collicular map. (B) For the same cell as in A, the firing frequency during plateaus is plotted against GPE. Mean firing frequency during gaze plateaus is zero at large GPE and increases as GPE → 0°. Points numbered 1 to 3 correspond to firing frequency during equivalently numbered plateaus in A. Other points are from plateaus in other multiple-step gaze traces. Solid line through points is a Gaussian fit giving a variance-accounted-for (R2) of 0.67. Open triangle at bottom indicates when discharge rises significantly above zero. (C and E). Same as A but for cells recorded in middle and caudal SC respectively. (D and F). Firing frequency versus GPE profiles for the cells in C and E respectively. See panel B for details. Dark triangle on abscissa indicates location of peak activity.

those reported by Krauzlis et al.,[49,50] in the head-fixed monkey, but in their analysis they focused on the peak discharge and were able to determine that it occurred at values of GPE close to but not equal to 0°.

Figures 10.4C and 10.4E show the discharge of two different neurons, cells K7 and K5, located on the representation of the horizontal meridian on the SC motor

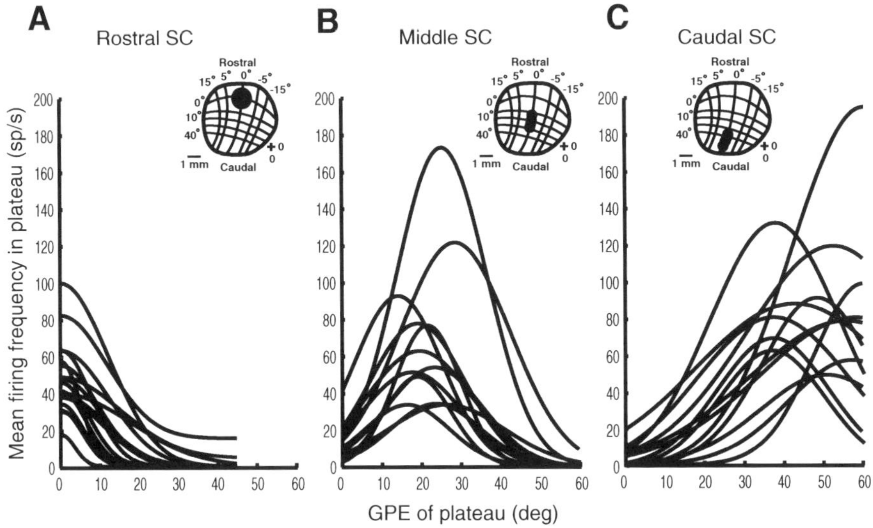

FIGURE 10.5 Dependence of firing frequency on GPE for groups of cells located in different zones on the SC map. The best-fit Gaussian (see Figure 10.4) is shown for each cell in our sample. (A) Fixation neurons (SCFNs) located in the rostral SC. (B and C) Neurons located in the middle and caudal SC respectively. See Reference 43 for details. Insets in each panel show SC motor map and location of relevant cell group. (From Bergeron, A., Matsuo, S., and Guitton, D. In multiple-step gaze shifts one site on the motor map of the superior colliculus encodes distance to target, not saccade amplitude, *Nat. Neurosci.*, 6, 404, 2003. With permission.)

map at points corresponding to 25° and 50° gaze shifts, respectively. As in panels A and B the associated multiple-step gaze shifts and firing frequency versus GPE plots are shown at the top and bottom, respectively. In each case a Gaussian curve was fitted to the points, but note that for each cell the peak discharge occurred at a different GPE. We found that the GPE encoded by a cell's peak discharge corresponded to the value of GPE encoded by the cell's location on the motor map. Figure 10.5B and 10.5C shows the population discharge profiles.

Our recordings from tecto-reticular and tecto-reticular-like neurons, summarized in Figures. 10.4 and 10.5, show that the classic *one-site one-saccade* view of the SC motor map does not hold during multiple-step gaze shifts (see Reference 43 for details). The arguments leading to this conclusion are summarized in Figure 10.6 (see Reference 51). Figure 10.6A illustrates our barrier paradigm, described above. Figure 10.6B shows a dorsal view of the collicular map with the numbered circles identifying recording sites that correspond to the values of GPE at each plateau position in the multiple-step gaze shift shown at the top of Figure 10.6C and repeated in Figure 10.6D. This example 40° multiple-step gaze shift is composed of an initial 20° saccade followed by two consecutive 10° saccades. Figure 10.6C shows the discharge pattern predicted by the one-site one-saccade hypothesis. Fixation cells at location 1 burst tonically and pause for every gaze saccade. The cells at 2 burst prior to and during the two 10° saccades. cells at 3 burst in relation to the 20° saccade. In contrast to the expected pattern, we found something very different (Figure

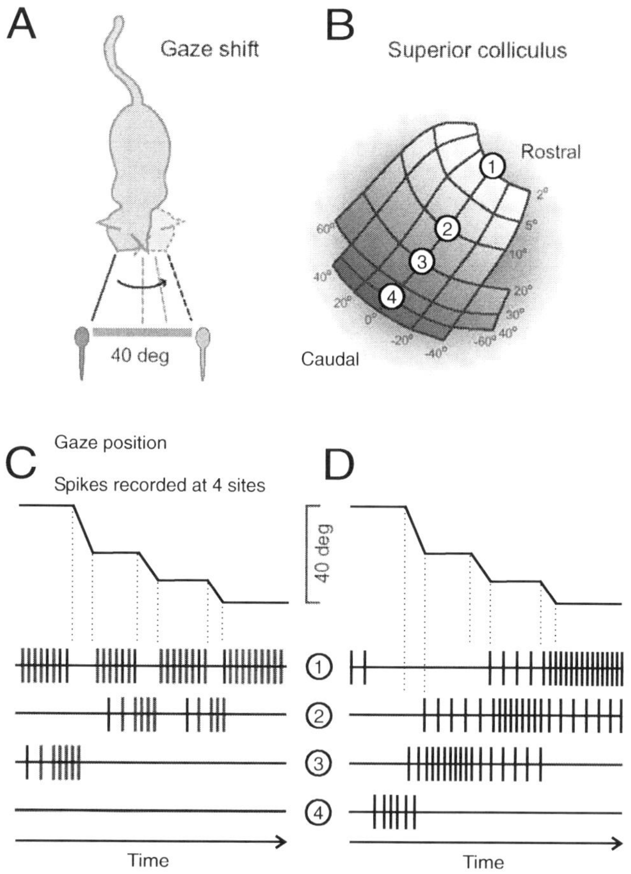

FIGURE 10.6 Schematized comparison of predictions made by the classic *one-site one-saccade* view of the SC motor map and actual recordings made during multiple-step gaze shifts. (A) Experimental, *barrier* paradigm. The head-unrestrained cat faces an opaque screen and a food target appears to either side. Just before the cat orients to the target the ambient lighting is extinguished, such that the gaze shift is made in complete darkness. (B) Dorsal view of SC motor map with iso-direction lines running rostro-caudally and iso-amplitude lines orthogonal. Numbered circles identify recording sites referred to in C and D. (C) Discharges at each identified location on the SC motor map predicted by the one-site one-saccade hypothesis during an example 40° multiple-step gaze shift comprised of three gaze saccades: 20°, 10°, 10°, respectively. (D) Discharges on the map observed by Bergeron et al.[43] encoding distance to goal. (Adapted from Figure 1 in Krauzlis R.J. and Carello C.D. Going for the goal, *Nat. Neurosci.*, 6, 332, 2003. With permission.)

10.6D). Fixation cells paused continuously until gaze arrived within about 10° from the target. Cells at 2, 3, and 4 fired maximally during plateaus at GPE = 10°, 20°, and 40° respectively, and the saccades that follow them.

The observations just reviewed are compatible with the moving hill hypothesis applicable to multiple-step gaze shifts. This is illustrated in Figure 10.7 for a hypothetical 60° multiple-step gaze shift composed of three 20° steps. Here we have used the data in Figure 10.5 to reconstruct the spatial activity on the SC map during this multiple-step gaze shift, at the different values of GPE after each step. In this example, a hill of activity begins at the 60° location and shifts forward on the map during each step, sequentially activating caudo-rostral sites until it reaches the SC fixation zone.

Further support for this mechanism was obtained by recording from SCFNs during gaze shifts that were mechanically halted (as in Figure 10.2B). When gaze shifts were interrupted during the pause in fixation neuron activity, the pause was prolonged beyond the control duration, until the corrective gaze saccade attained the cell's preferred GPE.[39,40] Furthermore, SC neurons at locations appropriate for encoding the remaining GPE, stayed tonically active until the gaze trajectory was allowed to complete its course.[35,39,40] We will summarize below analogous observations in primate collicular neurons.

10.3.5 Neurons in Monkey SC Are in a Gaze Feedback Loop

The primate SC has traditionally been thought of as an oculomotor structure, but it is only recently that it has been shown to encode gaze displacements.[30,52–55] Models of collicular involvement in the control of primate saccadic eye movements made head-fixed are divided into those that place the SC either upstream of, or within, a feedback loop. However, even models that place the SC within a feedback loop are not identical regarding the role of the feedback signal.

Our recent observations in *head-unrestrained* monkey[32–34] support gaze feedback to the primate SC. Indeed, in experiments illustrated in Figure 10.2, when *large* 50° gaze shifts were mechanically perturbed shortly after the onset of head motion (Figure 10.8A), activity in fixation cells remained low until the end of the second saccade that corrected for the perturbation. Thus, the pause in fixation cell activity was prolonged in relation to the longer overall gaze shift duration. This is in line with fixation cells in cat, reviewed above, that come on only when gaze reaches a specific GPE independent of the number of intervening plateaus.

Our recordings from buildup neurons (BUNs) (for review of BUNs see References 56 and 57) show discharge characteristics that complement those of fixation neurons (Figure 10.8B). For BUNs recorded at the 15° location on the motor map, the firing frequency that just preceded a perturbation-induced gaze plateau remained constant during the plateau until the end of the corrective saccade, which in the dark, brought gaze into the target zone.

We still do not know whether this gaze feedback to the monkey's SC implies that, like in cat, this animal's SC is encoding distance to target. For example, to deduce whether the tonic activity during the plateau encodes gaze position error we would need to know more about the distribution of activity on the motor map in this brake paradigm. However, contrary to what was suggested by Soetedjo et al.,[58] we believe it is premature to lay to rest the moving hill hypothesis *for the monkey whose head is unrestrained.* We emphasize that the moving hill model applies to tecto-reticular neurons in the head-unrestrained cat. As proposed by Bergeron et al.,[43] the moving hill in the cat may not actually steer the gaze saccade, but rather

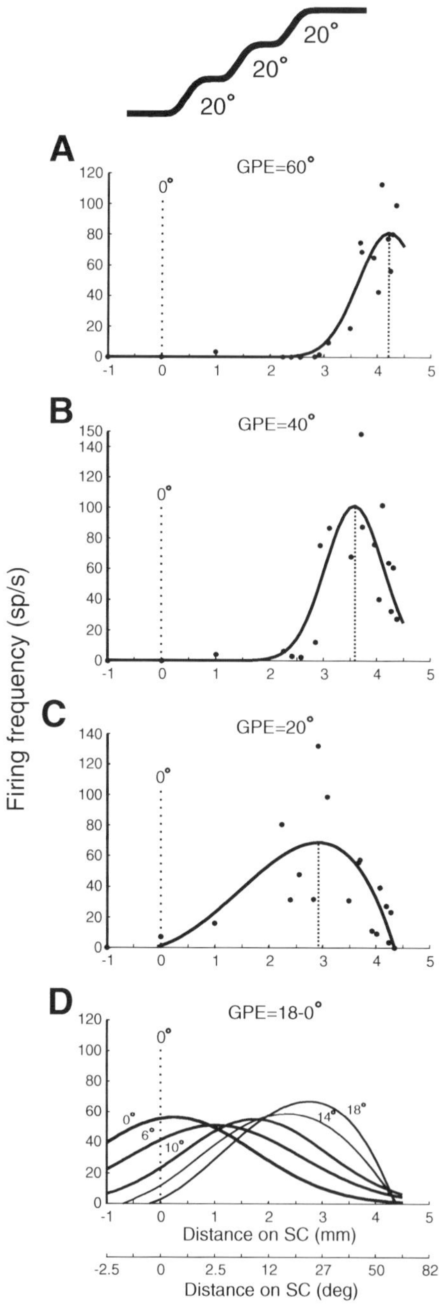
20°
20°
20°
A
GPE=60°
B
GPE=40°
C
GPE=20°
D
GPE=18-0°
0°
6°
10°
14°
18°
Firing frequency (sp/s)
Distance on SC (mm)
Distance on SC (deg)

FIGURE 10.7

represent *distance to the goal*. BUNs could play a similar role in the head-unrestrained monkey, a possibility not yet studied in detail, but in line with the Quaia et al.,[59] model for head-fixed saccade generation.

10.3.6 Neural Discharge Properties in Relation to Gaze Feedback Control: Role of Primate Burst and Omnipause Neurons

Cullen and Guitton[60–62] have analyzed quantitatively the discharge characteristics of inhibitory burst neurons (IBNs) in the head-fixed and head-unrestrained monkey. As a population IBNs carry the signal:

$$B(t) = 280 - 1.2\Delta E + 0.4\dot{E} + 0.2\dot{H}$$

where $B(t)$ is firing frequency as a function of time, ΔE is saccade amplitude, and $\dot{E}$, $\dot{H}$ are eye and head velocity. A similar signal is carried by excitatory burst neurons (EBNs)[38] (Sylvestre and Cullen, private communication). Thus, the burst generator carries a complex signal, part of which is related to gaze velocity. The head velocity term must be removed at the level of the ocular motoneurons.[63] This is most likely done via the residual activity carried by vestibular premotor neurons.[64,65] So too must the bias and eye amplitude terms be removed.

Cullen's group has examined the question of whether burst neurons and OPNs are in the dynamic gaze feedback loop.[38] To do this they transiently slowed monkey head motion during a gaze shift (Figure 10.9). This momentarily slowed the gaze movement and prolonged its duration. They found that the duration of both OPN pauses and EBN bursts increased. Furthermore, the same mathematical equation (preceding paragraph) described the burst discharge in both control and perturbed conditions. This demonstrates that EBNs are continuously updated dynamically during gaze shifts by a signal proportional to head velocity. Thus when head velocity decreased, so did EBN firing frequency, a finding that contradicts the assumption of the Freedman[66] model. These results, together with the discharge properties of SC neurons during perturbed gaze shifts, strongly support the hypothesis that the burst generator is in a gaze feedback loop. This proposal is further supported by

FIGURE 10.7 (See facing page.) Evolution of activity on the SC motor map at different epochs during a hypothetical multiple-step gaze shift of overall amplitude 60° and composed of three 20° saccades (top). Activity profiles expressed as firing frequency at different locations on the map given as distances — in both millimeters and degrees — from the map's zero at the rostral fixation zone. Data at a specific GPE for each cell taken from the curves in Figure 10.5. Each distribution of population activity on the SC map is fit by fourth order quadratic and is shown when gaze is at the following distances from the map's zero: (A) GPE = 60°, gaze shift start; (B) GPE = 40°, first gaze plateau; (C) GPE = 20°, second gaze plateau; and (D) GPE = 18 to 0°, during final gaze saccade. Panels A to C show that as gaze is moving towards the target and GPE is decreasing, the peak of population activity (vertical dotted line) is displaced rostrally on the map towards the fixation zone. See text for details.

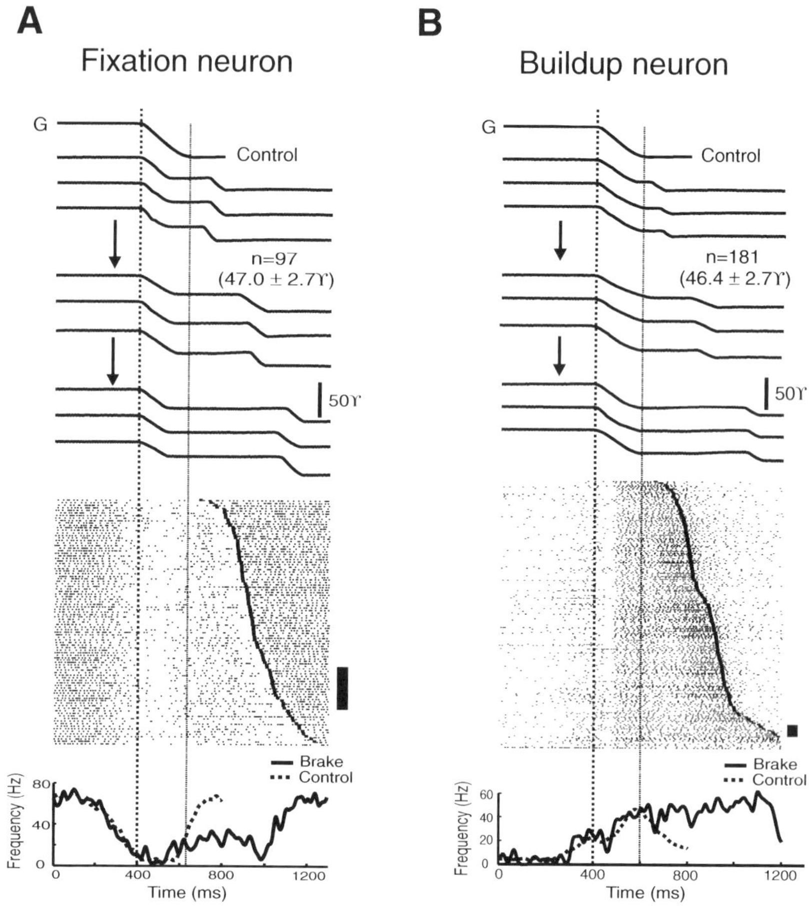

FIGURE 10.8 Discharge characteristics of primate SC neurons during perturbed gaze shifts. (A) Discharge characteristics of fixation neuron. *Top*, selected examples of gaze trajectories, ordered from top to bottom according to increasing gaze shift duration and aligned on gaze shift onset. *Middle*, rasters of action potentials for all trials. Each tick-mark identifies an action potential. Rasters are ordered from top to bottom according to increasing gaze shift duration and aligned on gaze shift onset. Thus, each line of action potentials corresponds to an equivalently place gaze trace above. Dark heavy tick marks on the raster plots indicate gaze shift end. *Bottom*, average firing frequency (spike-density) profiles for selected perturbed trials identified by short vertical bar near the bottom-right of the raster plot. For comparison, the average control gaze trajectory is shown at top and its associated firing frequency profile is shown dotted. (B) Discharge characteristics of buildup neuron. Same conventions and symbols as in A.

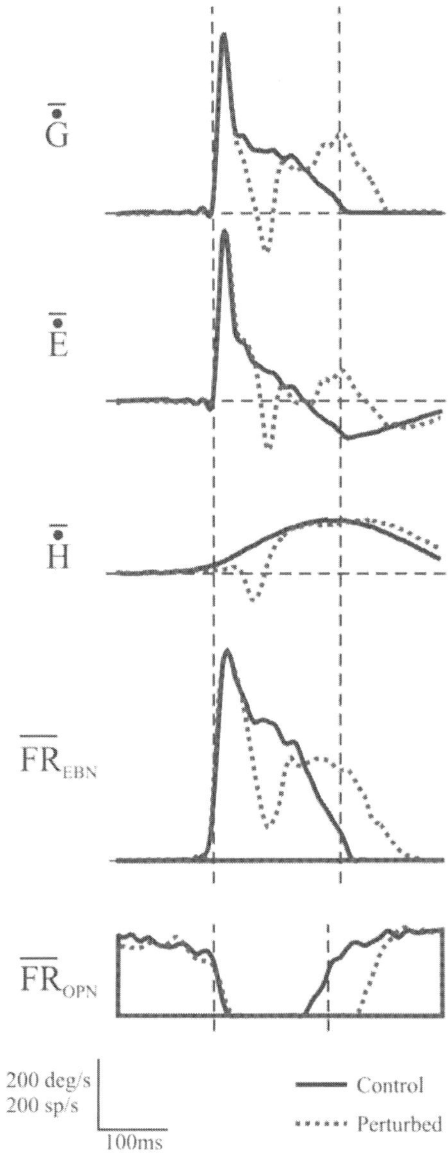

FIGURE 10.9 Gaze-related activity in burst and omnipause neurons revealed by mechanically perturbing head trajectory. Control and perturbed trajectories are shown as solid and dotted lines respectively. Traces, from top to bottom show: gaze velocity, eye velocity, head velocity, firing frequency profile of excitatory burst neuron (EBN), firing frequency profile of omnipause neuron (OPN). The control and perturbed movements and discharge profiles are aligned on start of the gaze shifts (left vertical dashed line). Right vertical dashed line shows end of control gaze trajectory. Perturbation to head trajectory shown as a transient reversal of head velocity that prolongs overall duration of perturbed gaze shift. Note associated prolongation of both the EBN burst activity and OPN pause. (This unpublished figure courtesy of P.A. Sylvestre and K.E. Cullen.)

the observation that OPN stimulation stops gaze shifts in the cat[36] (but see Ghandi and Sparks[67]).

10.3.7 On the Role of Cat Fastigial Nucleus in Gaze Feedback Control

Deficits in gaze shift accuracy due to subcortical lesions in the head-unrestrained cat have been reported to result from inactivation of the fastigial nucleus,[68–72] semicircular canals[73] and mesencephalic reticular formation.[74] We briefly consider the role of the fastigial nucleus in the control of head-unrestrained gaze shifts, which has been extensively studied in the cat by Pélisson and collaborators. In particular, Goffart et al.,[70,71] have studied whether the caudal fastigial nucleus is implicated in compensating for gaze shift perturbations generated by stimulating the SC. Since perturbations were compensated for, deficits caused by caudal fastigial nucleus do not compromise the function of the gaze feedback loop or of the summing junction, at least in the cat. This observation generates a problem regarding the extension, to the head-unrestrained condition, of the Quaia et al.,[59] model of eye saccade control wherein it is proposed that the cerebellar fastigial nucleus is within the feedback loop that generates the eye motor-error signal of the saccade burst generator.

10.4 FEEDBACK MODELS OF THE SUBCORTICAL GAZE-SACCADE GENERATOR

The experiments described above have shown that both the saccadic eye movement system (head-fixed) and the gaze control system (head-unrestrained) compensate with great efficacy for either natural or experimentally-induced trajectory modifications. The discharge properties of collicular, burst and pause neurons during perturbed gaze shifts suggest that gaze saccades are controlled by a feedback process, rather than by preprogrammed neural signals that are computed prior to saccade generation. In this section we will present an overview of some models that rely on feedback control. Many concepts contained in gaze control models originate in models of the head-fixed eye saccade system. Therefore, we first briefly consider some basic concepts in this system.

10.4.1 Head-Fixed

Most models of the saccade burst generator assume that the firing frequency of burst neurons encodes eye velocity. This is not strictly true[60–62] but is a useful simplification. Burst duration equals saccade duration. Accordingly, the neural code used by burst neurons is a temporal one, firing frequency and duration, as is that of motoneurons. By comparison, in the classic view the SC encodes saccades in a place code; a specific location on the map specifying the desired saccade vector. An important challenge to modelers is to *attach* the SC to the burst generator thereby accomplishing the enigmatic spatiotemporal transformation. A wealth of model configurations has been proposed to do this (reviewed in References 10 and 11). In all models the initially active site on the SC motor map encodes the desired saccade vector and amplitude. Models differ as to how this signal is treated. At the risk of excessive oversimplification, we place the model structures into three broad categories: (1) The SC receives no feedback on the actual accomplished saccade trajectory. A specific site on the motor map instructs a downstream local feedback loop with the desired saccade amplitude via either a site-specific weighted connection (e.g., see References 75 and 76) or a site specific fixed number of output spikes.[19] (2) The SC receives feedback on the ongoing saccade trajectory in order to either: (a) control the output spike count[21,58]; (b) generate instantaneous *motor error*[45,46,57,77–82]; or (c) encode current eye displacement, which is sent to a downstream feedback loop that produces motor-error.[83] (3) Finally there are models implicating both the SC and the cerebellum[84] some using feedback loops through both structures[59,85] or the cerebellum alone.[86,87]

10.4.2 Head Unrestrained

There are far fewer models of gaze control. Virtually all utilize gaze feedback — although in different model structures — to create a gaze error signal (that we also call gaze position error, GPE) that assures gaze accuracy in the face of trajectory variations.[2,24–26,28,45,46,79,88–90] In the open loop model of Freedman,[66] no gaze error signal is created. The initial desired gaze displacement signal is decomposed into separate desired eye and head displacement commands that become the inputs to downstream eye and head local feedback loops. The principal aim of the Freedman model is to replicate the experimentally observed complex eye velocity profiles in primate by assuming that head velocity subtracts from the eye burst generator output, a hypothesis not supported by the results (Figure 10.9) of Sylvestre et al.[38] Here we will consider only models that utilize gaze feedback because an open-loop model does not predict the responses of OPNs, BNs, and SC neurons to gaze trajectory perturbations (reviewed in previous sections). Neither can an open loop model predict the compensations for full gaze trajectory interruptions (Figure 10.2). In what follows we review two recent models that illustrate different conceptual approaches to modeling gaze control.

10.4.3 The Model of Goossens and Van Opstal (Figure 10.10)

Goossens and Van Opstal[89] investigated head-unrestrained gaze saccades made to continuously on visual or auditory targets presented in the two-dimensional frontal plane. They compared gaze shifts generated to a specific target when the eyes and head were initially either aligned straight ahead, or not aligned. In the latter condition they found that the head and eyes could move in different directions with different spatial trajectory profiles (see also References 91 and 92) but nevertheless, that the direction of initial eye motion could influence initial head motion. They argue that such observations are incompatible with the eyes and head being driven by a common gaze position error signal; but see Galiana model (below). Furthermore, they found that head movements were best described in craniocentric coordinates as if they were driven by a dynamic head motor error signal, as proposed in Vollé and Guitton[93]

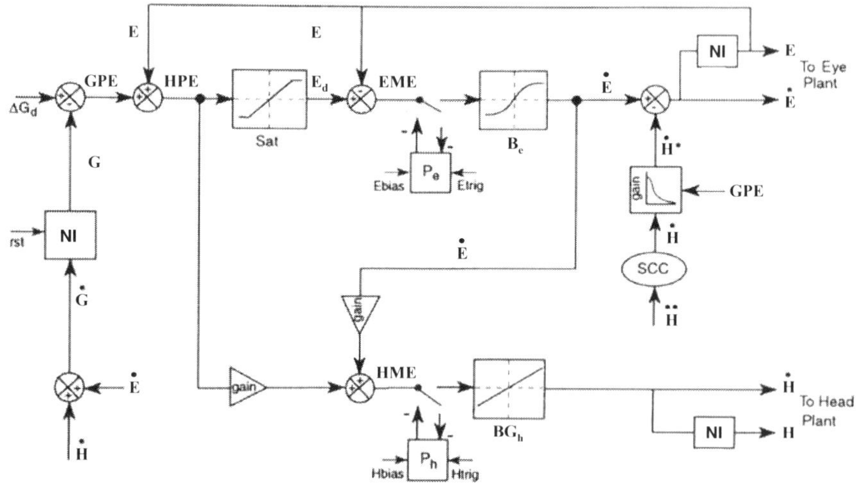

FIGURE 10.10 The two-dimensional gaze control model proposed by Goossens and Van Opstal.[89] To yield a dynamic vector gaze position error signal (GPE), the desired gaze displacement (fG_D) is compared with the current gaze displacement (fG). The latter is obtained by integration (NI) of gaze velocity $(\dot{G})(=\dot{E}+\dot{H})$ by a resettable integrator, which is reset to zero (rst) after each saccade. $\dot{H}$ is obtained from the semicircular canals. GPE is first converted into a dynamic head motor-error signal (HPE). A saturated version (E_D) of the latter signal drives the oculomotor system in a manner similar to Robinson's[8] local feedback model. The ocular burst generator (B_e) is driven until the dynamic eye motor-error signal (EME) is zero. Note that E_D is a dynamic signal. A neural estimate of $\dot{H}$ — with a gain that depends on GPE — interacts downstream of *Be* with the $\dot{E}$ signal. Thus, the VOR is partially suppressed during gaze shifts.[2,64] The head pulse generator is driven by both an attenuated version of HPE (where the gains are different for horizontal and vertical head movement components) and a collateral input from the oculomotor system. Partially independent control of the eye and head motor systems is achieved by functionally separated feedback loops and separate gating mechanisms. (From Goossens, H.H. and Van Opstal, A.J. Human eye-head coordination in two dimensions under different sensorimotor conditions, *Exp. Brain Res.*, 114, 542, 1997. With permission.)

(but see Reference 94). In their two-dimensional gaze-feedback model (similar in structure to that of Guitton and Volle[2]), the saturation in eye position frequently seen in large gaze shifts is accounted for by the saturation element, SAT. The head motor error signal (HME) that drives the head is a nonunity gain combination of the head position error signal (HPE) and the ocular burst signal. This latter input accounts for the fact that head movements are influenced by both HPE and eye saccades.[24,54,55] This model structure simulates accurate target acquisition in two dimensions, despite striking differences in the eye and head trajectories. However, although the model is a useful conceptual schema for exploring behavior, it does not account for key neurophysiological observations with regards to the burst generator. Thus, in the head unrestrained cat and primate BNs carry a $\dot{H}$ signal in addition to the classic $\dot{E}$ signal.[38,61] Furthermore, OPNs are not in an eye circuit, as proposed, but rather are gated by gaze duration[36,38] (see Figure 10.9). Note that Phillips et al.[95] argue that, in primate, the termination of the pause in OPN discharge is better linked to the onset of the VOR-induced eye counter-rotation than to gaze shift end. However, these two measures are equivalent in a gaze control system in which GPE is the controlling error signal.

10.4.4 Galiana Model (Figure 10.11)

This model is a successful predictor of many properties of horizontal saccadic gaze shifts made head-unrestrained,[24,79,88] with the added feature that its novel structure is conceptually quite different than that currently popular in oculomotor physiology. *It is important to note that the model uses feedback to the SC but, because of a reticular vestibular quick-phase generator, saccades can stop without this SC feedback.* Thus as proposed by Bergeron et al.,[43] the SC monitors progress towards the goal, but without necessarily controlling the details of a gaze saccade. It follows that both current views of the SC — within and outside the feedback loop — are accommodated by this model structure.

The detailed working model is incorporated as a bilateral system. This type of structure is too complicated to describe here and we show in Figure 10.11 only a collapsed conceptual schema. An earlier version, without the vestibular quick-phase generator, has been described extensively elsewhere.[24,79,88] Important model elements are: (1) The SC provides a gaze position error (GPE) signal to the reticular formation[96] eye and head motor networks. In the illustrated configuration, the SC acts as a resettable integrator[79] of efference copies of the eye velocity ($\dot{E}$ *) and head velocity ($\dot{H}$ *) inputs. For horizontal gaze shifts, this is easily implemented in a one-dimensional sheet of neurons[79]; a two-dimensional sheet can control oblique movements (P. Lefevre, personal communication) and permits a comparison of the desired angular change in gaze position (fG_D) with the actual change that is then sent, via tecto-reticular neurons, to both the eye and head motor machinery. Note that by simple parameter changes in the SC local neural network, the model implements either a moving hill encoding gaze-position-error spatially (discussed above) or a stationary hill that apparently collapses locally and jumps to the SC fixation zone at the end of the gaze saccade.[79] The GPE signal is conveyed directly to neck MNs and assures early slow head motion.[24,54,55] (2) The saturation element (SAT) in the GPE-BN pathway imposes the orbital eye position saturation or eye plateau seen in large gaze shifts (Figure 10.1).[88] This property is not intuitively obvious. (3) Burst neurons (BNs)[60–62,97,98] carry eye and head velocity signals and provide the main signal driving eye motoneurons (MNs) during saccades and contribute to driving neck MNs. Eye-head coupling[54,55,99,100] is assured by the combined sum of GPE and BN drives onto both eye and head motor systems. The synaptic weights of these common drives need not be equal. In a two-dimensional extension of this model, such different projections onto the head system would provide decoupled eye and head trajectories. (4) Inhibitory OPNs gate BNs[98] and both neuron types carry gaze-related signals (see Figure 10.9 and related text). OPNs are gated by combined signals from SC fixation cells,[101] inhibitory long-lead burst neurons[76] (iLLBN), $\dot{H}$[102] and E*,[36,103,104] the latter two for quick-phase generation. Such signals on OPNs may explain the decoupling between OPNs and collicular fixation cells (FIX) during

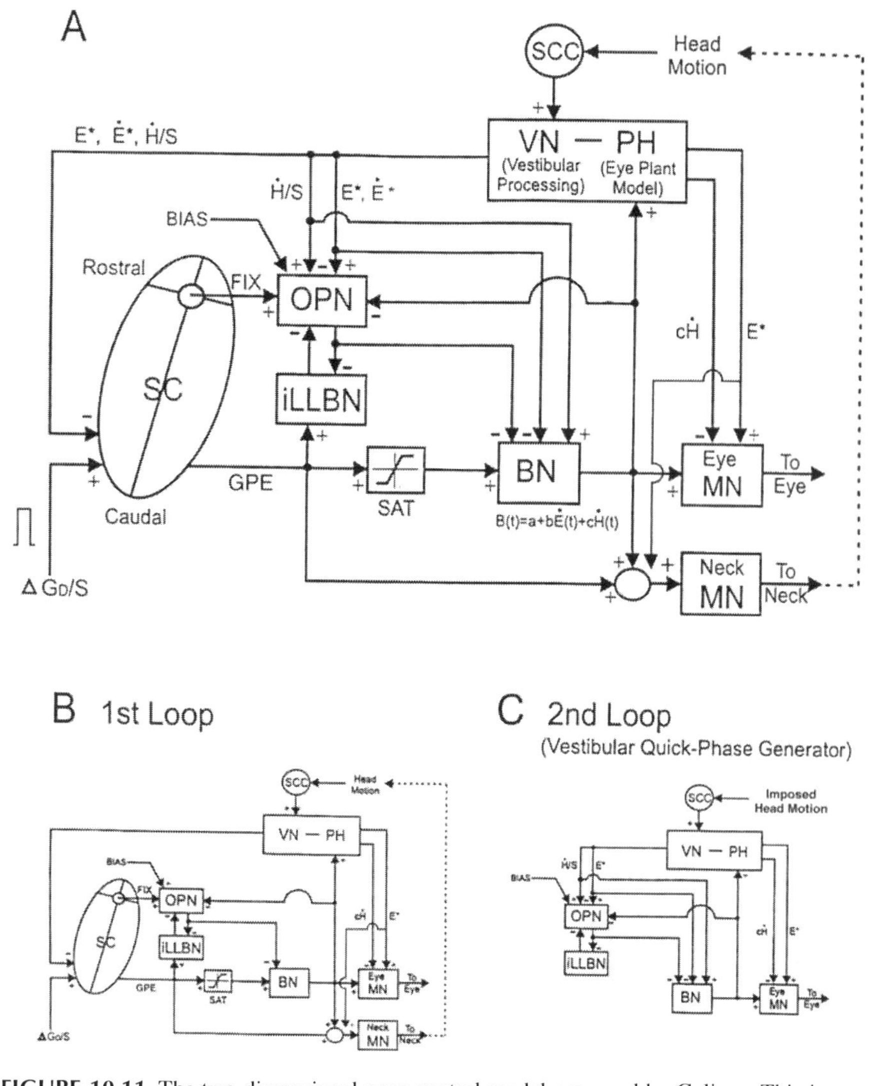

FIGURE 10.11 The two-dimensional gaze control model proposed by Galiana. This is normally implemented as a 2-sided bilateral model. (A) Collapsed one-sided version of the bilateral model. There are two main control loops: an outer loop (B) through the SC and an inner loop (C) through the burst generator (BN). The inner loop drives vestibular quick phases during passive head rotation, but it also contributes during active head movements. See text for details.

monkey eye saccades[105] and multiple-step gaze shifts.[42] (5) The vestibular nuclei (VN) provide head velocity $(\dot{H}/S)$[106] to SC for feedback; to OPNs[107] and BNs[104,108–111] for vestibular quick-phase generation (see below); and to MNs[104,112–116] to offset $\dot{H}/S$ signals on BNs.[61,64] (6) The prepositus hypoglossi nucleus (PH), which together with VN is shown as the VN-PH complex, is a system of loops acting as the oculomotor integrator in slow phases (large time constant). During saccades and fast phases (small time constant) the VN-PH provides E*, $\dot{E}$ * (also $\dot{H}$/S, if necessary) signals because it contains an eye plant model (and perhaps a head plant model).[117-123] The small extra eye saccade-related burst input onto eye motoneurons from VN-PH is required to boost the insufficient BN drive.[60,61] In Figure 10.11, no head plant model is used because the semicircular canals conveniently provide head velocity signals. (**7**) BNs are under the control of 2 loops: one outer through the SC producing the GPE input onto BN (Figure 10.11B), the other, inner, via $\dot{H}$/S and E* inputs to OPNs and BNs (Figure 10.11C). The inner loop generates vestibular quick-phases during both active and passive head rotations.[124] We propose that quick-phase generation during head rotations is an important mechanism in the gaze control circuitry. Thus, during passive head rotation, activation of the burst generator implies gaze motion.

Additional important predictions of the model are as follows. (1) It accounts for the head-velocity signal on BNs (Figure 10.9). (2) It explains the eye position saturation (*plateau*, Figures 10.1 and 10.2) and upper limits on eye velocity both resulting from the saturation element (SAT). In the cat, the plateau is also the *goal* to which head-fixed SC stimulation-evoked saccades are driven.[27,47,125] The model simulates such goal-directed saccades and associated OPN discharges, due to the SAT element and E* modulation of OPNs, respectively.[36] OPNs *change their allegiance* when the head is released, and now pause for evoked gaze shifts, rather than eye saccades. This is due to the $\dot{H}$ signal on OPNs. (3) The model simulates the effects of passive head rotations on BNs and OPNs. Our recording from OPNs and IBNs during *passive* head-on body rotations in the cat show that the pause and burst durations are both equal to gaze duration, and as predicted the IBN firing frequency ~ $\dot{G}$.[37,61,126] This also is due to model inner loop properties in which $\dot{H}$ and E* signals modulate OPNs (discussed in Reference 124). (4) The model explains, via quick-phase generation, multiple-step gaze shifts and associated collicular cell discharges.[41–43] Although the SC might call for a large, say 70° gaze shift, multiple-steps are caused by activation of the quick-phase generator via direct interactions between burst and OPN cells. (5) During gaze saccades, the model incorporates gaze-related suppression of the VOR[64] via the BN input to VN-PH.

One criticism of the model is that[89] it doesn't seem capable of generating head movements that are proportional to target-re-head, because the head is driven by a GPE signal. However, the model does replicate this property because, when gaze gets on target and GPE = 0, the system switches to the slow phase mode whose dynamics and initial conditions tend to keep the head moving until the eyes are centered in their orbits. This aligns the head on target.

10.5 CRITIQUES OF THE GAZE FEEDBACK HYPOTHESIS

10.5.1 Oculocentric Models

In early theories of eye-head coordination (reviewed in Reference 29) it was thought that the VOR acts throughout the entire eye saccade. It was proposed that head motion is added to that of the eye, because the head carries the eye, but that it is subtracted out by the VOR. In this oculocentric view of the gaze control system, the head motion is irrelevant to target acquisition; it is the eye saccade that gets gaze on target. This hypothesis is incorrect for large gaze shifts. Indeed, many experiments have shown that the VOR signal is not transmitted to ocular motoneurons during large gaze shifts (e.g., References 2 and 63 to 65). Furthermore, for the 120° gaze shifts in Figure 10.1, it is unrealistic to postulate that the brain programs 120° eye saccades. Thus, the contribution of the head must be taken into account in formulating a plausible gaze control model. However, specific model structures have been criticized on several grounds, which we will review below.

10.5.2 Single Comparator Models

Sparks[127] has argued that the finding that subjects compensate for perturbations to gaze trajectories does not unequivocally support the gaze feedback hypothesis whereby a single GPE comparator drives the eye and head motor circuits. He argues that compensation can also be explained by so-called *separate comparator* models, such as the ones by Freedman[66] and Phillips et al.,[90] which assume that the gaze displacement command is decomposed into separate eye and head displacement commands that drive separate eye and head generators, each controlled by its individual feedback circuit. Indeed, Freedman's open loop model, using feedback from neck reflexes, simulates gaze compensation for brief transient head motion perturbations. Although the latter scheme may be compatible with such specific experimental conditions, it is inadequate for controlling interrupted gaze movements in which the plateau period is long enough for the initial eye saccade to terminate (Figure 10.2). Consider the perturbed trajectory in Figure 10.2B. When head motion is unexpectedly halted, the eye completes its displacement and therefore, in the separate comparator model, *eye* position error has attained zero. Note that at the end of the gaze plateau, the initial eye position for the compensatory eye saccade differs from that at the end of the first saccade. Thus to get on target, a new eye saccade must be encoded whose amplitude is not related to that at the start of the overall gaze shift. A critical drawback of the separate comparator model is its inability to calculate, without knowledge of the remaining gaze position error, the new eye and head position displacement commands required by the independent eye and head loops. Furthermore, since gaze is only about half-way to the target during the plateau, the Phillips et al.,[90] model would have the VOR fully or partly turned *off*. This also is not compatible with experimental observations: in Figure 10.2B, when the head is released, the VOR moves the eye in a direction exactly opposite to head motion and gaze is perfectly stabilized in space. Some of these points have been discussed in Pélisson et al.,[4]

10.5.3 The Moving Hill Model

The hypothesis that places the SC within a feedback loop has been vigorously debated, particularly as to whether the moving hill hypothesis applies to the head-fixed monkey. We will briefly review some key points in this debate. Munoz and Wurtz[56,57] have described two cell types in the SC related to saccade generation in the head-fixed monkey; buildup neurons (BUNs) and burst neurons (BNs). Buildup neurons were reported to have a prelude of activity at a frequency > 30spikes/s, that preceded by about 100ms, eye saccades of the optimal amplitude and direction in the delayed saccade paradigm. BUNs also have *open-ended* movement fields; i.e., these cells discharge for all saccades larger than the amplitude of the *optimal* vector encoded by the cell's position on the motor map. By analogy with the moving hill, Munoz and Wurtz[57] described for monkey a moving front of activity that spread across the BUN layer towards the SC rostral pole as the saccade progressed. As in Munoz et al.,[45,46] the saccade was hypothesized to stop when the moving front activated fixation neurons. By comparison, collicular BNs showed a burst of activity that preceded a saccade, but negligible buildup activity. These neurons have a *closed* movement field: they discharge only during saccades that have a vector close to the *optimum*. According to Munoz and Wurtz,[57] activity in the BN layer remained topographically immobile during a saccade.

Further experiments, in which the activity of two buildup neurons was simultaneously recorded in the head-fixed monkey, have supported rostrally moving activity during saccades[128]: during a saccade activity in the rostral SC occurred later than in the caudal SC. However, Port et al.,[128] (see also Reference 58) argued that the time difference between the respective activations of rostral and caudal neurons was too small compared to that calculated from motor error. Put another way, they argued that the rostrally moving activity went faster than the movement itself and arrived in the rostral zone too early to stop saccades. A recent analysis by Optican (private communication) also showed moving activity but no encoding of eye motor error. A further argument against the moving hill hypothesis rests on the fact that one of its predictions — that saccades would be hypermetric if the rostral SC is deactivated[45,46,57] — has not been borne out experimentally.[129,130]

Other reports have looked specifically for moving activity on the collicular motor map, three in the *head-fixed* monkey and one in the head-fixed cat and all report no significant rostral movement of the locus of activity during eye saccades (imaging SC in monkey: see Reference 131; single cell recording in monkey: see References 58 and 132; single cell recording in cat: see Reference 133). One explanation for such discrepancies and the resulting debate on the applicability of the moving hill hypothesis to the monkey is that this idea has not been specifically tested in the head-unrestrained primate, a condition in which there appears to be feedback to the SC (Figure 10.8).

10.6 CONCLUSIONS

We have seen that in all experiments that have tested compensation for perturbations to eye or gaze trajectories — including complete interruption of the trajectory in cats, monkeys and humans — there is unanimous agreement that compensation is excellent and accuracy is virtually indistinguishable from control data. This supports feedback control, particularly in the head-unrestrained condition. The feedback concept is further supported by recordings of neural activity — in the SC during perturbed eye saccades[14,15] and in the SC and reticular formation during perturbed gaze saccades[32-34,38] — that show profound perturbation-induced changes in the neural discharge pattern.

The above considerations have focused on the horizontal gaze control system. However, in reality one must consider three-dimensional effects. The SC is organized in retinal coordinates and carries a two-dimensional gaze signal, but there is a three-dimensional quick-phase generator downstream.[134,135] This suggests that the hypothesized feedback signal to the SC (Figures 10.8 and 10.11) may have to be encoded in three dimensions.[136,137]

The challenge of understanding the neural mechanisms underlying gaze feedback control is daunting. The complexities of controlling together two mobile platforms, each rotating in three-dimensions, have just been emphasized. Even in the one-dimensional horizontal gaze control system, eye, head and gaze motion are coupled, making it difficult to assign neural signals to a particular platform.[127] Furthermore, as shown by even simple (by neural standards) models that use distributed processing, it is difficult to use single-cell discharges to distinguish what signals are being processed. For example, in the models of Arai et al.,[77] and Lefèvre and Galiana,[79] no cells explicitly encode motor error. These signals are an ensemble property. The problem is exacerbated when one realizes that many computations that could be done with a neural network can also be performed by single cells (e.g., see Reference 138) and that cell and network properties may be matched.

ACKNOWLEDGMENTS

Funded by the Canadian Institutes of Health Research (CIHR). A.B. and W.Y.C. held studentships from respectively the CIHR and Le Fonds de la Recherche en Santé du Québec. We thank P.A. Sylvestre and K. Cullen for permission to quote and illustrate their unpublished data.

REFERENCES

1. Todorov, E. and Jordan, M.I., Optimal feedback control as a theory of motor coordination, *Nat. Neurosci.*, 5, 1226, 2002.
2. Guitton, D. and Vollé, M., Gaze control in humans: eye head coordination during orienting movements to targets within and beyond the oculomotor range, *J. Neurophysiol.*, 58, 427, 1987.
3. Herter, T.M. and Guitton, D., Human head-free gaze saccades to targets flashed before gaze-pursuit are spatially accurate, *J. Neurophysiol.*, 80, 2785, 1998.
4. Pélisson, D., Guitton, D., and Munoz, D.P., Compensatory eye and head movements generated by the cat following stimulation-induced perturbations in gaze position, *Exp. Brain Res.*, 78, 654, 1989.
5. Sparks, D.L. and Mays, L.E., Spatial localization of saccade targets. I. Compensation for stimulation-induced perturbations in eye position, *J. Neurophysiol.*, 49, 45, 1983.
6. Sparks, D.L. and Porter, J.D., Spatial localization of saccade targets. II. Activity of superior colliculus neurons preceding compensatory saccades, *J. Neurophysiol.*, 49, 64, 1983.
7. Vossius, G., Das system der augenbewegung. (I.), *Z. Biol.*, 112, 27, 1960.

8. Robinson, D.A., Oculomotor control signals, in *Basic Mechanisms of Ocular Motility and their Clinical Implications,* Bach, Y., Rita, P., and Lennerstrand, G., Eds., Pergamon, Oxford, 1975, p. 337.
9. Zee, D.S. et al., Slow saccades in spinocerebellar degeneration, *Arch. Neurol.*, 33, 243, 1976.
10. Moschovakis, A.K., Scudder, C.A., and Highstein, S.M., The microscopic anatomy and physiology of the mammalian saccadic system, *Progr. Neurobiol.*, 50, 133, 1996.
11. Scudder, C.A., Kaneko, C.S., and Fuchs, A.F., The brainstem burst generator for saccadic eye movements. A modern synthesis, *Exp. Brain Res.*, 142, 439, 2002.
12. Quaia, C. et al., Extent of compensation for variations in monkey saccadic eye movements, *Exp. Brain Res.*, 132, 39, 2000.
13. Keller, E.L., Gandhi, N.J., and Shieh, J.M., Endpoint accuracy in saccades interrupted by stimulation in the omnipause region in monkey, *Vis. Neurosci.*, 13, 1059, 1996.
14. Keller, E.L. and Edelman, J.A., Use of interrupted saccade paradigm to study spatial and temporal dynamics of saccadic burst cells in superior colliculus in monkey, *J. Neurophysiol.*, 72, 2754, 1994.
15. Munoz, D.P., Waitzman, D.M., and Wurtz, R.H., Activity of neurons in monkey superior colliculus during interrupted saccades, *J. Neurophysiol.*, 75, 2562, 1996.
16. Keller, E.L., Gandhi, N.J., and Sekaran, S.V., Activity in deep intermediate layer collicular neurons during interrupted saccades, *Exp. Brain Res.*, 130, 227, 2000.
17. Raybourn, M.S. and Keller, E.L., Colliculoreticular organization in the primate oculomotor system, *J. Neurophysiol.*, 40, 861, 1977.
18. Goossens, H.H. and Van Opstal, A.J., Blink-perturbed saccades in monkey. I. Behavioral analysis, *J. Neurophysiol.*, 83, 3411, 2000.
19. Goossens, H.H. and Van Opstal, A.J., Blink-perturbed saccades in monkey. II. Superior colliculus activity, *J. Neurophysiol.*, 83, 3430, 2000.
20. Kaneko, C.R., Effect of ibotenic acid lesions of the omnipause neurons on saccadic eye movements in rhesus macaques, *J. Neurophysiol.*, 75, 2229, 1996.
21. Soetedjo, R., Kaneko, C.R., and Fuchs, A.F., Evidence that the superior colliculus participates in the feedback control of saccadic eye movements, *J. Neurophysiol.*, 87, 679, 2002.
22. Fuller, J.H., Maldonado, H., and Schlag, J., Vestibular oculomotor interaction in cat eye head movements, *Brain Res.,* 271, 241, 1983.
23. Guitton, D., Douglas, R.M., and Volle, M., Coordinated eye head movements in the cat, *J. Neurophysiol.,* 52, 1030, 1984.
24. Guitton, D., Munoz, D.P., and Galiana, H.L., Gaze control in the cat: Studies and modelling of the coupling between orienting eye and head movements in different behavioral tasks, *J. Neurophysiol.*, 64, 509, 1990.
25. Laurutis, V.P. and Robinson, D.A., The vestibulo ocular reflex during human saccadic eye movements, *J. Physiol.*, 373, 209, 1986.
26. Pélisson, D., Prablanc, C., and Urquizar, C., Vestibuloocular reflex inhibition and gaze saccade control characteristics during eye-head orientation in humans, *J. Neurophysiol.*, 59, 997, 1988.
27. Roucoux, A., Guitton, D., and Crommelinck, M., Stimulation of the superior colliculus in the alert cat. II. Eye and head movements evoked when the head is unrestrained, *Exp. Brain Res.*, 39, 75, 1980.
28. Tomlinson, R.D., Combined eye head gaze shifts in the primate. III. Contributions to the accuracy of gaze saccades, *J. Neurophysiol.*, 64, 1873, 1990.
29. Guitton, D., Control of eye head coordination during orienting gaze shifts, *Trends Neurosci.*, 15, 174, 1992.

30. Freedman, E.G. and Sparks, D.L., Eye-head coordination during head-unrestrained gaze shifts in rhesus monkeys, *J. Neurophysiol.*, 77, 2328, 1997.
31. Castet, E. and Masson, G., Motion perception during saccadic eye movements, *Nat. Neurosci.*, 3, 177, 2000.
32. Choi, W.Y. and Guitton, D., Discharge characteristics of saccade-related neurons in primate superior colliculus during head-perturbed gaze shifts, *Soc. Neurosci. Abstr.*, 11.7, 2002.
33. Choi, W.Y. and Guitton, D., Discharge characteristics of fixation neurons in primate superior colliculus during head-perturbed gaze shifts, *Soc. Neurosci. Abstr.*, 11.8, 2002.
34. Choi, W.Y. and Guitton, D., Superior colliculus monitors unexpected perturbations to gaze shifts in darkness, submitted.
35. Guitton, D. et al., On the feedback control of orienting gaze shifts made with eye and head movements, *Prog. Brain Res.*, 142, 55, 2003.
36. Paré, M. and Guitton, D., Brain stem omnipause neurons and the control of combined eye-head gaze saccades in the alert cat, *J. Neurophysiol.*, 79, 3060, 1998.
37. Cullen, K.E. and Guitton, D., Inhibitory burst neuron activity encodes gaze, not eye, metrics and dynamics during passive head on body rotation. Evidence that vestibular signals supplement visual information in the control of gaze shifts, *Ann. N. Y. Acad. Sci.*, 781, 601, 1996.
38. Sylvestre, P.A., Roy, J.E., and Cullen, K.E., The brainstem burst generator is updated online in response to head perturbations during gaze shifts, *Soc. Neurosci. Abstr.*, 27, 784.4, 2001.
39. Matsuo, S., Bergeron, A., and Guitton, D., Further evidence for gaze-feedback to the superior colliculus: discharges on motor map reflect gaze trajectory perturbations, *Soc. Neurosci. Abstr.*, 71:17, 2001.
40. Matsuo, S., Bergeron, A., and Guitton, D., Activity of superior colliculus neurons during gaze trajectory perturbations in head-unrestrained cat, *Soc. Neurosci. Abstr.*, 266:8, 2002.
41. Bergeron, A. and Guitton, D., Fixation neurons in superior colliculus encode gaze-position-error, the distance between current and desired gaze positions, *Nat. Neurosci.*, 3, 932, 2000.
42. Bergeron, A. and Guitton, D., Discharge of superior colliculus fixation neurons (SCFNs) and brain stem omnipause neurons (OPNs) during multiple step gaze shifts in cat: OPNs, not SCFNs, gate saccades; both cells encode gaze position error, *J. Neurophysiol.*, 88, 1726, 2002.
43. Bergeron, A., Matsuo, S., and Guitton, D., In multiple-step gaze shifts one site on the motor map of the superior colliculus encodes distance to target, not saccade amplitude, *Nat. Neurosci.*, 6, 404, 2003.
44. Pélisson, D., Guitton, D., and Goffart, L., On-line compensation of gaze shifts perturbed by micro-stimulation of the superior colliculus in the cat with unrestrained head, *Exp. Brain Res.*, 106, 196, 1995.
45. Munoz, D.P., Pélisson, D., and Guitton, D., Movement of neural activity on the superior colliculus motor map during gaze shifts, *Science*, 251, 1358, 1991.
46. Munoz, D.P., Guitton, D., and Pélisson, D., Control of orienting gaze shifts by the tectoreticulospinal system in the head-free cat. III. Spatiotemporal characteristics of phasic motor discharges, *J. Neurophysiol.*, 66, 1642, 1991.
47. Paré, M., Crommelinck M., and Guitton D., Gaze shifts evoked by stimulation of the superior colliculus in the head-free cat conform to the motor map but also depend on stimulus strength and fixation activity, *Exp. Brain Res.*, 101, 123, 1994.

48. Paré, M. and Guitton, D., The fixation area of the cat superior colliculus: effects of electrical stimulation and direct connection with brainstem omnipause neurons, *Exp. Brain Res.*, 101, 109, 1994.
49. Krauzlis, R.J., Basso, M.A., and Wurtz, R.H., Shared motor error for multiple eye movements, *Science*, 276, 1693, 1997.
50. Krauzlis, R.J., Basso, M.A., and Wurtz, R.H., Discharge properties of neurons in the rostral superior colliculus of the monkey during smooth-pursuit eye movements, *J. Neurophysiol.*, 84, 876, 2000.
51. Krauzlis R.J. and Carello C.D., Going for the goal, *Nat. Neurosci.*, 6, 332, 2003.
52. Freedman, E.G., Stanford, T.R., and Sparks, D.L., Combined eye head gaze shifts produced by electrical stimulation of the superior colliculus in rhesus monkeys, *J. Neurophysiol.*, 76, 927, 1996.
53. Freedman, E.G. and Sparks, D.L., Activity of cells in the deeper layers of the superior colliculus of the rhesus monkey evidence for a gaze displacement command, *J. Neurophysiol.*, 78, 1669, 1997.
54. Corneil, B.D., Olivier, E., and Munoz, D.P., Neck muscle responses to stimulation of monkey superior colliculus. I. Topography and manipulation of stimulation parameters, *J. Neurophysiol.*, 88, 1980, 2002.
55. Corneil, B.D., Olivier, E., and Munoz, D.P., Neck muscle responses to stimulation of monkey superior colliculus. II. Gaze shift initiation and volitional head movements, *J. Neurophysiol.*, 88, 2000, 2002.
56. Munoz, D.P. and Wurtz, R.H., Saccade-related activity in monkey superior colliculus. I. Characteristics of burst and buildup cells, *J. Neurophysiol.*, 73, 2313, 1995.
57. Munoz, D.P. and Wurtz, R.H., Saccade-related activity in monkey superior colliculus. II. Spread of activity during saccades, *J. Neurophysiol.*, 73, 2334, 1995.
58. Soetedjo, R., Kaneko, C.R., and Fuchs, A.F., Evidence against a moving hill in the superior colliculus during saccadic eye movements in the monkey, *J. Neurophysiol.*, 87, 2778, 2002.
59. Quaia, C., Lefèvre, P., and Optican, L., Model of the control of saccades by superior colliculus and cerebellum, *J. Neurophysiol.*, 82, 999, 1999.
60. Cullen, K.E. and Guitton, D., Analysis of primate IBN spike trains using system identification techniques, I Relationship To eye movement dynamics during head-fixed saccades, *J. Neurophysiol.* 78, 3259, 1997.
61. Cullen, K.E. and Guitton, D., Analysis of primate IBN spike trains using system identification techniques. II. Relationship to gaze, eye, and head movement dynamics during head-free gaze shifts, *J. Neurophysiol.*, 78, 3283, 1997.
62. Cullen, K.E. and Guitton, D., Analysis of primate IBN spike trains using system identification techniques. III. Relationship To motor error during head-fixed saccades and head-free gaze shifts, *J. Neurophysiol.*, 78, 3307, 1997.
63. Cullen, K.E., Galiana, H.L., and Sylvestre P.A., Comparing extraocular motoneuron discharges during head-restrained saccades and head-unrestrained gaze shifts, *J. Neurophysiol.*, 83, 630, 2000.
64. Roy, J.E. and Cullen, K.E., A neural correlate for vestibulo ocular reflex suppression during voluntary eye head gaze shifts, *Nat. Neurosci.*, 5, 404, 1998.
65. Roy, J.E. and Cullen K.E., Vestibuloocular reflex signal modulation during voluntary and passive head movements, *J. Neurophysiol.*, 87, 2337, 2002.
66. Freedman, E.G., Interactions between eye and head control signals can account for movement kinematics, *Biol. Cybern.*, 84, 453, 2001.
67. Gandhi, N.J. and Sparks, D.L., Accuracy of head-unrestrained gaze shifts interrupted by stimulation of the omnipause neurons in monkey, *Soc. Neurosci. Abstr.*, 784.3, 2001.

68. Goffart, L. and Pélisson, D., Changes in initiation of orienting gaze shifts after muscimol inactivation of the caudal fastigial nucleus in the cat, *J. Physiol. (Lond.)*, 503, 657, 1997.
69. Goffart, L. and Pélisson, D., Orienting gaze shifts during muscimol inactivation of caudal fastigial nucleus in the cat. I. Gaze dysmetria, *J. Neurophysiol.*, 79, 1942, 1998.
70. Goffart, L., Pélisson, D., and Guillaume, A., Orienting gaze shifts during muscimol inactivation of caudal fastigial nucleus in the cat. II. Dynamics and eye-head coupling, *J. Neurophysiol.*, 79, 1959, 1998.
71. Goffart, L., Guillaume, A., and Pélisson, D., Compensation for gaze perturbation during inactivation of the caudal fastigial nucleus in the head-unrestrained cat, *J. Neurophysiol.*, 80, 1552, 1998.
72. Pélisson, D., Goffart, L., and Guillaume, A., Contribution of the rostral fastigial nucleus to the control of orienting gaze shifts in the head-unrestrained cat, *J. Neurophysiol.*, 80, 1180, 1998.
73. Fakhri, S., Pélisson, D., and Guitton, D., Compensation for perturbation of gaze and role of vestibular signals in gaze control, in *Information Processing underlying Gaze Control,* Delgado Garcia, J.M., Godaux, E., Vidal, P.P., Eds., Elsevier Science, Oxford, U.K., 1994, p.53.
74. Waitzman, D.M. et al., Contribution of the superior colliculus and the mesencephalic reticular formation to gaze control, *Ann. N. Y. Acad. Sci.*, 956, 111, 2002.
75. Bozis, A. and Moschovakis, A.K., Neural network simulations of the primate oculomotor system. III. An one-dimensional, one-directional model of the superior colliculus, *Biol. Cybern.*, 79, 215, 1998.
76. Scudder, C.A. A new local feedback model of the saccadic burst generator, *J. Neurophysiol.*, 59, 1455, 1988.
77. Arai, K. et al., A distributed model of the saccade system: simulations of temporally perturbed saccades using position and velocity feedback, *Neural Networks,* 12, 1359-1375, 1999.
78. Grossberg, S. et al., A neural model of multimodal adaptive saccadic eye movement control by superior colliculus, *J. Neurosci.*, 17, 9706, 1997.
79. Lefèvre, P. and Galiana, H.L., Dynamic feedback to the superior colliculus in a neural network model of the gaze control system, *Neural Networks*, 5, 871, 1992.
80. Moschovakis, A.K., Neural network simulations of the primate oculomotor system. I. The vertical saccade burst generator, *Biol. Cybern.*, 70, 291, 1994.
81. Waitzman, D.M. et al., Superior colliculus neurons provide the saccadic motor error signal, *Exp. Brain Res.*, 72, 649, 1988.
82. Waitzman, D.M. et al., Superior colliculus neurons mediate the dynamic characteristics of saccades, *J. Neurophysiol.*, 66, 1716, 1991.
83. Optican, L., A field theory of saccade generation: temporal to spatial transform in the superior colliculus, *Vision Res.*, 35, 3313, 1995.
84. Dean, P., Modeling the role of the cerebellar fastigial nuclei in producing accurate saccades: the importance of burst timing, *Neuroscience*, 68, 1059, 1995.
85. Optican, L.M. and Quaia, C., Distributed model of collicular and cerebellar function during saccades, *Ann. N. Y. Acad. Sci.*, 956, 164, 2002.
86. Schweighofer, N., Arbib, M.A., and Dominey, P.F., A model of the cerebellum in adaptive control of saccadic gain. I. The model and its biological substrate, *Biol. Cybern.*, 75, 19, 1996.
87. Schweighofer, N., Arbib, M.A., and Dominey, P.F., A model of the cerebellum in adaptive control of saccadic gain. II. Simulation results, *Biol. Cybern.*, 75, 29, 1996.

88. Galiana, H.L. and Guitton, D., Central organization and modeling of eye-head coordination during orienting gaze shifts, *Ann. N. Y. Acad. Sci.,* 656, 452, 1992.
89. Goossens, H.H. and Van Opstal, A.J., Human eye-head coordination in two dimensions under different sensorimotor conditions, *Exp. Brain Res.*, 114, 542, 1997.
90. Phillips, J.O. et al., Rapid horizontal gaze movement in the monkey, *J. Neurophysiol.*, 73, 1632, 1995.
91. Glenn, B. and Vilis, T., Violations of Listing's law after large eye and head gaze shifts, *J. Neurophysiol.*, 68, 309, 1992.
92. Tweed, D., Glenn, B., and Vilis, T., Eye-head coordination during large gaze shifts, *J. Neurophysiol.*, 73, 766, 1995.
93. Vollé, M. and Guitton, D., Human gaze shifts in which the head and eye are not initially aligned, *Exp. Brain Res.*, 94, 463, 1993.
94. Delreux, V. et al., Eye-head coordination: influence of eye position on the control of head movement amplitude, in *Brain and Space*, Paillard, J., Ed., Oxford University Press, Oxford, 1991, p. 38.
95. Phillips, J.O., Ling, L., and Fuchs, A.F., Action of the brain stem saccade generator during horizontal gaze shifts. I. Discharge patterns of omnidirectional pause neurons, *J. Neurophysiol.*, 81, 1284, 1999.
96. Izawa, Y., Sugiuchi, Y., and Shinoda Y., Neural organization from the superior colliculus to motoneurons in the horizontal oculomotor system of the cat, *J. Neurophysiol.*, 81, 2597, 1999.
97. Büttner-Ennever, J.A. and Büttner, U., The reticular formation, in *Neuroanatomy of the Oculomotor System*, Büttner-Ennever, J.A., Ed., Elsevier, Amsterdam, 1988, p. 119.
98. Fuchs, A.F, Kaneko, C.R.S., and Scudder, C.A., Brainstem control of saccadic eye movements, *Ann. Rev. Neurosci.*, 8, 307, 1985.
99. André-Deshays, C., Berthoz, A., and Revel, M., Eye-head coupling in humans, *Exp. Brain. Res.*, 69, 399, 1988.
100. Vidal, P.P., Roucoux, A., and Berthoz, A., Horizontal eye position related activity in neck muscles of the alert cat, *Exp. Brain Res.*, 46, 448, 1982.
101. Büttner-Ennever, J.A. et al., Projections from the superior colliculus motor map to omnipause neurons in monkey, *J. Comp. Neurol.*, 413, 55, 1999.
102. Grantyn, A.A. et al., Neuronal mechanisms of two-dimensional orienting movements in the cat. I. A quantitative study of saccades and slow drifts produced in response to the electrical stimulation of the superior colliculus, *Brain Res. Bull.*, 41, 65, 1996.
103. Belknap, D.B. and McCrea, R.A., Anatomical connections of the prepositus and abducens nuclei in the squirrel monkey, *J. Comp. Neurol.*, 268, 13, 1988.
104. McCrea, R.A., The nucleus prepositus, in *Neuroanatomy of the Oculomotor System*, Büttner-Ennever, J.A., Ed., Elsevier, Amsterdam, 1988, p. 203.
105. Everling, S. et al., Comparison of the discharge characteristics of brain stem omnipause neurons and superior colliculus fixation neurons in monkey: implications for control of fixation and saccade behavior, *J. Neurophysiol.*, 79, 511, 1998.
106. Wilson, V.J. and Melvill Jones G., *Mammalian Vestibular Physiology*, Plenum Press, New York, 1979.
107. Ito, J., Markham, C.H., and Curthoys, I.S., Projections to eye movement-related pause neuron region in cat using HRP, *Exp. Neurol.*, 86, 93, 1984.
108. Henn, V., Baloh, R.W., and Hepp, K., The sleep-wake transition in the oculomotor system, *Exp. Brain Res.*, 54, 166, 1984.

109. Ito, J., Markham, C.H., and Curthoys, I.S., Direct projection of type II vestibular neurons to eye movement-related neurons in the cat pontine reticular formation, *Exp. Neurol.*, 91, 331, 1986.
110. McCrea, R.A., Strassman, A., and Highstein, S.M., Anatomical physiological characteristics of vestibular neurons mediating the vertical vestibulo-ocular reflexes of the squirrel monkey, *J. Comp. Neurol.*, 264, 571, 1987.
111. Ohki, Y., Shimazu, H., and Suzuki, I., Excitatory input to burst neurons from the labyrinth and its mediating pathway in the cat: location and functional characteristics of burster driving neurons, *Exp. Brain Res.*, 72, 457, 1988.
112. Büttner-Ennever, J.A., Patterns of connectivity in the vestibular nuclei, in *Sensing and Controlling Motion. Vestibular and Sensorimotor Function*, Cohen, B., Tomko, D.L., Guedry, F., Eds., The New York Academy of Sciences, New York, 1992, p. 363.
113. Cullen, K.E. and McCrea, R.A., Firing behavior of brain stem neurons during voluntary cancellation of the horizontal vestibuloocular reflex. I. Secondary vestibular neurons, *J. Neurophysiol.*, 70, 828, 1993.
114. Cullen, K.E., Chen-Huang, C., and McCrea, R.A., Firing behavior of brain stem neurons during voluntary cancellation of the horizontal vestibuloocular reflex. II. Eye movement related neurons, *J. Neurophysiol.*, 70, 844, 1993.
115. Highstein, S.M. and McCrea, R.A., The anatomy of the vestibular nuclei, in *Neuroanatomy of the Oculomotor System*, Büttner-Ennever, J.A., Ed., Elsevier, Amsterdam, 1988, p. 177.
116. Scudder, C.A., Fuchs, A.F., and Langer, T.P., Characteristics and functional identification of saccadic inhibitory burst neurons in the alert monkey, *J. Neurophysiol.*, 59, 1430, 1988.
117. Corvisier, J. and Hardy, O., Topographical characteristics of preposito-collicular projections in the cat as revealed by Phaseolus vulgaris-leucoagglutinin technique. A possible organisation underlying temporal-to-spatial transformations, *Exp. Brain Res.*, 114, 461, 1997.
118. Escudero, M., de la Cruz, R.R., and Delgado-García, J.M., A physiological study of vestibular and prepositus hypoglossi neurones projecting to the abducens nucleus in the alert cat, *J. Physiol.*, 458, 539, 1992.
119. Hardy, O. and Corvisier, J., Firing behaviour of anticompensatory neurones in the prepositus hypoglossi nucleus of alert cat, *Neurosci. Lett.*, 139, 178, 1992.
120. Hardy, O. and Corvisier, J., Firing properties of preposito-collicular neurones related to horizontal eye movements in the alert cat, *Exp. Brain Res.*, 110, 413, 1996.
121. Hartwich-Young, R., Nelson, J.S., and Sparks, D.L., The perihypoglossal projection to the superior colliculus in the rhesus monkey, *Vis. Neurosci.*, 4, 29, 1990.
122. Stechison, M.T., Saint-Cyr, J.A., and Spence, S.J., Projections from the nuclei prepositus hyploglossi and intercalatus to the superior colliculus in the cat: an anatomical study using WGA-HRP, *Exp. Brain Res.*, 59, 139, 1985.
123. Stechison, M.T. and Saint-Cyr, J.A., Organization of spinal inputs to the perihypoglossal complex in the cat, *J. Comp. Neurol.*, 246, 555, 1986.
124. Galiana, H.L., A nystagmus strategy to linearize the vestibulo-ocular reflex, *IEEE Trans. Biomed. Eng.*, 38, 532, 1991.
125. Guitton, D., Crommelinck, M., and Roucoux, A., Stimulation of the superior colliculus in the alert cat. I. Eye movements and neck EMG activity evoked when the head is restrained, *Exp. Brain Res.*, 39, 63, 1980.
126. Cullen, K.E. et al., Gaze-related activity of putative inhibitory burst neurons in the head-free cat, *J. Neurophysiol.*, 70, 2678, 1993.

127. Sparks, D.L., Conceptual issues related to the role of the superior colliculus in the control of gaze, *Curr. Opin. Neurobiol.*, 9, 698, 1999.
128. Port, N.L., Sommer, M.A., and Wurtz, R.H., Multielectrode evidence for spreading activity across the superior colliculus movement map, *J. Neurophysiol.*, 84, 344, 2000.
129. Aizawa, H. and Wurtz, R.H., Reversible inactivation of monkey superior colliculus. I. Curvature of saccadic trajectory, *J. Neurophysiol.*, 79, 2082, 1998.
130. Quaia, C. et al., Reversible inactivation of monkey superior colliculus. II. Maps of saccadic deficits, *J. Neurophysiol.*, 79, 2097, 1998.
131. Moschovakis, A.K., Gregoriou, G.G., and Savaki, H.E., Functional imaging of the primate superior colliculus during saccades to visual targets, *Nat. Neurosci.*, 4, 1026, 2001.
132. Anderson, R.W. et al., Two dimensional saccade related population activity in superior colliculus in monkey, *J. Neurophysiol.*, 80, 798, 1998.
133. Kang, I. and Lee, C., Properties of saccade-related neurons in the cat superior colliculus: patterns of movement fields and discharge timing, *Exp. Brain Res.*, 131, 149, 2000.
134. Klier, E., Wang, H., and Crawford, J.D., The superior colliculus codes gaze commands in retinal coordinates, *Nat. Neurosci.*, 4, 627, 2001.
135. Klier, E., Wang, H., and Crawford, J.D., Three-dimensional eye-head coordination is implemented downstream from the superior colliculus, *J. Neurophysiol.*, 89, 2839, 2003.
136. Medendorp, W.P. et al., Rotational remapping in human spatial memory during eye and head motion, *J. Neurosci.*, 22, 196, 2002.
137. Smith, M.A. and Crawford, J.D., Implications of ocular kinematics for the internal updating of visual space, *J. Neurophysiol.*, 86, 2112, 2001.
138. Egorov, A.V. et al., Graded persistent activity in entorhinal cortex neurons, *Nature*, 420, 173, 2002.

11 Distributed Neural Processing in the Saccadic System

Edward L. Keller

CONTENTS

11.1 INTRODUCTION

This chapter is focused on models of the primate saccadic system that attempt to explain the role played by superior colliculus (SC) in the generation of these specialized eye movements. In particular, it concentrates on reviewing more recent attempts to include a distributed representation of the operation of the deeper layers of the SC and their role in saccade generation, in contrast to earlier models, which used lumped systems representations of the colliculus that provided a single output signal from the SC to the brainstem saccadic burst generator. In the context of the present chapter, distributed representations are those that represent the SC as a dynamic network that processes visuomotor information with a spatial grid of neuron-like elements interconnected with excitatory and inhibitory connections. Lumped representations use single-input/single-output, dynamic engineering elements to represent signal processing in the SC. The chapter begins with a brief

0-8493-0097-5/04/$0.00+$1.50

summary of relevant anatomical and physiological data on the SC that the new distributed models attempt to encapsulate. A further brief summary reviews work that has made estimates of the spatiotemporal population discharge in the SC during saccades. Knowledge of this population activity is a crucial guide to the development of distributed models with biological realism. New experimental results that need to be incorporated in distributed models are also discussed.

The review will concentrate on a description of the functional organization of the deeper layers of the primate SC and saccade-rated cells recorded there. Various classes of saccade-related cells have been described in the deeper layers. In particular, some authors have described prelude or buildup cells (cells with considerable presaccadic discharge in a delayed saccade paradigm) and burst cells (cells without prelude discharge) as separate classes of cells with possibly different functions.[1–3] In the delayed saccade paradigm a visual target appears in the periphery, but the subject is not allowed to saccade to this target until the fixation point goes off after a variable interval from the appearance of the target. In several recent papers it has been argued that the purported separate classes of cells, burst neurons and buildup neurons, are instead a continuum of cell type with a common function, at least with respect to saccade generation.[4–8] Whenever possible this review will use the more generic term, saccade-related burst neurons (SRBNs) to refer to all types of SC neurons located in the intermediate and deeper layers of the SC and that discharge prior to or during saccades.

11.2 SPATIAL PROPERTIES ENCODED BY SINGLE COLLICULAR NEURONS AND SPATIAL ORGANIZATION OF THE COLLICULUS FOR SACCADES

Several early studies[9–11] recorded single neurons in the alert monkey colliculus and showed that the superficial layers of this structure contain cells that respond after a short latency to the appearance of visual stimuli in a restricted region of the visual field. Cells in the deeper layers responded with a burst of discharge with an onset that precedes saccades moving the eyes to a restricted region in the visual field, called the motor field. Some cells in the deeper layers have both a visual and saccade-related burst of activity. The centers of the visual and the motor fields are similarly located in space, and these locations of the centers of the response fields were shown to be organized in a precise retinotopically organized map of the contralateral visual field over the surface of the SC.[11,12] Cells with field centers near the representation of the fovea were located rostrally in the SC and cells with centers in the eccentric visual field were located caudally in SC. Cells with centers along the horizontal meridian of the visual field were located along a middle, rostral-caudal meridian of the contralateral colliculus, while cells with centers in the upper portion of the visual field were located medially and cells with down preferences were found laterally on each colliculus. Here and elsewhere in the text the words *collicular meridians* refer to rostrocaudal or mediolateral displacements on the SC surface, which represent actual meridians in the visual field.

The details of the visuomotor map on the SC were further confirmed with electrical microstimulation techniques in alert monkeys.[13,14] In particular, Robinson[13] systematically explored the SC with closely spaced microelectrode penetrations and plotted a response (a saccade vector) at each stimulated location. He then fit curves through this vector field map to provide a smooth, continuous representation of saccadic vectors over the surface of the SC. Robinson's[13] study and that of Cynader and Berman[12] clearly showed a systematic distortion in the visual and motor maps on the SC such that the central region of visual or motor space received a much larger representation, with the result that the central 10° of visual space occupies more than 30% of the surface of the SC. Ottes et al.,[15] used the map proposed by Robinson along with additional single-unit recordings in the SC to generate a mathematical formulation of the motor map in the SC. The essential feature of this mathematical formulation is a logarithmic compression of visual space along the central meridian of the SC representing horizontal saccades into the opposite hemifield. Compression of vertical visual space is accomplished by an arctan function along the medial-lateral meridians of the SC.

The shape of the response field of individual neurons in the SC varies, but for most cells the response gradient within the field declines more rapidly for locations toward the fovea than for more eccentric positions.[16] The declining gradient for nonoptimal directions is symmetrical on both sides of the optimal vector.[16] Smoothed, two-dimensional estimates of the shape of the motor fields of SC cells as a function of saccade amplitude and direction have usually assumed that the discharge data measured for a distribution of saccade vectors over the field of the cell can be fit by a log-Gaussian function in amplitude and a Gaussian function in direction.[15–17] However, many cells, particularly those in the more ventral SC, have broad spatially distributed responses that produce poor fits or fits that do not converge for the Gaussian type of model estimator.[1,18] These cells have been called open-field cells because they discharge for all saccades with a contralateral component and an amplitude larger than some minimum that varies systematically with the cell's location on the map.[1] Munoz and Wurtz[1] used spline estimates to provide one-dimensional fits to the saccade amplitude component of these cells' movement fields for cells with horizontal preferred directions. In a more general approach, Anderson et al.,[18] used a radial basis function technique to estimate the two-dimensional shape of the movement fields of SC cells (Figure 11.1). The advantages of model free approaches, e.g., splines or radial basis functions, is that no assumptions need to be made about the particular mathematical function that describes the movement field shape; the method converges to an estimate for all cells and the goodness of fit is usually better than with Gaussian fits for many cells. The disadvantage of these techniques is that they contain a large number of parameters, and no concise description of the fit (as summarized by a few parameters for log-Gaussian fits) can be made, except to show it graphically.

In addition to the topographical map of visual and motor space that exists in the SC, it has been found that the size of SC cells' movement fields tend to increase as a function both of the cell's location (distance) from the rostral pole of the SC and of the depth of the cell from the dorsal surface of the SC.[16,19] Also some studies have suggested that the intensity of the discharge in individual cells, for movements

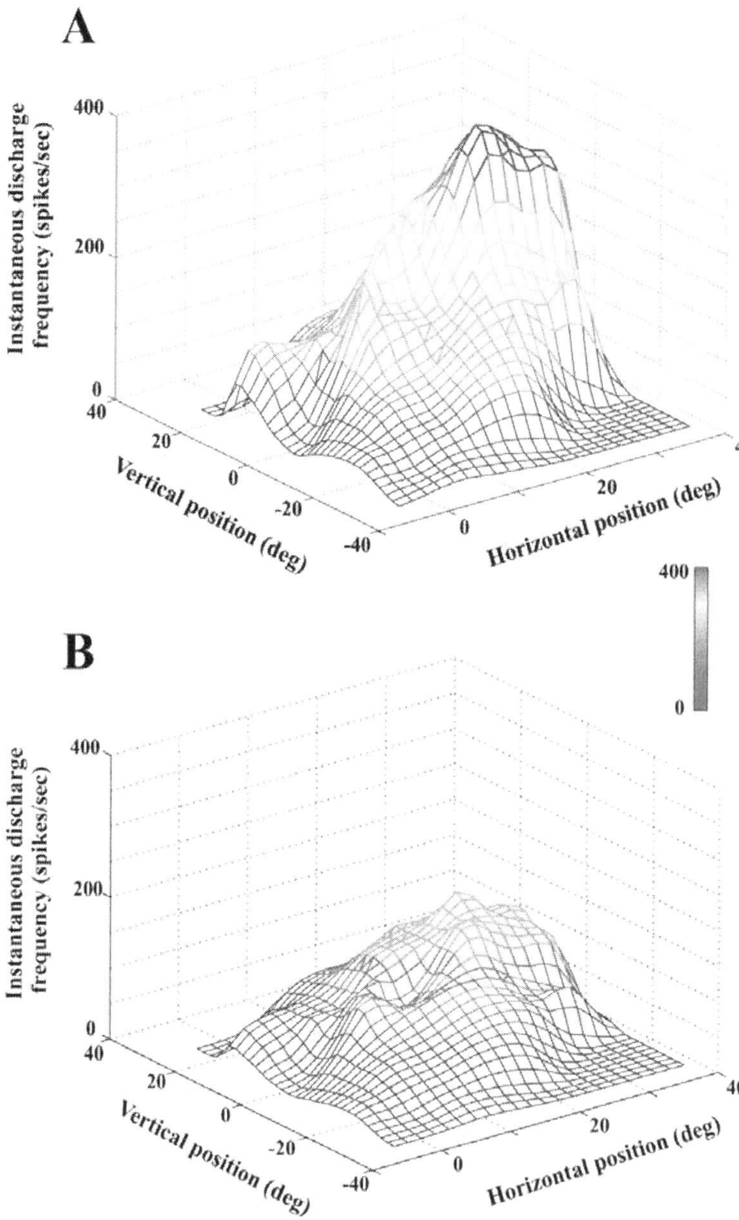

FIGURE 11.1 (See color insert following page 176.) Movement field (MF) fit for a sample SRBN in the left SC. The estimation of the two-dimensional shape of the field was made at two peri-saccadic time epochs using a model-free radial basis function technique.[18] The MF fit was computed from discharge data recorded while the animal made saccades to over 100 locations scattered in the right visual hemifield. (A) Discharge at 8 ms before saccade onset when activity was maximal. (B) Discharge at saccade end. An attempt to fit the two-dimension discharge with a log-Gaussian model in amplitude and a Gaussian in direction did not converge.

toward the center of the response field, does not vary systematically as a function of rostral/caudal location[1,16] (but see Reference 7 for evidence that the intensity of discharge at the center of SRBNs' movement fields decreases as recordings are made more caudally in the SC).

Taken together the set of spatial properties of movement fields described above shows that a particular saccadic eye movement cannot be specified by a single cell in the SC. Indeed, the data indicate that a large number of collicular cells are active for any saccade. Early theoretical estimates based on very limited spatial sampling by Ottes and his colleagues[15] suggested that the activated area of the SC for any saccade was invariant in size and covered about a third of the entire area of one colliculus.

In conclusion, early work emphasized the spatial properties of the population discharge in the SC. This work also suggested that the colliculus supplied a purely place-coded, distributed signal to downstream elements of the saccadic system.

11.3 INDICATIONS THAT TEMPORAL CODING IS ALSO IMPORTANT IN COLLICULAR FUNCTION

The concept of a place-coded output from the SC assumed a static operation of the SC. That is, it was thought that the population activity present in this structure was read out and decoded by downstream structures just before movement onset. The result of this decoding process would provide a static, desired saccade vector (the goal) for the ongoing saccade. This concept is no longer tenable. First of all, several studies have shown that the intrasaccadic discharge intensity of SC neurons is correlated with saccade speed.[20–22] Furthermore, when saccades are evoked by electrical microstimulation in the SC, the parameters of the electrical stimulation (at least within a limited range) affect the peak velocity and other kinematic characteristics of the saccade evoked at a particular site in the SC.[21,23] Finally, the decay in discharge rate of many SC neurons during a saccade into the cell's movement field is well correlated with a decline in dynamic motor error (the distance at each time point remaining in the movement to reach the intended saccadic goal).[24] The later dynamic characteristic for SC discharge has also been shown to be the case for the population discharge in the SC during saccades.[7]

In summary, it is now generally agreed that the population spatiotemporal pattern of discharge in the SC provides an important control signal to downstream elements of the saccadic system.

11.4 ESTIMATING THE SPATIOTEMPORAL POPULATION DISCHARGE IN SC DURING SACCADES

Because the population spatiotemporal discharge in the SC provides an important functional output of this structure during saccades, the computation or estimation of this distributed signal is necessary for understanding the exact role of the SC in saccade generation. Ideally, one would have an array of recording microelectrodes at evenly spaced sites in the deeper layers of the SC and would record the activity of neurons at these sites during saccades with a variety of different vectors. From this distributed information one could then estimate the population activity for any arbitrary saccade. Technically this is not yet possible. Therefore, all current studies have attempted to estimate population discharge in the SC from data collected from single-cell recordings.

An initial estimate of the population discharge sacrificed temporal resolution and averaged the number of spikes in the saccade-related burst over the duration of the saccade.[15] The data were obtained from the activity of a small sample of SC cells for a variety of saccade vectors. Each of the recorded cells was located on the SC motor map using a nonlinear mapping technique derived from Robinson's[13] microstimulation study of this map. Finally, the sampled data from all cells were fit with a two-dimensional, bivariate Gaussian function to give a continuous spatial representation of population activity over the SC for any particular saccade vector. Furthermore, the Gaussian function was hypothesized to be rotation-symmetric on the SC surface. Therefore, the gradient describing the decline of population activity from the center and most intense locus of activity is the same in their estimates in the rostro-caudal direction on the SC as in the mediolateral. The conclusion was that a *mound of activity*, invariant in size and intensity existed in the SC for saccades of different sizes and directions of movement. Only the locus of this mound of activity shifted over the SC map as a function of saccade amplitude and direction.

More recently a new spatiotemporal hypothesis has been proposed in which a rostrally spreading wave of activity exists within the SC during saccades, and it is this dynamic spread of population activity that controls saccades.[25,26] In order to test this hypothesis Munoz and Wurtz[26] recorded from a distribution of cells that were located along the horizontal meridian of the SC during horizontal saccades of a wide range of amplitudes. The activity of cells they called burst neurons, which tended to be located more dorsally in the intermediate layers, increased rapidly just before saccade onset and then decayed rapidly as the saccade progressed, but remained at the same locus on the SC motor map. On the other hand, the activity of buildup cells, which tend to be located more ventrally in the intermediate layers and even the deep layers of the SC, showed a rostrally directed spread as the saccade progressed. Finally, their hypothesis included the notion that fixation cells, which are located in the rostral pole of the SC and discharge during fixation, were reactivated by the rostrally spreading activity in buildup cells when it reached the rostral pole. The renewed activity of the fixation cells was proposed to contribute to the termination of the saccade.

Anderson et al.,[7] recorded the activity of a sample of SC cells located at irregularly spaced intervals over a majority of the spatial extent of the motor map. They divided their cells into the same two classes–the burst and buildup cells–based on presaccadic discharge level in a delayed saccade paradigm, although they argued that the cells more accurately fell along a continuum with respect to presaccadic discharge. They developed a new, model-free method of estimating the population discharge at 2-ms time intervals in the SC for an arbitrary saccade in either a horizontal or oblique (up to 45° from horizontal) direction. Examples of these population activity estimates for burst neurons at three discrete peri-saccadic time points for a small and a large horizontal saccade are shown in Figure 11.2.

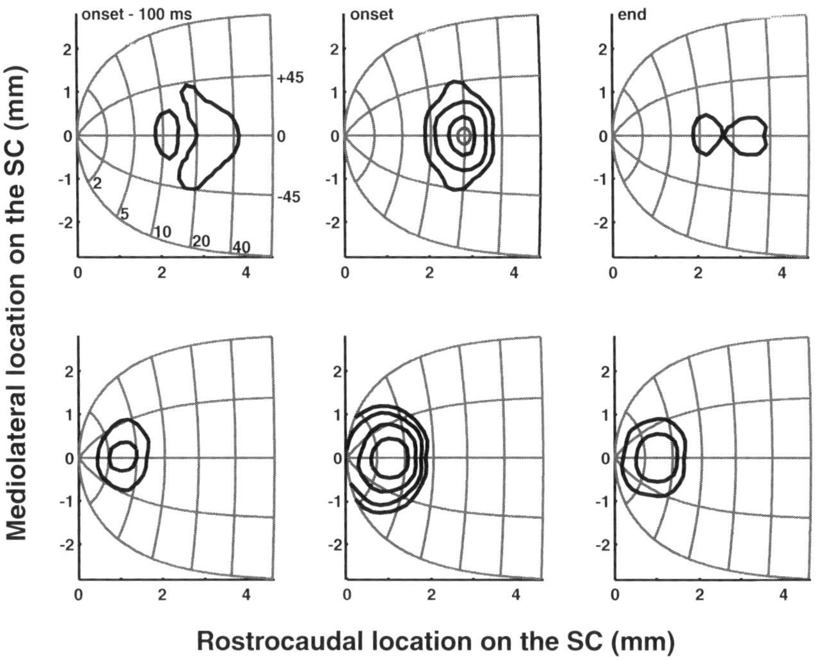

FIGURE 11.2 Estimated population activity in burst neurons in the SC for a 20° (upper sections) and a 2.5° (lower sections) horizontal saccade. From left to right activity at 100 ms before saccade onset, at saccade onset and at saccade end. Contour curves of constant discharge are 30, 60, 120, and 240 spikes/sec. (Modified from Anderson, R.W. et al., Two-dimensional saccade-related population activity in superior colliculus in monkey, *J. Neurophysiol.*, 80, 798, 1998. With permission.)

On the basis of the results obtained in their study, they concluded that the population activity in the more dorsally situated SC neurons (burst cells) produced an extensive mound of activity that was rather invariant in size and amplitude for any saccade. This mound of activity peaked just before saccade onset and then decayed rapidly during saccades. The center and most intense locus of activity in these cells remained in place on the SC map throughout the saccade. The population activity for the more ventrally situated buildup cells was qualitatively similar to that shown for burst cells in Figure 11.1 except that the locus of activity was even larger, occupying a full half or more of the SC at saccade onset. The spatial extent of the buildup neuron activity expanded during the saccade in the rostral as well as the mediolateral directions, but there was only a small rostrally directed shift in the centroid of this activity during saccades. Moreover, there was no evidence of an organized rostral spread of activity with temporal characteristics that could be used to control the end of saccades. Instead, the spread proceeded at about the same rate and to the same extent in the mediolateral as in the rostral direction. Further, the population discharge in both burst and buildup cells declined more rapidly for small saccades than for large saccades. Thus, the spatial size of the population discharge at saccade onset appears to be invariant with saccade amplitude (and therefore location on the SC), but this discharge declines more rapidly in time for saccades with smaller amplitudes.

Both the study of Munoz and Wurtz[26] and Anderson et al.,[7] suffered from the necessity of using averaged data obtained from several monkeys recorded over extended periods of time to obtain sufficient data to compute their estimated population spatiotemporal activity profiles within the SC. A more recent study has succeeded in recording from two cells simultaneously at separate locations on the SC.[27] Thus, the latter authors were able to directly observe, in individual trials, simultaneously occurring activity from a caudal and a rostral cell. They conclude that some cell pairs show a caudal-to-rostral sequence of activation, but others do not, and some even showed the opposite (rostral-to-caudal) sequence of activation. Overall, they agreed with the conclusion of Anderson et al.[7] that the activation patterns they observed could not serve as a distributed neural mechanism for controlling the dynamics of saccades. The latter conclusion was supported by another study that showed large saccades were not affected by lesions placed in the rostral SC that would have interrupted any rostrally spreading wave of activity.[28] Another study using functional imaging also showed that activity in the monkey SC was confined to a circumscribed region during saccades.[29]

11.5 USING SACCADE PERTURBATIONS TO TEST THEORIES OF SC FUNCTION

Because saccades normally have very stereotyped durations and velocity trajectories that vary systematically with saccade size, it has been difficult to establish quantitatively whether SC population activity codes dynamic motor error, eye velocity or current saccade displacement, and thus, might be part of the local feedback loop that controls these movements. The question is often phrased: Is the SC inside the local feedback loop controlling saccades? It has been difficult to establish causality in recording studies of single SC neurons due to the stereotyped nature of saccades. In addition, as pointed out by Quaia et al.,[8] the individual cells in highly distributed neural control systems do not necessarily reflect the simple dynamic signals that would be expected to appear in man-made, lumped element systems. Thus, there may be no one place or even any one group of neurons that *close the local feedback loop* for saccade control. Nevertheless, it is clear that the intermediate layers of the SC receive significant input from structures (e.g., the nucleus prepositus hypoglossi, the fastigial nuclei of the cerebellum and the mesencephalic reticular formation) that have been shown to contain neurons whose discharge is correlated with eye position (or change of eye position) or eye velocity or both,[30–34] and it would be surprising if the saccadic system did not make some use of this information to control the discharge of SC cells during saccades. By definition this would mean that the SC is inside a feedback loop, but not necessarily in the sense of a single-variable, negative feedback system.

Perhaps the most productive way that has been used to shed light on how feedback affects SC discharge during saccades has been to perturb saccade trajectory in mid-flight, and thus, break the normally stereotyped relationship of saccade characteristics (e.g., duration, peak velocity, decline of motor error) to saccade size.

When the activity of SC cells has been recorded during these perturbed trajectories, it has been possible to infer additional information about the relationship of this activity and saccade kinematics and dynamics.

Saccades can be arrested in midflight by stimulation in either the omnipause neuron (OPN) region in the brain stem[35] or the rostral part of the SC.[36,37] In these paradigms the eyes resume their movement toward the target (which had been removed at initial saccade onset) shortly after the electrical stimulation was turned off and end near the location of the now extinguished target. However, the dynamics of the overall interrupted movements were dramatically altered compared to the unstimulated saccades, which occurred with a fixed duration and a single-peaked velocity profile. Keller and Edelman[35] recorded the activity of SRBNs in the SC during these interrupted saccades and found that the burst activity of most SC cells was totally inhibited shortly after the onset of stimulation. After the stimulation ended, the same population of SC cells that were active before the stimulation resumed their discharge even though the resumed movement was too small to belong to the movement field of these cells. Only sporadic activity was found at the SC site that would have normally discharged for single movements with the same smaller magnitude vector as the resumed movement. Discharge in the active SC neurons showed a similar relationship with the remaining dynamic motor error to that computed for the final phase of a normal saccade to the same target location, although there were quantitative differences. This was taken as strong evidence of feedback to the colliculus about the time course and kinematics of the ongoing movement, but was not sufficient to establish that the SC computed dynamic motor error and was therefore *inside the local feedback loop*. This issue will be discussed more fully below in the section on distributed models of saccadic control.

In addition, in these experiments the interrupted SC discharge did not maintain a representation of the remaining motor error during the interrupted movement because most collicular cells were totally shut off during the stimulation, which could last as long as 30 to 40 ms. In more recent papers, two groups of investigators reported that a minority of saccade-related SC cells located in the deeper layers of the SC, and that showed the characteristics of buildup cells, were not turned off during the perturbing OPN or rostral SC stimulation.[4,36] It is possible that a few such cells with maintained coherent discharge at the center of the desired target location in the SC are sufficient to rekindle the discharge at this previously active region. One model described more fully below shows resumed activity at this site coding the original saccade goal after the end of the interrupting stimulus.[38]

The interrupted saccade paradigms discussed so far produced no (for OPN stimulation) or only minimal directional deviations (for SC stimulation). Furthermore, the perturbations were produced by electrical stimulation in the brain a technique, which always adds the potential confounding factor of stimulation of unknown paths and regions in addition to the assumed site of primary effect. Two more recent experiments have produced perturbations with alternative methods and have added additional information about the question of dynamic control of saccades by the SC. In the first, Goossens and Van Opstal[5] induced ocular blinks with airpuffs delivered near the onset of visually triggered saccades. The visual target was turned off before the onset of the targeting saccade. The blinks induced large spatiotemporal perturbations in the trajectories of saccades, including large direction changes, greatly slowed speeds and associated increases in the durations of saccades that still ended near the former location of the target. The authors speculated that the directional changes were produced by input from the brainstem blink centers to the saccadic burst generators and the slowing of saccades was produced by inhibitory inputs to the SC from these centers. These perturbations were associated with large changes in the discharge of SC neurons when compared to normal saccades with the same final displacement vector. Shortly after the onset of the blink, SC activity was transiently suppressed, but then resumed at a lower rate (when compared with the expected discharge for unperturbed saccades) as the perturbed saccade trajectory continued and eventually was redirected to land near the target. In further analysis of these events the authors found that dynamic motor error, now a vector function because of the perturbation of saccade direction, was no longer well correlated with instantaneous firing rate in SC cells. However, the increased durations of the burst of discharge in SC cells were rather well correlated with the increased saccade durations in the perturbed trajectories. Interestingly, the number of spikes in the total burst remained very similar in the perturbed and normal movements. On the basis of all their observations, the authors argue that the SC is not included in the local feedback loop controlling saccades, but that the lowered rate of discharge in SC cells for perturbed saccades causes a lowered rate in brainstem burst cells, and not the other way around. Finally, they suggest that a spike counting mechanism downstream from the SC accounts for the encoding of the actual displacement vector for the saccade.

Soetedjo et al.[39] slowed saccades by injecting muscimol into the OPN region of the brain stem. These saccades still ended accurately on the target. When they recorded SC neuron activity during the slowed movements, they found that the discharge of many of these cells was also slowed proportionally to the slowing of the movements and that the increased duration of the saccades was correlated with the duration of the burst activity in SC cells. The number of spikes in the burst remained constant for a given movement displacement. Although the results appear similar to those of Goossens and Van Opstal,[5] Soetedjo and colleagues interpreted their results as proof that the SC is inside the local feedback loop.

What are we to conclude on the basis of these perturbation studies about the conjecture that the SC is included in the local feedback loop? Overall the observations seem to show that the SC receives rather instantaneous feedback about the ongoing saccadic movement, but fall short of proving that the SC is the site where feedback exclusively controls saccade accuracy.

11.6 THE WINNER-TAKE-ALL CONCEPT OF SC FUNCTION

Summaries presented above suggest that a single, albeit large, locus of activity is present on the SC motor map by the onset of any saccadic movement. The center of this invariantly sized, distributed locus of activity shifts over the map as a function of the desired saccade vector. This type of activity in a computational neural map is commonly called a *winner-take-all* type of behavior.[40] Most conceptual and computational models of the SC assume that the SC operates as this type of map. Winner-take-all type of behavior fits well with the notion that the SC is close to the motor output of the saccadic system, and that this motor system, by definition, cannot carry out more than one saccade at a time.

A number of reports have recently challenged this traditional view of the operation of the SC. When two equally salient visual targets are presented simultaneously, the saccadic system often executes a movement that is directed in between the spatial locations of the two targets–a result that is called an averaging saccade.[41,42] Edelman and Keller[17] recorded SC activity during normal and averaging saccades and reported that the pattern of activity in the motor layers of the SC when two targets were present better represented the two separate locations of the visual stimuli than the single vector of the actual averaging saccade being executed. Glimcher and Sparks[43] produced conflicting data, but in a delayed-saccade paradigm.

Port and Wurtz[44] recorded simultaneously from two sites on the SC using a double-step saccadic paradigm. In the double-step paradigm two possible saccade targets are presented with variable inter-stimulus presentation times. The subject's task is to make sequential saccades to each target in the order that the targets were presented. When the temporal separation of the two visual targets in this paradigm was very short, monkeys sometimes executed saccades with highly curved trajectories. In these curved movements the trajectory initially pointed toward one target, but then curved around into the direction of the second target, and landed close to it at movement end. The activity initially was stronger at the site of the first target, but then activity declined at that site as activity increased at the site coding the other target. Considerable activity co-existed simultaneously during the movement at the two target sites on the SC. In another experiment McPeek and Keller[45] recorded activity at distractor sites in the SC during a visual search paradigm. In their paradigm an array of visual stimuli appeared simultaneously at equal eccentricity, but separated by 90° of visual angle. One of these stimuli was of an odd color. The animal's task was to saccade to the odd colored stimulus (the target) and not to one of the other stimuli (the distractors). They found that this paradigm also produced many curved saccades that were preceded by an accelerating discharge in SC neurons located at distractor sites. This discharge was terminated at, or just after, saccade onset. There was little or no discharge at distractor sites for the more frequently observed straight saccades. In order to confirm that saccade curvature was caused by presaccadic activity at SC sites representing distractors, they next presented a single visual target to elicit a saccade. On randomly selected trials they delivered an electrical stimulation at an SC site that would have coded the location of a distractor in the search task (Figure 11.3). This stimulus began just before saccade onset, increased in frequency and terminated at saccade onset, and thus, resembled the previously recorded distractor site activity for curved saccades during search. In those trials in which stimulation was delivered, the initial saccade direction was in between the direction of the visual target and the direction coded by the stimulated site on the SC (arrow 1). When the stimulation was terminated near saccade onset, the direction of the ongoing saccade changed over time (arrow 2) and the movement eventually ended near the location of the visual target. These authors interpreted their results to show that activity initially existed at two sites on the SC–the site representing the visual

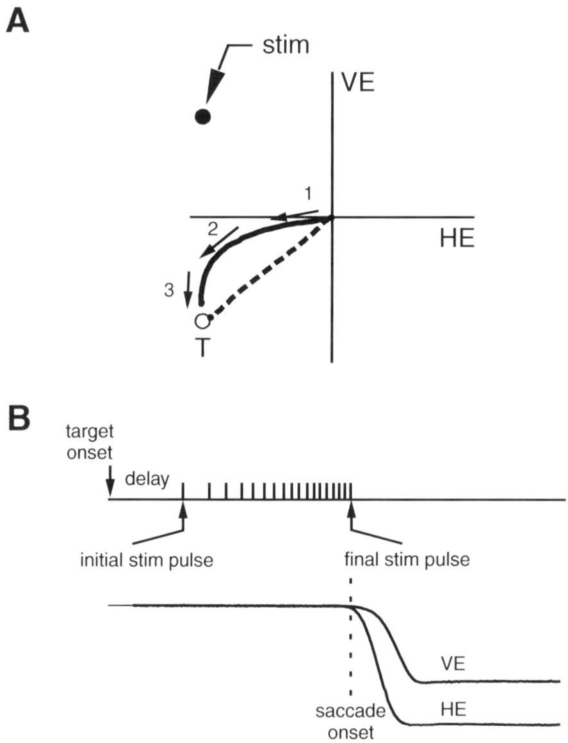

FIGURE 11.3 Curved saccade produced by presaccadic, subthreshold stimulation in the SC. (A) Arrangement in visual space of the target (open circle at T) that evoked the saccade and the image in visual space of the SC stimulation site (stim). Solid curve shows a curved saccade for a stimulated trial. Dashed curve shows an unstimulated control. Numbered arrows indicate the trajectory phases described in the text. (B) Schematic of the stimulus timing. Downward arrow on left indicates the onset of the visual target. The vertical dashed line indicates saccade onset. HE, horizontal eye position; VE, vertical eye position.

target and the site where the electrical stimulation was applied or the distractor appeared. Neither of these two experiments could account, on the basis of the activity they recorded in the SC, for the final direction of the saccade (arrow 3) as it approached the target. The final saccade direction would not normally be coded by activity at either of the two active SC sites (the site representing the visual target and the site of stimulation). Both Port and Wurtz[44] and McPeek and Keller[45] suggested that either a third SC site became active for the terminal phase of the movement, or that circuits downstream from the colliculus controlled the final trajectory of the movement.

Visual search paradigms in which an array of similar visual stimuli appear simultaneously, but the subject or animal is cued by the odd color or shape of one stimulus in the array to saccade to its location, produce sequential saccades with extremely short or no intersaccadic intervals.[46,47] McPeek and Keller[6] recorded the activity of cells in the SC during this paradigm and reported that activity could occur concurrently at the site that represented the target of the second, rapidly following saccade, even during the initial saccade to a distractor. They interpreted this result to mean that saccade planning for the second movement could proceed in parallel at a separate site in the SC even during the execution of a saccade coded by full-blown activity at another site.

11.7 DISTRIBUTED MODELING OF THE SUPERIOR COLLICULUS

Robinson's[48] concept that saccades are controlled by local feedback of efferent motor activity remains a cornerstone of all present models of the saccadic system. Initially these models used a lumped engineering systems approach to explain saccadic behavior.[49–52] Moschovakis[53] has reviewed the development of these types of models. A generic version of one form of these models is shown in Figure 11.4A to illustrate what is meant by the term *lumped.* All the dynamic elements in these models have single-input, single-output signals and well defined variables appear at discrete locations. For example, dynamic motor error is computed discretely in a comparator of the desired change in eye position (the goal) versus the instantaneous change in eye position. These models have the advantage of predicting the appearance of precise variables at single junctions in the model, but unfortunately, probably do not represent the reality of the distributed signal processing that is present in the brain. Figure 11.4B shows a partially distributed representation of the system in which processing in the SC is represented in a distributed fashion carried out with intracollicular excitatory and inhibitory connections. The input to this model is a distributed representation of visual target location represented spatially by which input lines are active (the inputs with solid lines and arrow heads). The desired saccade goal is computed by the distributed activity that is engendered by the dynamic neural network in the SC. Once the saccade begins, this distributed dynamic network receives feedback about current eye velocity and displacement from the SBG and outputs a distributed spatiotemporal signal that is decoded to yield a single input command to the SBG. Note that single command lines in all these schematics do not represent single neurons, but rather a group of neurons carrying a single signal. An even more realistic schematic of the saccadic system is shown in Figure 11.4C in which some of the multiple distributed feedback loops that exist in the system are shown. In this schematic the boxes represent dynamic neural networks existing in the SC, the posterior vermis/fastigial nucleus (CEREB) and the SBG. Experimental authentication of a model of this latter form is much more complex when function is not only distributed within the topographic motor maps present in the SC and possibly in the CEREB, but also over other brain regions. In this situation control function (e.g., the computation of dynamic motor error) is a system property that cannot be pinpointed to any one structure in the brain.

Nevertheless, interest in the development of a model of the saccadic system utilizing a distributed representation of the SC (Figure 11.4B) has grown rapidly because this type of model may be able to help unify and explain the explosion of experimental data collected in the SC. One common feature of many currently existing distributed SC models is that they utilize intracollicular connections and laterally distributed feedback of collicular output neurons to shape the model's population activity profiles present in the SC just before and during saccadic movements to match that present on the actual SC. This makes the model of the motor map a recurrent, dynamic network.[38,54–62] Some distributed models have focused on dynamic map updating to ensure spatially accurate sequences of saccades.[55,56,63–65] Due to space limitations this is a topic beyond the scope of the present review.

Depending on the symmetry and the parameters used to describe the intracollicular connections, two basic types of behavior can be obtained with these distributed

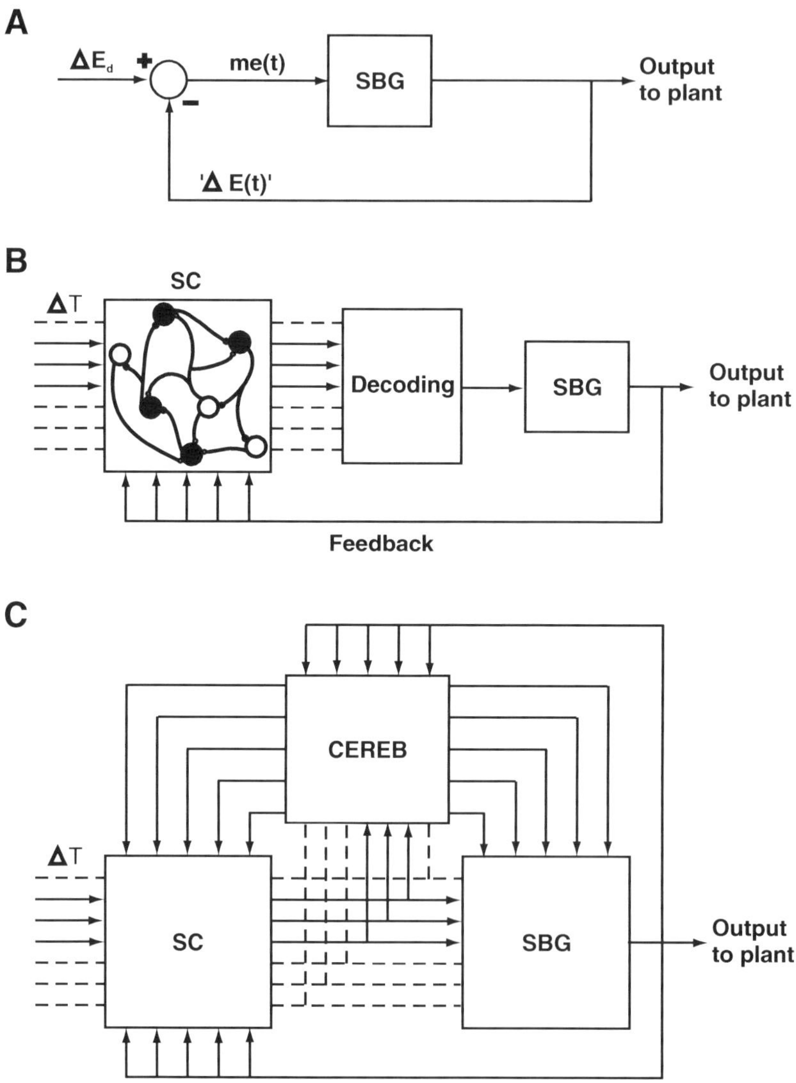

FIGURE 11.4

dynamic maps. In one class of models presaccadic activity is initially centered in the rostral colliculus in the so called fixation neurons. As saccade preparation proceeds activity in these rostral SC neurons declines and a separate locus of distributed activity begins to build up at the caudal SC site that codes the desired saccade vector. During the saccade this latter locus (mound of activity) or its leading edge travels rostrally across the colliculus, and when this spreading wave of activity reaches the rostral pole of the colliculus, the saccade is terminated.[57,59,61] Some of these models also utilize three distinct types of SC neurons, the fixation neurons, which are located in the rostral pole of the SC,[66] the burst neurons, and the buildup neurons. There are reciprocal inhibitory connections between the fixation neurons and the buildup neurons so that only one group can be fully turned on at any one time. Another feature of the moving wave class of models is that the rate and direction of motion on the SC map during a saccade is driven by the feedback of an eye velocity signal to the SC from the brainstem saccadic burst generator (SBG). The dynamic network in the SC map performs a spatial integration of this feedback such that the forward motion of the wave and its remaining distance to the rostral pole encodes the dynamic motor error. A number of investigations have cast doubt on the validity of this moving mound or wave hypothesis.[7,27–29,67] The remainder of this review will concentrate on dynamic models of the SC that do not ascribe this specific type of control function to the spatiotemporal pattern of peri-saccadic activity of SRBNs.

The other class of dynamic SC models also uses intracollicular connections and recurrent feedback from output cells to shape the SC network response in space and time for saccades of arbitrary size.[38,54,55,62,68] The fixation cells in the rostral SC play a secondary role in these models, and it is the activity in the more caudal cells–both burst neurons and buildup neurons at the site coding the desired saccade vector that determine saccade dynamics. In particular, two groups have taken quite different approaches to generate the spatiotemporal activity present in the SC in the peri-

FIGURE 11.4 (See facing page.) Schematic models of the saccadic system. (A) Single-signal path (lumped) model of the local feedback loop. ΔE_d; desired eye position at saccade end (*the saccade goal*); $\Delta E(t)$; brain's estimate of the current eye displacement (*efference copy*); me(t), dynamic motor error; SBG, saccadic burst generator. (B) Similar schematic with a distributed signal representation of the SC. A visual input (ΔT) indicates the initial excited region of the map. Excitatory cells in the SC network are shown with open circles and inhibitory cells with solid circles. Efferent feedback goes to the SC, but is distributed. Error signals must be represented in the distributed output of the SC. This output is readout or decoded below the level of the colliculus before input to the lumped SBG. Sets of parallel lines are meant to represent spatiotemporal coding. Active elements are shown with solid lines and arrow heads, inactive elements with dashed lines. Efferent feedback from the SBG (possibly eye displacement and velocity) is considered to be temporally coded so that all signal lines are active. (C) Realistic representation of the actual saccadic system in which each box is a dynamic, distributed network similar to SC in B. Here there are multiple feedback loops and the *feedback loop* is not closed at any one point in the system. CEREB, cerebellum. The output of the cerebellum is shown with temporal coding, but this fact is not yet established.

saccadic interval.[38,54,55] Both groups use local recurrent excitatory feedback from tecto-reticular burst neurons to help generate a large region of active neurons surrounding the locus of the visual input to the motor layer. The existence of this local excitation is supported anatomically[69,70] and physiologically.[71] Further, both models use inhibitory connections to limit the spatial extent of the burst activity. In both, the excitatory burst is gated by inhibitory input from the substantia nigra. All these mechanisms have extensive experimental support.[70,72–74]

An important distinguishing feature of the many SC models is how the distributed motor map is connected to the brainstem saccadic burst generator (SBG) cells. This process is labeled "decoding" in Figure 11.3B. This figure shows the distributed connections by which tectoreticular cells in the SC project through the decoding box to the SBG. Since the cells that constitute the SBG are very stereotyped in their behavior, all current models use a lumped representation at this level of the system with various alterations of a topology originally proposed by Robinson and his colleagues.[49] There are several existing hypotheses about how the distributed SC input is decoded in the SBG to yield a saccadic command.[75] Most experimental and theoretical evidence favor an analytic technique called vector averaging.[22] According to this mechanism each active site in the colliculus represents a potential vector displacement command — the location of the site determining direction, and the level of activity at the site, the magnitude of this potential vector. All the potential vectors are then averaged by circuits in the SBG to yield a single desired displacement vector.

Another hypothesis called vector summation assumes that each active site represents a minivector specifying that site's contribution to the desired displacement command. These minivectors are simply summed up in the SBG to yield the overall command. Since the amount of active colliculus and the magnitude of this distributed activity is rather constant for saccades of varied size and direction,[7] the vector summation model requires that each labeled line activity descending from the SC to the SBG be weighted at the level of the latter to reflect origin on the SC motor map. For example, projecting cells in the caudal colliculus should have greater synaptic effect on the SBG than rostral SC cells. Experimental evidence that this pattern of weighted connection exists has been provided by Moschovakis et al.[76] Both the model of Bozis and Moschovakis[55] and that of Arai and colleagues[38] use weighted vector summation to produce accurate saccadic behavior over a wide range of saccades just as vector averaging models do. It has been argued that one definitive way to determine which type of decoding process better represents the operation carried out in the saccadic system is to examine the saccadic behavior after the placement of a discrete lesion in its distributed motor map in the SC. Experimental results in such situations show a pattern of saccadic dysmetria that seem to be accounted for by a vector averaging, but not a vector summation, operation.[22] In particular, when a lesion is placed at a particular location on the SC map and a monkey is asked to make a saccade to a target whose image on the SC map is more caudal than the site of the lesion, the movement is found to be hypermetric. This result would not be predicted by a vector summation model because some of the SC minivectors specifying the desired saccade have been removed by the lesion. Only hypometric behavior is predicted by such models. However, such discriminative predictions are only valid for static SC motor maps. That is, systems in which the

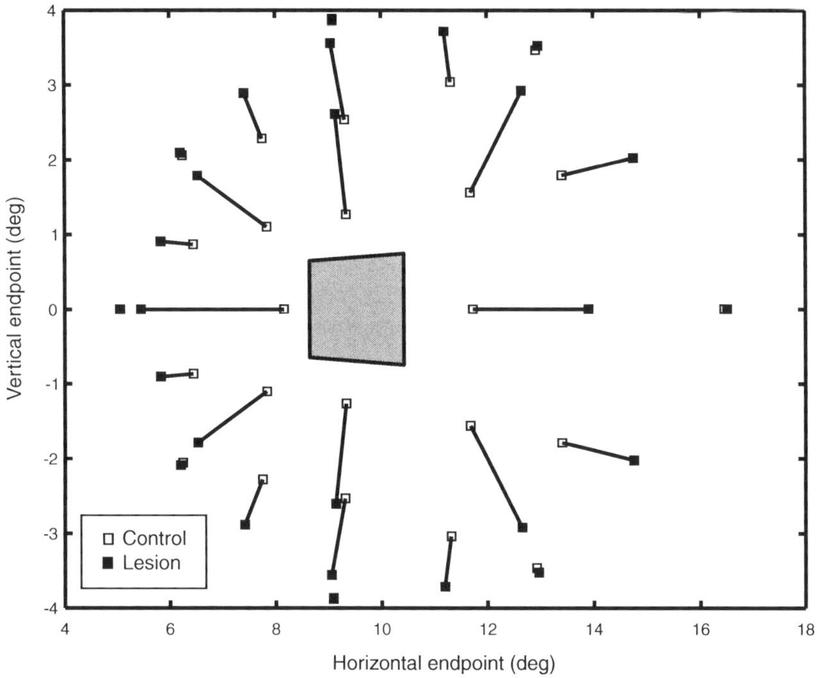

FIGURE 11.5 Map of errors in saccadic endpoints produced in a distributed vector summation model of the system by a lesion placed in the dynamic SC network. The open squares show the saccadic endpoints produced by the model before the lesion and the filled squares after the lesion. Straight lines connect the before and after symbols for the same target location. Where only filled symbols appear, these cover up identically located filled symbols. The gray region shows the extent of the simulated SC lesion transformed into visual space. (From Badler, J.B. and Keller, E.L., Decoding of a motor command vector from distributed activity in superior colliculus, *Biol. Cybern.*, 86, 179, 2002. With permission.)

distributed activity in the SC is read out once at some time just before saccade onset and this activity is then decoded to yield a desired saccade displacement vector (goal) that does not change with progression of the movement. Bozis and Moschovakis[55] first showed with a one-dimensional, linear SC model map and Badler and Keller[77] confirmed with a two-dimensional, logarithmically warped map that the center of the locus of activity on the SC does not remain in place when a lesion is placed in a dynamic map. The normal reciprocal and symmetrically balanced lateral connections in such maps are upset by the lesion, and the center of the activity moves radially away from the site of the lesion even before the saccade starts. As a result a new movement is coded by the time of saccade onset. The two-dimensional results of the latter authors' model shown in Figure 11.5 illustrate a pattern of saccadic dysmetria very similar to the experimental results of Lee et al.[22] In particular, saccades made to visual targets whose image on the SC map lies caudal to the site of the lesion produce hypermetric saccades. This emergent property of distributed, dynamic motor maps shows that one needs to be careful in interpreting experimental results with reasoning based on static SC maps. The saccade averaging results cited above are also predicted by intracollicular interactions in distributed dynamic network models.[54,55,60] When simulated visual stimuli appear at two, spatially separate locations in these models, the symmetrically arranged pattern of local excitation and distant inhibition are disturbed. Units located at SC sites in between the loci of the two visual stimuli are activated by the excitatory connections from both sites, but the units located at the two sites of visual input mutually inhibit each other. The result is a saccade that lands in between the two visually stimulated locations — the so-called averaging saccade. Some experimental data recorded from SC neurons in the monkey support this model demonstration of an increased activity locus at the in between site that appears over time following the appearance of the two visually activated loci.[43,78] Another experimental study does not support this model derived notion,[17] and shows instead that the activity profile on the SC remains close to the spatial distribution expected from the two visual stimuli even during the execution of the averaging saccade. This latter study also shows that the saccades that result from the simultaneous presentation of two visual stimuli at equal eccentricity, but separated in direction, are seldom in the true averaging direction, but instead land with a rather uniform distribution between the two targets. This behavior has been approximated in a dynamic network model in which random spatial and temporal noise was added to the distributed network activity generated by the two simulated visual stimuli.[79]

The feature that separates distributed SC models to the greatest extent is the role that each assigns to feedback from the SBG to the network portion of the model. In some dynamic models feedback about ongoing eye velocity is supplied in a distributed fashion to the SC network.[54,57,59,61,62] In others this feedback takes the form of a change in eye position signal.[54,56] In another, both eye displacement and eye velocity signals from the SBG are used to modulate activity in the SC network layer.[38] Other models do not explicitly use feedback to the SC network, but do have a change in eye position signal as an input to it.[8,55]

All models have to explain the fact that SC population discharge declines rapidly during saccades, and this decline is on average more rapid in the rostral colliculus (which is active for small, shorter duration saccades).[7] The decline in activity could be modeled as a preprogrammed feature built into the connection patterns of the dynamic SC network,[55] but this would not explain the SC activity recorded in many of the perturbed trajectory experiments. This activity remains tightly coupled to saccade duration, velocity and dynamic motor error even for saccades with widely varied characteristics. Unrealistic preprogramming of burst profile could be partially overcome by the feedback of an inhibitory signal roughly coupled with eye displacement to the dynamic network.[8] It is unproven whether this latter arrangement could reproduce the pattern of SC discharge that occurs in the perturbed trajectory experiments, although such an arrangement can produce accurate saccadic behavior.[8] The two-dimensional model of Arai et al.,[38] produces realistic distributed collicular activity and accurate saccadic behavior, when the dynamic SC network is supplied with both velocity and position feedback about the ongoing movement.

A number of additional arguments have been put forward to exclude the possibility that the SC is inside the local feedback loop.[8] These arguments include results obtained with prolonged electrical stimulation of the SC or with large lesions of the SC. This subject remains a contentious issue among oculomotor researchers and modelers and is likely to remain so due to the difficulty encountered in the interpretation of experimental data produced in highly distributed systems.

The experimental evidence that the SC motor layer does not always function as a winner-take-all network was presented above. One attempt to explain this with a distributed model divided the deeper portion of the SC into two layers containing burst cells dorsally and buildup cells ventrally.[56] Only the burst cells implement a winner-take-all scheme, while the buildup cells could display a much broader spatial locus or even multiple loci of activity. A similar approach has been used to model cellular discharge observed in the parietal cortex during closely spaced, sequential saccades. In this approach the model map is separated into two layers, one of which implements strong winner-take-all behavior, and therefore codes the present saccade, while a separate group of cells maintained a memory of the second saccade during the execution of the first.[80] McPeek and Keller[6] found evidence against both these schemes, at least for the SC, because they recorded concurrent activity coding the second saccade during the first saccade in all types of SRBNs in the SC. Strong winner-take-all behavior in a distributed neural map is implemented by the strength and extent of the lateral inhibitory connections. Until recently the experimental evidence for the existence in the SC of such lateral inhibitory connections seemed quite strong.[70,74] Recently, evidence has been presented that the long-range inhibitory connections needed to implement a strong winner-take-all type of network behavior may not be present in the SC motor layers.[81] These authors have used photostimulation and whole-cell patch clamp recording techniques to show this intriguing result. If these preliminary results are proved to be correct, distributed SC network models would have to abandon the usually represented strong lateral connections, and perhaps achieve a limited extent of map activation by the use of other inhibitory connections from the substantia nigra, for example.[54,55] Another approach would be to use the patchy cholinergic inputs from the pedunculopontine nucleus[82] to modify the excitability properties of local sites in the SC map to permit local regions where visual stimuli had appeared to be relatively immune to the inhibitory influence of feedback and lateral inhibition during the ongoing saccade. A focal increase in cell excitability has been demonstrated in the rat SC following the application of nicotine.[83]

Up to the present time most distributed models of the saccadic system have concentrated on the operation of the SC. However, evidence has been accumulating that the cerebellum plays an equally important, if not even more dominant role in controlling saccade trajectory than the SC.[84,85] The portions of the cerebellum contributing to this saccade function include pathways through the posterior vermis and fastigial nucleus.[86–88] This saccade-related circuitry will be referred to as just the cerebellum for simplicity. Two particularly significant models have appeared that attempt to include a role for the cerebellum in a more comprehensive scheme of operation of the saccadic system.[8,89] Quaia et al.[8] present a distributed model that produces saccades by the parallel cooperation of the SC and the cerebellum. Both structures are influenced by feedback information about the ongoing saccade. A spatial integrator located in the bilateral fastigial nucleus (FN) is hypothesized to play a dominate role in terminating saccades on target and to also correct directional errors during the movement. In this model activity in the FN would advance from one location at saccade onset to another mirror image location in the opposite FN at saccade end with a wave-like motion similar to that previously hypothesized to occur in the SC during saccades.[26] No direct evidence for this hypothesized wave-like motion in the FN exists,[88] and additional recordings with multiple electrodes are required to test the idea quantitatively, just as is the case for similar hypothesized activity in the SC. The model is able to reproduce accurate saccades in the face of prolonged electrical microstimulation of the SC, and it reproduces the effects of placing lesions in the FN. Furthermore, it has the very important property of explaining the curved saccadic trajectories that result from the placement of focal lesions in SC, a behavior that has not been possible to reproduce with distributed SC models alone.[90] The model has not been tested on the interrupted saccade paradigm[33] or the highly perturbed saccades observed in the blink paradigm.[5] The ability to produce accurate saccadic endpoint with realistic SC discharge for perturbed movements would seem to be an important test for this and other distributed models, because precise data on the kinematics and discharge in the SC during such perturbed movements exist. The distributed model of Arai et al.[38] was able to produce orthometric saccades with realistic discharge patterns in a model SC when saccade interruptions were simulated by brief reactivation of the model OPNs. This feature was an emergent property of the model after it was trained to make accurate slow saccades. The authors argued that the ability to make accurate saccades of different speeds is a natural condition faced by the saccadic system. During the training of their model, both position and velocity feedback to the SC were optimized to produce accurate saccades as saccadic speed was randomly varied by alterations of SBG gain. Apparently, the system was then able to compensate for interrupted movements as an extreme case of lowered speeds encountered in the training set.

No current distributed models have attempted to explain the curved trajectories that result when multiple visual targets exist, and no single winner for the desired saccadic displacement vector has emerged by the onset of the first saccade. In this situation the pattern of activity on the SC may no longer be located at the site normally associated with the actual saccade that is produced.[44,45] Similarly, during averaging movements that are directed between the locations of two equally salient targets, the discharge on the SC also does not represent the actual accomplished saccade vector, but better represents the locations coded by the two targets.[17] Past attempts to model the latter situation have used winner-take-all models of the SC distributed network, and have therefore predicted a resolution of the two visual mounds of activity originally present on the SC map into a single mound at the location of the averaged saccade vector by saccade onset.[54,55]

In conclusion, it appears that the use of distributed models that represent several separate functional anatomical structures in the saccadic system will become more common in the future. Progress in the validation of such models will be highly dependent on data obtained from more distributed methods of neural data collection in the alert behaving brain (e.g., multielectrode arrays, voltage sensitive dyes or functional imaging).

ACKNOWLEDGMENT

Supported by NIH grant R01 EY06860.

REFERENCES

1. Munoz, D.P. and Wurtz, R.H., Saccade-related activity in monkey superior colliculus. I. Characteristics of burst and buildup cells, *J. Neurophysiol.*, 73, 2313, 1995.
2. Glimcher, P.W. and Sparks, D.L., Movement selection in advance of action in the superior colliculus, *Nature*, 355, 542, 1992.
3. Basso, M.A. and Wurtz, R.H., Modulation of neuronal activity in superior colliculus by changes in target probability, *J. Neurosci.*, 18, 7519, 1998.
4. Keller, E.L., Gandhi, N.J., and Sekaran, S.V., Activity in deep intermediate layer collicular neurons during interrupted saccades, *Exp. Brain Res.*, 130, 227, 2000.
5. Goossens, H.H.L.M. and Van Opstal, A.J., Blink-perturbed saccades in monkey. II. Superior colliculus activity, *J. Neurophysiol.*, 83, 3430, 2000.
6. McPeek, R.M. and Keller, E.L., Superior colliculus activity related to concurrent processing of saccade goals in a visual search task, *J. Neurophysiol.*, 87, 1805, 2002.
7. Anderson, R.W. et al., Two-dimensional saccade-related population activity in superior colliculus in monkey, *J. Neurophysiol.*, 80, 798, 1998.
8. Quaia, C., Lefèvre, P., and Optican, L.M., Model of the control of saccades by superior colliculus and cerebellum, *J. Neurophysiol.*, 82, 999, 1999.
9. Wurtz, R.H. and Goldberg, M.E., Activity of superior colliculus in behaving monkey. III. Cells discharging before eye movements, *J. Neurophysiol.*, 35, 575, 1972.
10. Goldberg, M.E. and Wurtz, R.H., Activity of superior colliculus in behaving monkey: I. Visual receptive fields of single neurons, *J. Neurophysiol.*, 35, 542, 1972.
11. Schiller, P.H. and Koerner, F., Discharge characteristics of single units in superior colliculus of the alert rhesus monkey, *J. Neurophysiol.*, 34, 920, 1971.
12. Cynader, H. and Berman, N., Receptive-field organization of monkey superior colliculus, *J. Neurophysiol.*, 35, 187, 1972.
13. Robinson, D.A., Eye movements evoked by collicular stimulation in the alert monkey, *Vision Res.*, 12, 1795, 1972.
14. Schiller, P.H. and Stryker, M.P., Single-unit recording and stimulation in superior colliculus of the alert rhesus monkey, *J. Neurophysiol.*, 35, 915, 1972.
15. Ottes, F.P., Van Gisbergen, J.A.M., and Eggermont, J.J., Visuomotor fields of the superior colliculus: a quantitative model, *Vision Res.*, 26, 857, 1986.
16. Sparks, D.L., Holland, R., and Guthrie, B.L., Size and distribution of movement fields in the monkey superior colliculus, *Brain Res.*, 113, 21, 1976.
17. Edelman, J.A. and Keller, E.L., Dependence on target configuration of express saccade-related activity in the primate superior colliculus, *J. Neurophysiol.*, 80, 1407, 1998.
18. Anderson, R.W., Das, S., and Keller, E.L., Estimation of spatiotemporal neural activity using radial basis function networks, *J. Comput. Neurosci.*, 5, 421, 1998.
19. Mohler, C.W. and Wurtz, R.H., Organization of monkey superior colliculus: Intermediate layer cells discharging before eye movements, *J. Neurophysiol.*, 39, 722, 1976.
20. Berthoz, A., Grantyn, A., and Droulez, J., Some collicular efferent neurons code saccadic eye velocity, *Neurosci. Lett.*, 72, 289, 1986.

21. Van Opstal, A.J. et al., Comparison of saccades evoked by visual stimulation and collicular electrical stimulation in the alert monkey, *Exp. Brain Res.*, 79, 299, 1990.
22. Lee, C., Rohrer, W.H., and Sparks, D.L., Population coding of saccadic eye movements by neurons in the superior colliculus, *Nature*, 332, 357, 1988.
23. Stanford, T.R., Freedman, E.G., and Sparks, D.L., Site and parameters of microstimulation: evidence for independent effects on the properties of saccades evoked from the primate superior colliculus, *J. Neurophysiol.*, 76, 3360, 1996.
24. Waitzman, D.M. et al., Superior colliculus neurons mediate the dynamic characteristics of saccades, *J. Neurophysiol.*, 66, 1716, 1991.
25. Munoz, D.P., Pélisson, D., and Guitton, D., Movement of neural activity on the superior colliculus motor map during gaze shifts, *Science*, 251, 1358, 1991.
26. Munoz, D.P. and Wurtz, R.H., Saccade-related activity in monkey superior colliculus. II. Spread of activity during saccades, *J. Neurophysiol.*, 73, 2334, 1995.
27. Port, N.L., Sommer, M.A., and Wurtz, R.H., Multielectrode evidence for spreading activity across the superior colliculus movement map, *J. Neurophysiol.*, 84, 344, 2000.
28. Aizawa, H. and Wurtz, R.H., Reversible inactivation of monkey superior colliculus. I. Curvature of saccadic trajectory, *J. Neurophysiol.*, 79, 2082, 1998.
29. Moschovakis, A.K., Gregoriou, G.G., and Savaki, H.E., Functional imaging of the primate superior colliculus during saccades to visual targets. *Nature Neurosci.*, 4, 1026, 2001.
30. Moschovakis, A.K., Karabelas, A.B., and Highstein, S.M., Structure-function relationships in the primate superior colliculus. II. Morphological identity of presaccadic neurons, *J. Neurophysiol.*, 60, 263, 1988.
31. Edwards, S.B. et al., Sources of subcortical projections to the superior colliculus in the cat, *J. Comp. Neurol.*, 184, 309, 1979.
32. Hartwich-Young, R., Nelson, J.S., and Sparks, D.L., The perihypoglossal projection to the superior colliculus in the rhesus monkey, *Vis. Neurosci.*, 4, 29, 1990.
33. May, P.J. et al., Cerebellotectal pathways in the macaque: implications for collicular generation of saccades, *Neuroscience*, 36, 305, 1990.
34. Corvisier, J. and Hardy, O., Topographical characteristics of preposito-collicular projections in the cat as revealed by Phaseolus vulgaris-leucoagglutinin technique. A possible organisation underlying temporal-to-spatial transformations. *Exp. Brain Res.* 114, 461, 1997.
35. Keller, E.L. and Edelman, J.A., Use of interrupted saccade paradigm to study spatial and temporal dynamics of saccadic burst cells in superior colliculus in monkey, *J. Neurophysiol.*, 72, 2754, 1994.
36. Munoz, D.P., Waitzman, D.M., and Wurtz, R.H., Activity of neurons in monkey superior colliculus during interrupted saccades, *J. Neurophysiol.*, 75, 2562, 1996.
37. Gandhi, N.J. and Keller, E.L., Comparison of saccades perturbed by stimulation of the rostral superior colliculus, the caudal superior colliculus, and the omnipause neuron region, *J. Neurophysiol.*, 82, 3236, 1999.
38. Arai, K. et al., A distributed model of the saccade system: simulations of temporally perturbed saccades using position and velocity feedback, *Neural Netw.*, 12, 1359, 1999.
39. Soetedjo, R., Kaneko, C.R.S., and Fuchs, A.F., Evidence that the superior colliculus participates in the feedback control of saccadic eye movements, *J. Neurophysiol.*, 87, 679, 2002.
40. Didday, R., A model of visuomotor mechanisms in the frog optic tectum, *Math. Biosciences*, 30, 169, 1976.
41. Findlay, J.M., Global visual processing for saccadic eye movements, *Vision Res.*, 22, 1033, 1982.

42. Ottes, F.P., Van Gisbergen, J.A.M., and Eggermont, J.J., Metrics of saccade responses to visual double stimuli: two different modes, *Vision Res.*, 24, 1169, 1984.
43. Glimcher, P.W. and Sparks, D.L., Representation of averaging saccades in the superior colliculus of the monkey, *Exp. Brain Res.*, 95, 429, 1993.
44. Port, N.L. and Wurtz, R.H., Two electrode recordings in monkey superior colliculus during curved saccades, *Soc. Neurosci. Abstr.*, 26, 297, 2000.
45. McPeek, R.M. and Keller, E.L., Competition between saccadic goals in the superior colliculus produces saccade curvature, *J. Neurophysiol.*, 89, 2577, 2002.
46. McPeek, R.M., Skavenski, A.A., and Nakayama, K., Concurrent processing of saccades in visual search, *Vision Res.*, 40, 2499, 2000.
47. McPeek, R.M. and Keller, E.L., Short-term priming, concurrent processing, and saccade curvature during a target selection task in the monkey, *Vision Res.*, 41, 785, 2001.
48. Robinson, D.A., Oculomotor control signals, in *Basic Mechanisms of Ocular Motility and Their Clinical Implications*, Lennerstrand, G. and Bach-Y-Rita, P., Eds., Pergamon Press, Oxford, 1975,
49. Van Gisbergen, J.A.M., Robinson, D.A., and Gielen, S., A quantitative analysis of generation of saccadic eye movements by burst neurons, *J. Neurophysiol.*, 45, 417, 1981.
50. Tweed, D. and Vilis, T., A two dimensional model for saccade generation, *Biol. Cybern.*, 52, 219, 1985.
51. Scudder, C.A., A new local feedback model of the saccadic burst generator, *J. Neurophysiol.*, 59, 1455, 1988.
52. Jürgens, R., Becker, W., and Kornhuber, H.H., Natural and drug-induced variations of velocity and duration of human saccadic eye movements: Evidence for a control of the neural pulse generator by local feedback, *Biol. Cybern.*, 39, 87, 1981.
53. Moschovakis, A.K., Neural network simulations of the primate oculomotor system I. The vertical saccadic burst generator, *Biol. Cybern.*, 70, 291, 1994.
54. Arai, K., Keller, E.L., and Edelman, J.A., Two-dimensional neural network model of the primate saccadic system, *Neural Netw.*, 7, 1115, 1994.
55. Bozis, A. and Moschovakis, A.K., Neural network simulations of the primate oculomotor system. III. A one-dimensional, one-directional model of the superior colliculus., *Biol. Cybern.*, 79, 215, 1998.
56. Grossberg, S. et al., A neural model of multimodal adaptive saccadic eye movement control by superior colliculus, *J. Neurosci.*, 17, 9706, 1997.
57. Lefèvre, P. and Galiana, H.L., Dynamic feedback to the superior colliculus in a neural network model of the gaze control system, *Neural Netw.*, 5, 871, 1992.
58. Lefèvre, P., Quaia, C., and Optican, L.M., Distributed model of control of saccades by superior colliculus and cerebellum, *Neural Netw.*, 11, 1175, 1998.
59. Massone, L.E. and Khoshaba, T., Local dynamic interactions in the collicular motor map: a neural network model, *Network.*, 6, 1, 1995.
60. Van Opstal, A.J. and Van Gisbergen, J.A.M., A nonlinear model for collicular spatial interactions underlying the metrical properties of electrically elicited saccades, *Biol. Cybern.*, 60, 171, 1989.
61. Optican, L.M., A field theory of saccade generation: temporal-to-spatial transform in the superior colliculus, *Vision Res.*, 35, 3313, 1995.
62. Van Opstal, A.J. and Kappen, H., A two-dimensional ensemble coding model for spatial-temporal transformation of saccades in monkey superior colliculus, *Network.*, 4, 19, 1993.
63. Dominey, P.F. and Arbib, M.A., A cortico-subcortical model for generation of spatially accurate saccades, *Cereb. Cortex.*, 2, 153, 1992.

64. Droulez, J. and Berthoz, A., A neural network model of sensoritopic maps with predictive short-term memory properties, *Proc. Natl. Acad. Sci. U.S.A.*, 88, 9653, 1991.
65. Krommenhoek, K.P. et al., Remapping of neural activity in the motor colliculus: A neural network study, *Vision Res.*, 33, 1287, 1993.
66. Munoz, D.P. and Wurtz, R.H., Fixation cells in monkey superior colliculus. I. Characteristics of cell discharge, *J. Neurophysiol.*, 70, 559, 1993.
67. Sparks, D.L., Are gaze shifts controlled by a "moving hill" of activity in the superior colliculus?, *TINS*, 16, 214, 1993.
68. Van Opstal, A.J. and Van Gisbergen, J.A.M., A model for collicular efferent mechanisms underlying the generation of saccades, *Brain Behav. Evol.*, 33, 90, 1989.
69. Moschovakis, A.K., Karabelas, A.B., and Highstein, S.M., Structure-function relationships in the primate superior colliculus. I. Morphological classification of efferent neurons, *J. Neurophysiol.*, 60, 232, 1988.
70. Olivier, E. et al., Evidence for glutamatergic tectotectal neurons in the cat superior colliculus: a comparison with GABAergic tectotectal neurons, *Eur. J. Neurosci.*, 12, 2354, 2000.
71. Pettit, D.L. et al., Local excitatory circuits in the intermediate gray layer of the superior colliculus, *J. Neurophysiol.*, 81, 1424, 1999.
72. Karabelas, A.B. and Moschovakis, A.K., Nigral inhibitory termination on efferent neurons of the superior colliculus: an intracellular horseradish peroxidase study in the cat, *J. Comp. Neurol.*, 239, 309, 1985.
73. Hikosaka, O. and Wurtz, R.H., Visual and oculomotor functions of monkey substantia nigra pars reticulate V. Relationship of substantia nigra to superior colliculus, *J. Neurophysiol.*, 49, 1285, 1983.
74. Munoz, D.P. and Istvan, P., Lateral inhibitory interactions in the intermediate layers of the monkey superior colliculus, *J. Neurophysiol.*, 79, 1193, 1998.
75. Groh, J.M., Converting neural signals from place codes to rate codes, *Biol. Cybern.*, 85, 159, 2001.
76. Moschovakis, A.K. et al., An anatomical substrate for the spatiotemporal transformation, *J. Neurosci.*, 18, 10219, 1998.
77. Badler, J.B. and Keller, E.L., Decoding of a motor command vector from distributed activity in superior colliculus, *Biol. Cybern.*, 86, 179, 2002.
78. Van Opstal, A.J. and Van Gisbergen, J.A.M., Role of monkey superior colliculus in saccade averaging, *Exp. Brain Res.*, 79, 143, 1990.
79. Das, S., Keller, E.L., and Arai, K., A distributed model of the saccadic system: The effects of internal noise, *Neurocomputing*, 11, 245, 1996.
80. Mazzoni, P. et al., Motor intention activity in the macaque's lateral intraparietal area. I. Dissociation of motor plan from sensory memory, *J. Neurophysiol.*, 76, 1439, 1996.
81. Hall, W.C., Helms, M.C., and Özen, G., Horizontal intrinsic circuitry of the superior colliculus, *Soc. Neurosci. Abstr.*, 31, 36, 2001.
82. Graybiel, A.M., A stereometric pattern of distribution of acetylcholinesterase in the deep layers of the superior colliculus, *Nature*, 272, 539, 1978.
83. Aizawa, H. et al., Injection of nicotine into the superior colliculus facilitates occurrence of express saccades in monkeys, *J. Neurophysiol.*, 82, 1642, 1999.
84. Robinson, F.R., Role of the cerebellum in movement control and adaptation, *Curr. Opin. Neurobiol.*, 5, 755, 1995.
85. Takagi, M., Zee, D.S., and Tamargo, R.J., Effects of lesions of the oculomotor vermis on eye movements in primate: saccades, *J. Neurophysiol.*, 80, 1911, 1998.

86. McElligott, J.G. and Keller, E.L., Neuronal discharge in the posterior cerebellum: Its relationship to saccadic eye movement generation, in *Functional Basis of Ocular Motility Disorders*, Lennerstrand, G., Zee, D.S., and Keller, E.L., Eds., Pergamon Press, Oxford, 1982,
87. Fujikado, T. and Noda, H., Saccadic eye movements evoked by microstimulation of lobule VII of the cerebellar vermis of macaque monkeys, *J. Physiol. Lond.*, 394, 573, 1987.
88. Fuchs, A.F., Robinson, F.R., and Straube, A., Role of the caudal fastigial nucleus in saccade generation. I. Neuronal discharge patterns, *J. Neurophysiol.*, 70, 1723, 1993.
89. Dean, P., Modelling the role of the cerebellar fastigial nuclei in producing accurate saccades: the importance of burst timing, *Neuroscience*, 68, 1059, 1995.
90. Keller, E.L., unpublished data, 2002.

12 Commands for Coordinated Eye and Head Movements in the Primate Superior Colliculus

David L. Sparks

CONTENTS

12.1 INTRODUCTION

Results of recent experiments have modified concepts of how neurons in the intermediate and deeper layers of the superior colliculus (SC) represent commands for

0-8493-0097-5/04/$0.00+$1.50

orienting movements of the eyes and head. This chapter explores the implications of these new findings for three important problems associated with collicular function: the spatial-to-temporal transform problem, population coding, and the types of feedback signals received by collicular neurons.

12.2 KEY QUESTIONS

12.2.1 Spatial-to-Temporal Transform Problem (STTP)

Electrical stimulation of the intermediate and deeper layers of SC of monkeys with their heads restrained produces conjugate, contralateral saccades with amplitudes and directions that depend on the site of collicular stimulation.[1] When different regions of the colliculus are activated, the direction and amplitude of the stimulation-evoked movements change systematically,[1] revealing a motor map that is in anatomical register with the retinotopic map found in the overlying superficial layers.[2] Early studies reported that variations in the parameters of stimulation (intensity, frequency, pulse width, and train duration) had no effect on the direction, amplitude, or velocity of the stimulation-evoked movements.[1,2] Thus, it was concluded that the spatial distribution of activity within the SC encodes information about the characteristics of a saccade, and that neither the level nor duration of collicular activity plays an important role in saccade execution.

Downstream from the SC, both the level and duration of motor command signals are important. Excitatory burst neurons (EBNs) in the pons generate activity that is primarily responsible for the horizontal component of a saccade. Similarly, EBNs located in the midbrain produce activity responsible for vertical and torsional components of saccades. The temporal characteristics of EBN activity are related to the temporal features of the saccade. Burst rate is highly correlated with saccade velocity and saccade duration is tightly coupled to burst duration.[3–6] EBN signals are conveyed monosynaptically to motoneurons innervating extraocular muscle fibers. The motoneurons display a pulse-step discharge pattern. The transient increase in muscle tension needed to move the eye at saccadic velocity results from a pulse of innervation, and the steady-state tension needed to hold the eye at its new location in the orbit is due to a step change in the firing rate of motoneurons.[7–12] The amplitude and duration of the pulse affect the size and speed of saccades.

How are the place coded command signals observed in SC translated into the temporally coded signals observed in premotor neurons located in pons and midbrain? This question,[1] called the spatial-to-temporal transform problem (STTP), remains one of the most important problems in the neural control of gaze. The related problem of extracting separate signals for the horizontal and vertical burst generator circuits is known as the vector decomposition problem. The transformed signals must control both saccade direction and amplitude.

The amplitude of saccades evoked by collicular stimulation increases as the stimulation site moves more caudally. A rostrocaudal gradient of the number of tectal neurons projecting to pontine and midbrain burst neurons could be an important aspect of controlling saccade amplitude. Such a gradient in the number of neurons projecting to the region surrounding the abducens nucleus was described in an early experiment,[13] but was not observed in subsequent studies.[14–16] In retrospect, the failure to replicate the early observation is understandable because, in fact, a simple rostrocaudal gradient is not the pattern of connectivity required. Collicular neurons coding for movements having a 5° horizontal component reside in many different locations in the motor map. Cells discharging before 5° purely horizontal movements do reside in rostral SC, but those discharging maximally before movements with a 30° vertical component and a 5° horizontal component reside in caudal SC. The amplitude lines on Robinson's motor map represent vectorial amplitude, not the amplitude of the horizontal or vertical components.

Contemporary anatomical methods were used to reexamine the rostrocaudal gradient of SC connections to abducens nucleus using an approach that takes the details of the collicular map into account. Grantyn and colleagues[17] injected rabies virus into the lateral rectus muscle of rhesus monkeys and measured the retrograde transneuronal transport of the virus to SC to determine the density of neurons projecting to abducens nucleus. If the relative density of projection neurons is an important part of the STTP, the number of horizontal eye movement-related SC neurons should increase with distance from the rostral pole of the SC and decrease as a function of distance from the representation of the horizontal meridian. Counts of the number of labeled neurons in the intermediate SC layers located inside a 1-mm-wide band that matched the horizontal meridian of the collicular motor map were plotted against distance from the rostral SC pole. Caudal sites contained more neurons, but the experimentally observed density gradients were not as steep as the gradient produced by models in which the density of collicular output neurons was the critical factor accounting for the increase in saccade amplitude. The distributions of cell densities in the intermediate SC layers representing saccades of equal vectorial amplitudes but different directions were also examined. Concentrations of labeled cells were highest in the vicinity of the horizontal meridian but decreased more sharply toward the periphery than predicted by models. Thus, the authors conclude that the density gradients of cells projecting to motoneurons cannot fully account for the amplitude of a pure horizontal saccade or for the amplitude of the horizontal component of oblique saccades. The retrograde labeling method permits a description of the relative number of cells projecting to the abducens nucleus, but possible differences in the synaptic efficacy of each labeled cell's output are more difficult to quantify.

Moschovakis and colleagues[18] stimulated different regions of the collicular map, noted the amplitudes of the horizontal and vertical components of the stimulation-evoked movement, and then made small injections of biocytin into the stimulated areas. They counted the number of boutons labeled in the paramedian pontine reticular formation (PPRF) after anterograde transport of the biocytin for each injection. A significant positive correlation was observed between the normalized number of labeled boutons and the amplitude of the horizontal component of movements evoked by stimulation. This differential weighting of anatomical projections from different parts of the collicular map to PPRF must be a critical aspect of the STTP. However, additional processing is probably required to account for the site-specific maximal amplitudes observed with collicular stimulation. With a simple weighting of synaptic connections, the effects of strong stimulation at a rostral site could be equivalent to weaker stimulation at a more caudal site.

12.2.2 Population Coding

The receptive and movement fields of collicular neurons are large and coarsely tuned.[19–21] Initially, it was assumed that neurons with such coarsely tuned receptive and movement fields could not generate accurate saccades and, therefore, that the SC was not involved in coding the exact location of a visual target or in specifying the exact distance and direction of the saccade required to look to the target. Instead, it was proposed that the activity of collicular neurons merely facilitated movements that directed gaze to a broad area of the visual field.[21] Later, researchers realized that information about the precise parameters of an impending saccade could be coded spatially — the location of the active population of neurons within the SC could specify, precisely, saccade metrics. A number of schemes by which this information could be extracted from a spatial or anatomical code were suggested.[20,22–25] In these schemes, each member of the active population contributes to the movement but the exact trajectory of a saccade is determined by the average or sum of the population activity.

Results of experiments in which small subsets of the active population were reversibly inactivated or activated,[26,27] supported a population averaging hypothesis. The vector averaging model of Sparks and colleagues,[20] assumes that the region of collicular neurons active before a given saccade occupies a symmetrical area within the motor map. Only the neurons in the center of the active population discharge maximally before the programmed movement. But, for each subset of active neurons producing a movement tendency with a direction and amplitude other than the programmed movement, a second subset of neurons producing an opposing movement tendency will be active. The resultant of the two movements will have the programmed direction and amplitude. The vector-averaging hypothesis does not require neurons with sharply tuned movement fields in order to generate accurate saccades.

According to the vector-averaging hypothesis, a predictable pattern of errors should be observed when a small region of colliculus is inactivated and animals are required to make saccades of different directions and amplitudes. The predicted pattern of errors was observed after small injections produced a local reversible inactivation of collicular neurons.[27] Further evidence for a vector-averaging model was obtained in experiments in which saccades were evoked by microstimulation of the frontal eye field (FEF) before and after local inactivation of neurons in SC. When SC inactivation affected only a subset of the neurons that would normally be active during the FEF evoked saccades, the direction and amplitude of the evoked saccade shifted in a manner that was consistent with the SC inactivation removing a component of a vector average.[28]

12.2.3 The Role of Feedback

Most current models assume that saccades, once considered to be ballistic or preprogrammed movements, are under feedback control. What is the source of the putative feedback signal? It is not visual. A saccade ends before signals from the retina reach central visual areas that could provide feedback about accuracy. Necessary feedback signals are not derived from sensory receptors in extraocular muscle fibers. Saccadic accuracy is not affected significantly by elimination of proprioceptive feedback.[29] Zee and colleagues[30] hypothesized that a copy of a motor command (corollary discharge) could be used as a feedback signal for controlling saccade amplitude and almost all contemporary models of the saccadic system use a corollary discharge signal for feedback. This is possible because the load confronting the control system is relatively constant. Changes in the mechanical properties of the plant associated with trauma or aging would not prevent the use of this computational strategy if the gain of the critical signals is modifiable.

Some models of the saccadic system assume that the SC is within a dynamic feedback loop so that the locus[31,32] or level[33] of collicular activity continuously changes as a saccade is executed. Munoz and colleagues hypothesized that during a large saccade, a population of neurons in the caudal colliculus will discharge first. As the eye moves (reducing the difference between the desired displacement of the eye and the current eye position), the site of neural activity migrates rostrally in the SC until, finally, neurons at the rostral pole of the colliculus encoding a motor error of zero become active. If the distribution of activity within the SC is represented as a relief map, the nested contours of increasing activity can be thought of as a *hill* that, according to the hypothesis, migrates rostrally during the course of the movement. According to this hypothesis, motor error is mapped dynamically by the location of active cells within the motor map found in the deeper layers of the SC.

Initial evidence offered in support of this hypothesis was evaluated as either inconsistent with the model or subject to plausible alternative explanations.[34] Subsequent quantitative analyses of the distribution of activity within SC during a saccade have not been consistent with the notion of a systematic and sequential shift from caudal to rostral SC.[35–37]

Waitzman and colleagues hypothesized that saccade dynamics are controlled by the discharge profiles of SC neurons.[33] In this model, the locus of activity in the SC encodes the desired eye movement and the level of activity represents the dynamic motor error. The evidence supporting this hypothesis is meager. The hypothesis was based upon a correlation (observed for a subset of collicular cells) between the decay in spike discharge at the end of a burst and the reduction in vectorial motor error occurring during a saccade. There is no direct evidence that the decay in collicular activity causes the decay in motor error. Moreover, the methods used to obtain this correlation are problematic and other necessary lines of supporting evidence are missing.

Waitzman et al.[33] obtained a heavily filtered estimate of the decay in spike frequency of collicular cells with saccade-related activity by obtaining a spike density function that was the average of several trials. The spike density function was generated by substituting a Gaussian pulse of variable width for each action potential. The temporal waveform of this average spike density function was compared to the time course of dynamic motor error by shifting one of the waveforms to optimize the fit between the curves, rather than shifting the function by a fixed time. One concern with this analysis is based upon the fact that saccades are highly stereotyped and the bursts generated by neurons in different regions of the SC are quite similar.[19] Thus, the relationship between dynamic motor error and the decay in spike discharge could be an epiphenomenon that emerges when two stereotyped decay functions are filtered and time-shifted to maximize the optimal fit. Large gaze shifts that involve coordinated movements of the eyes and head extend the time in which a relationship between frequency of cell discharge and motor error can be examined. The activity of cells in caudal SC of monkeys that discharge during combined eye/head gaze shifts is poorly correlated with dynamic motor error.[38] It should also be noted that Bozis and Moschovakis developed a model of the saccadic system that generated a monotonic relationship between motor error and the instantaneous discharge of collicular burst cells without placing the SC within a feedback loop.[39]

A second concern is that the analysis of Waitzman et al. ignores all members of the active population, except those in the center. Compared to the entire active population, only a few cells reside in the center of the active population and they may contribute very little to the total population output. Although surrounding cells may discharge less vigorously and with a different time course, because the number of cells is so large, their contribution to the population output may actually be greater than the contribution of the few cells in the middle. Because the activity of cells on the periphery of the active population contribute to the each movement,[27] what needs to be computed to test the Waitzman et al. hypothesis is the time course of the population output — not the time course of the activity of the cells in the middle of the population. The relationship between the total population output and dynamic motor error is unknown.

Also unknown is whether the subset of cells displaying the putative monotonic decay in spike discharge, rather than the cells that do not display this relationship, are those that project to downstream premotor circuits. Equally critical to the hypothesis is a demonstration that the necessary feedback signal actually reaches the putative collicular dynamic motor error cells. As Waitzman et al. point out, this requires an inverse temporal-to-spatial transform in which the inhibitory feedback from the resettable integrator is arranged topographically and weighted according to the termination point in SC.

Finally, in a number of recent experiments a variety of methods have been used to modify the time course of saccades while recording the activity of neurons in SC. Convincing evidence that the activity of SC neurons codes dynamic motor error by instantaneous firing rate has not been obtained in any of these experiments.[37,40,41]

12.3 RECENT FINDINGS

12.3.1 The Level and Duration of Collicular Activity

The traditional view that the direction and amplitude of a saccade are determined solely by the site of collicular activity and that the level or duration of collicular activity plays no role in saccade execution is not supported by the results of recent experiments. Stanford and colleagues[42] repeated the early microstimulation experiments but systematically varied the duration and frequency of the stimulation trains over a wider range than used in the early experiments. In contrast to the results of the earlier studies, they found that both the site of stimulation and the parameters of stimulation were important factors in determining the properties of stimulation-induced eye movements. Discrepancies with earlier results are probably due to differences in the ranges over which stimulation parameters were varied. Consistent with early findings, the site of stimulation within the collicular motor map determines the largest movement that can be produced. But, for any stimulation site, movement amplitude depends upon the duration of the stimulation train. As train duration increases, movement amplitude increases monotonically until it reaches the site-specific limit. Within the range over which amplitude can be modulated, movement offset is tightly linked to the offset of the stimulation train. Truncation of the movement is detectable if collicular stimulation ends prematurely. This finding strongly suggests that collicular activity must be sustained throughout the execution of the movement if the amplitude specified by the site of collicular stimulation is to be achieved. Additionally, the peak velocity of an evoked movement is influenced by the frequency of stimulation; higher frequencies produce movements with higher velocities. This finding is consistent with the hypothesis that the peak velocity of a saccade is determined, at least in part, by the level of collicular activity. The effects of train duration and frequency can be traded to produce movements that have comparable amplitudes but different velocity profiles. Movements with high velocity and short duration can be produced by high-frequency, short duration stimulation trains. Stimulation with low frequency, long duration trains produces low-velocity movements of long duration. Thus, the parameters of collicular stimulation can be adjusted to provide independent control over saccade amplitude and saccade velocity.

The findings of Stanford and colleagues, summarized above, are remarkably similar to observations made in several nonprimate species.[43–46] Consequently, the older view that the functional organization of the primate SC is fundamentally different from the organization of the SC or optic tectum of other species is no longer warranted. Moreover, these findings require a revised view of the function of the primate SC — a view in which, in addition to the site of collicular activity, the duration and level of activity contributes to determining the amplitude, velocity and duration of saccadic eye movements. Ongoing collicular activity sustains the movement until the desired displacement is accomplished while the level of collicular activity influences movement velocity. Stanford and colleagues[42] concluded that at least three independent signals are derived from the spatial and temporal pattern of collicular activity — one specifying the desired displacement, another related to saccadic velocity, and a third involved in the initiation of a saccade.

Several other lines of evidence also challenge the notion of a purely topographic motor representation of saccades within the primate SC. Pharmacological inactivation of collicular regions by injection of lidocaine[27,47] or muscimol[48] causes dramatic decreases in the velocity of visually guided saccades without necessarily modifying the amplitude of the movement.[27] In a systematic study extending earlier microstimulation results, Van Opstal and colleagues reported movement amplitude to be an increasing, but saturating, function of current strength.[49] Assuming that variations in current strength influence the size and vigor of the active collicular population, they suggested that the computation of desired displacement does, in fact, depend on the level of collicular activity. They also noted that differences in the effects of inactivation and current strength might be reconciled by considering the possibility that the duration of collicular activity is important. In agreement with the inactivation studies, Van Opstal and colleagues reported decreases in movement velocity with decreases in current strength and suggested that reductions in amplitude might reflect a failure to compensate for decreases in velocity with suitable increases in the duration of collicular activation.[49]

12.3.2 Gaze, Not Eye or Head, Commands?

12.3.2.1 Microstimulation Experiments

How many command signals are generated by collicular neurons? Microstimulation of the optic tectum/superior colliculus in a number of nonprimate species produces coordinated movements of the eyes, head, pinnae, vibrissae, and body. But the results of early experiments in which coordinated movements of the eyes and head were not observed following microstimulation of the SC in monkeys[50] were interpreted as evidence that the monkey SC is not involved in the coordination of eye and head movements. This conclusion was reinforced by the failure to find single cell activity in the monkey SC related to head movements.[51] Thus, the primate SC was thought to be involved in the production of saccadic movements of the eyes but not in the implementation or coordination of combined eye-head movements that occur during large gaze shifts.

Freedman and colleagues stimulated the SC of rhesus monkeys with completely unrestrained heads while systematically varying the site and parameters of microstimulation.[52] In contrast to earlier primate experiments that failed to observe coordinated orienting movements of the eyes and head, they found that collicular stimulation produces high velocity, combined eye-head gaze shifts that are remarkably similar to naturally occurring visually guided gaze shifts of comparable amplitude and direction. The failure of earlier workers to obtain similar findings can be attributed to the restricted range of stimulation parameters and/or stimulation sites used in those studies.

The amplitude and velocity of stimulation-induced gaze shifts depends on the site of stimulation and on the parameters (frequency, current level, and duration of the stimulation train) of stimulation. Increases in the duration of the stimulation train systematically increase the amplitude of evoked gaze shifts until a site-specific maximal amplitude is reached. The frequency of stimulation affects the velocity of evoked gaze shifts; peak velocity increases when higher frequencies are used. The head contribution to stimulation-induced gaze shifts depends on the positions of the eyes in the orbit at the onset of stimulation. The head contribution increases and the latency to head movement onset decreases when the eyes are deviated in the direction of the ensuing gaze shift. The head contribution to stimulation-induced gaze shifts also depends upon the direction of the gaze shift, contributing less for gaze shifts with more vertical components.

Stimulation drives the eyes to approximately the same orbital position regardless of whether the head is restrained or unrestrained. Movements produced when the head is restrained are reduced in amplitude by approximately the amount that the head would have contributed if free to move. Consequently, when the head is restrained, there may be a dissociation between the *desired* gaze amplitude specified by the locus of collicular activity and the stimulation-induced movement that actually occurs. Also, during head restrained stimulation, the observed movement may be only a portion of the movement encoded by the locus of collicular activity. Thus, the collicular *motor map* defined using microstimulation in head restrained subjects is grossly distorted.

Which neuronal signals are being activated by collicular stimulation? Stimulation of a particular site using constant stimulation parameters produces gaze shifts with relatively constant amplitudes and directions. This does not depend upon a particular eye movement always being coupled with a particular head movement. Rather, stimulation-induced gaze shifts of similar directions and amplitudes are accomplished with many combinations of eye and head components, depending on the initial positions of the eyes in the orbits and the direction of the gaze shift. The most parsimonious interpretation of this finding is that microstimulation is activating neurons generating a command for a change in gaze angle — signals that are separated into eye and head components below the level of the SC. It is more difficult to explain this pattern of results if one assumes that the combined eye-head movements produced by microstimulation result from activating one subset of cells generating an eye command and another subset of cells generating a command to move the head.

12.3.2.2 Recording Data

Freedman and Sparks also recorded the activity of single neurons in the intermediate and deep layers of the monkey SC during combined eye-head gaze shifts made to visual targets.[53] The cells were studied in conditions in which: (1) the amplitude and direction of gaze movements was relatively constant, but the eye and head components varied over a wide range; and (2) either the eye or head contribution was fixed but the direction and amplitude of gaze changed over a large range. Thus, the activity of collicular neurons was studied under conditions in which gaze and the eye and head contributions to gaze were dissociable.

For all of the cells in their sample for which these analyses could be performed, motor-related activity was best correlated with the amplitude and direction of the gaze shift, and only weakly correlated with eye or head components of gaze. Gaze shifts having similar amplitudes and directions were associated with similar motor-related bursts. They conclude that cells in the superior colliculus generate commands related to the displacement of the line of sight, and that this activity is not directly related to the individual components of the orienting gaze shifts.[53] These findings provide further support for the hypothesis that a desired gaze displacement signal is derived from the locus of collicular activity, and is decomposed into separate eye and head signals downstream from the SC.

12.4 IMPLICATIONS OF RECENT FINDINGS

12.4.1 Population Coding

Previous studies of population coding of motor commands in SC have examined animals with their heads restrained (for example, see References 26 and 27). As a consequence, little is known about how large populations of broadly tuned SC neurons encode precise gaze shifts and population coding of motor commands in SC needs to be re-examined under more natural, head unrestrained conditions.

If the activity of collicular neurons specifies a change in the direction of gaze rather than a change in eye position, then population coding schemes must be modified to account for coordinated movements of the eyes and head. Currently, no published studies have determined whether or not the vector averaging scheme of Sparks and colleagues[20] can be extended to the more natural, head-unrestrained condition. A simple extension of the vector averaging hypothesis to coordinated eye-head movements predicts that local inactivation of collicular neurons should cause the SC to output an erroneous signal for desired gaze displacement (except when the area affected by the injection coincides with the center of the activated region of SC) that would also affect the velocity of the gaze shift. Other than the velocity effect, the injection should have no direct effect upon ability of the animal to execute eye or head movements of any direction or amplitude. Thus, the relative contributions of the eyes and head to these erroneous gaze shifts should match the ratio observed for similar (but normometric), naturally occurring gaze shifts. As mentioned above, the starting position of the eyes influences this ratio. Systematically manipulating initial orbital position would make it possible to test the population coding hypothesis over a wide range of behavioral conditions.

12.4.2 STTP and Feedback

12.4.2.1 Gaze Direction

In retrospect, one of the reasons the STTP has been so intractable is that the question is ill-posed. For the primate, the original statement of the problem is based upon erroneous conclusions about both *which* signals are represented in the SC and *how* the signals are represented. New findings indicate that the location of collicular activity encodes a signal of desired gaze displacement rather than a signal of desired eye displacement. While the site of activity in the SC sets the limit on the maximal gaze shift that will occur, the time course and level of activity are also important. Collicular activity must be sustained for a movement to continue and the level of activity is a determinant of the speed of a movement.

This new perspective has important consequences for experimental attempts to solve the STTP. For example, more experiments are needed to provide additional detail about how projections from different regions of the collicular map differ in terms of the ratio of synaptic drive to neurons controlling the horizontal and vertical movements of the eyes. But parallel experiments must also examine the ratio of synaptic drive to neurons involved in the control of head movements.[54] In this context, a recent study of the patterns of neck muscle activity following SC stimulation in head-restrained monkeys[55] provides insights into how the locus of collicular activity could specify which neck muscles are activated and influence the temporal pattern of neck muscle activity. Corneil and colleagues found that the drive to neck muscle motoneurons is widely distributed throughout almost the entire SC and that there is a clear topography to the stimulation-evoked neck muscle EMG responses.

EMG activation became faster, stronger, and more prevalent for more caudal stimulation locations.[55]

12.4.2.2 Initial Eye Position

Gaze shifts of the same direction and amplitude are accomplished by many different combinations of eye and head contributions, depending upon the initial positions of the eyes in the orbits and the direction of the gaze shift.[56] Even when the head is immobilized, the tonic component of the activity of several dorsal neck muscles is highly correlated with the horizontal position of the eyes during spontaneous eye movements and during vestibular nystagmus.[57–59] The relationship between neck muscle activity and horizontal eye position is nonlinear. Very little, if any, neck muscle activity is observed until the eye reaches a threshold position in the orbit. Then, neck muscle activity increases linearly as a function of eye position until the recruitment of additional motor units makes it difficult to isolate individual units. The relationship between tonic levels of neck muscle activity and eye position is greater when the head is fixed. When the head is free, neck muscle activity is usually related to head position, but muscle activity is phasically modulated during saccadic eye movements and head accelerations. In head restrained animals, neck EMG responses evoked by collicular stimulation are systematically influenced by the positions of the eyes in the orbits at the onset of the stimulation train.[55]

12.4.2.3 Gaze Amplitude

Most models of the coordinated eye-head movements occurring during large gaze shifts assume that gaze direction and amplitude are under feedback control. Results of experiments in which head movements were perturbed by the application of an external force during a large gaze shift have been interpreted as evidence that gaze is controlled by feedback mechanisms. In a variety of conditions[60–65] gaze amplitude is unaffected by perturbations in head trajectory. However, one model[66] of the gaze system compensates for mechanical perturbations of the head without using dynamic feedback control of gaze amplitude.

The microstimulation experiments summarized above[42,43,45,52] demonstrate that collicular activity must be sustained for a movement to continue (see previous discussion). Accordingly, one possible way of controlling the duration (and indirectly, the amplitude) of a gaze shift is to control the duration of collicular activity. For large gaze shifts, the cessation of the movement-related activity of collicular neurons is highly correlated with the end of a gaze shift[53] (because the end of the gaze shift and the end of the eye movement are coupled tightly, the end of the motor burst also is highly correlated with the end of the eye movement). The observation that this correlation does not depend on the amplitude and direction of movements relative to the movement field center suggests that there is a general cessation of motor activity over large regions of the collicular motor map associated with the end of the gaze shift. These findings seem consistent with models[60,67,68] in which gaze, rather than eye or head, displacement is under feedback control and the SC is located within the feedback loop. However, a causal relationship between the cessation of collicular activity and the end of a gaze shift has not been established and other experimental findings are difficult to explain using such a model.

If the eye and head components of large gaze shifts are under feedback control and if neurons in SC are located inside the feedback loop, then electrical stimulation of output fibers will produce a nonzero output signifying a mismatch between actual gaze displacement and the desired gaze displacement. Gaze position should continue to change for as long as the stimulation train continues to activate the output neurons or until the displacement of the eyes and/or head reaches a physiological limit. This is what occurs for eye movements with stimulation of the MNs or PPRF (unpublished observations), but not what happens with stimulation of SC. One possible explanation for the failure to see continued changes in gaze position with sustained collicular stimulation is that the stimulation becomes ineffective, through fatigue or otherwise. However, over the range of stimulation train durations that have been studied, the head continues to move as long as the stimulation train continues, even though gaze stops when it reaches the site specific limit and the eyes begin to counter-rotate at that time to compensate for the continued movement of the head.

If the comparator resides downstream from the SC, then collicular stimulation produces a signal that specifies a particular desired displacement, depending upon which site in the motor map is activated. This signal serves as a reference signal to the downstream comparator and when the actual displacement matches the desired displacement, the movement stops, even if the stimulation train continues. This is what appears to happen when collicular neurons are stimulated with long trains.

12.5 SUMMARY

Despite much progress in understanding the role of the SC in the control of orienting movements of the eyes and head, much remains to be learned. Important issues have not been resolved. Is gaze under feedback control? If so, how and where are signals of the eye and head components of the gaze shift combined? Where are signals of desired and actual gaze displacement compared? How is the result of this comparison used to control the direction and amplitude of eye and head movements? How and where is the desired gaze displacement command observed at the level of SC decomposed into signals appropriate to control the eye and head components? Is the head component of the gaze shift under feedback control? What is the status of the VOR during large gaze shifts? Is the VOR equally active during the entire gaze shift? When does the reduction in VOR gain end? How are reductions in VOR gain implemented? These and other questions about the neural control of coordinated eye and head movements bring a different perspective to studies of the role of the SC in the control of gaze.

ACKNOWLEDGMENTS

Work in the author's laboratory is supported by NIH grant EY01189. Many of the ideas presented in this paper emerged from extensive discussions with Terry Stanford, Ed Freedman, Jamie Nichols, and Raj Gandhi.

REFERENCES

1. Robinson, D.A., Eye movements evoked by collicular stimulation in the alert monkey, *Vision Res.,* 12, 1795–1808, 1972.
2. Schiller, P.H. and Stryker, M., Single-unit recording and stimulation in superior colliculus of the alert rhesus monkey, *J. Neurophysiol.,* 35, 915–924, 1972.
3. Keller, E.L., Participation of medial pontine reticular formation in eye movement generation in monkey, *J. Neurophysiol.,* 37, 316–331, 1974.
4. King, W.M. and Fuchs, A.F., Reticular control of vertical saccadic eye movements by mesencephalic burst neurons, *J. Neurophysiol.,* 42, 861–876, 1979.
5. Luschei, E.S. and Fuchs, A.F., Activity of brain stem neurons during eye movements of alert monkeys, *J. Neurophysiol.,* 35, 445–461, 1972.
6. van Gisbergen, J.A.M., Robinson, D.A., and Gielen, S.A., A quantitative analysis of generation of saccadic eye movements by burst neurons, *J. Neurophysiol.* 45, 417–442, 1981.
7. Sparks, D.L. and Mays, L.E., Signal transformations required for the generation of saccadic eye movements, *Ann. Rev. Neurosci.,* 13, 309–336, 1990.
8. Scudder, C.A., Kaneko, C.R. S., and Fuchs, A.F., The brainstem burst generator for saccadic eye movements: A modern synthesis, *Exp. Brain Res.,* 142, 439–462, 2002.
9. Moschovakis, A.K. and Highstein, S.M., The anatomy and physiology of primate neurons that control rapid eye movements, *Ann. Rev. Neurosci.,* 17, 465–488, 1994.
10. Moschovakis, A.K., Scudder, C.A., and Highstein, S.M., The microscopic anatomy and physiology of the mammalian saccadic system, *Progress in Neurobiology* 50, 133–254, 1996.
11. Fuchs, A.F., Kaneko, C.R.S., and Scudder, C.A., Brainstem control of saccadic eye movements, *Ann Rev. Neurosci.,* 8, 307–337, 1985.
12. Sparks, D.L., The brainstem control of saccadic eye movements, *Nat. Rev. Neurosci.,* 3, 952–964, 2002.
13. Edwards, S.B. and Henkel, C.K., Superior colliculus connections with the extraocular motor nuclei in the cat, *J. Comp. Neurol.,* 179, 451–468, 1978.
14. Kawamura, K. and Hashikawa, T., Cell bodies of origin of reticular projections from the superior colliculus in the cat: an experimental study with the use of horseradish peroxidase as a tracer, *J. Comp. Neurol.,* 182, 1–16, 1978.
15. Olivier, E., Chat, M., and Grantyn, A., Rostrocaudal and lateromedial density distributions of superior colliculus neurons projecting in the predorsal bundle and to the spinal cord: a retrograde HRP study in the cat, *Exp. Brain Res.,* 87, 268–282, 1991.
16. Stanton, G.B. and Greene, R.W., Brain stem afferents to the periabducens reticular formation (PARF) in the cat. An HRP study, *Exp. Brain Res.,* 44, 419–426, 1981.
17. Grantyn, A., Brandi, A.-M., Dubayle, D., Graf, W., Ugolini, G., Hadjidimitrakis, K., and Moschovakis, A., Density gradients of trans-synaptically labeled collicular neurons after injections of rabies virus in the lateral rectus muscle of the rhesus monkey, *J. Comp. Neurol.,* 451, 346–361, 2002.
18. Moschovakis, A.K., Kitama, T., Dalezios, Y., Petit, J., Brandi, A.M., and Grantyn, A.A., An anatomical substrate for the spatiotemporal transformation, *J. Comp. Neurol.,* 18, 10219–10229, 1998.
19. Sparks, D.L. and Mays, L.E., Movement fields of saccade-related burst neurons in the monkey superior colliculus, *Brain Res.,* 190, 39–50, 1980.
20. Sparks, D.L., Holland, R., and Guthrie, B.L., Size and distribution of movement fields in monkey superior colliculus, *Brain Res.,* 113, 21–34, 1976.

21. Wurtz, R.H. and Goldberg, M.E., Activity of superior colliculus in behaving monkey. III. Cells discharging before eye movements, *J. Neurophysiol.,* 35, 575–586, 1972.
22. McIlwain, J.T., Large receptive fields and spatial transformations in the visual system, *Int. Rev. Physiol.,* 10, 223–248, 1976.
23. Tweed, D. and Vilis, T., A two-dimensional model for saccade generation, *Biol. Cybern.,* 52, 219- 227, 1985.
24. van Gisbergen, J.A.M., van Opstal, A.J., and Tax, A.A.M., Collicular ensemble coding of saccades based on vector summation, *Neuroscience,* 21, 541–555, 1987.
25. van Opstal, A.J. and van Gisbergen, J.A. M., A nonlinear model for collicular spatial interactions underlying the metrical properties of electrically elicited saccades, *Biol. Cybern.,* 60, 171–183, 1989.
26. Sparks, D.L., Lee, C., and Rohrer, W.H., Population coding of the direction, amplitude, and velocity of saccadic eye movements by neurons in the superior colliculus, *Cold Spring Harbor Symposia on Quantitative Biology* LV, 805–811, 1990.
27. Lee, C., Rohrer, W.H., and Sparks, D.L., Population coding of saccadic eye movements by neurons in the superior colliculus, *Nature,* 332, 357–360, 1988.
28. Hanes, D.P. and Wurtz, R.H., Interaction of the frontal eye field and superior colliculus for saccade generation, *J. Neurophysiol.,* 85, 804–815, 2001.
29. Guthrie, B.L., Porter, J.D., and Sparks, D.L., Corollary discharge provides accurate eye position information to the oculomotor system, *Science,* 221, 1193–1195, 1983.
30. Zee, D.S., Optican, L.M., Cook, J.D., Robinson, D.A., and King, W., Slow saccades in spinocerebellar degeneration, *Arch. Neurol.,* 33, 243–251, 1976.
31. Munoz, D.P., Guitton, D., and Pélisson, D., Control of orienting gaze shifts by the tectoreticulospinal system in the head-free cat. III. Spatiotemporal characteristics of phasic motor discharges, *J. Neurophysiol.,* 66, 1642–1666, 1991.
32. Munoz, D.P. and Wurtz, R.H., Saccade-related activity in monkey superior colliculus. II. Spread of activity during saccades, *J. Neurophysiol.,* 73, 2334–2348, 1995.
33. Waitzman, D.M. et al., Superior colliculus neurons mediate the dynamic characteristics of saccades, *J. Neurophysiol.,* 66, 1716–1737, 1991.
34. Sparks, D.L., Are gaze shifts controlled by a 'moving hill' of activity in the superior colliculus? *Trends Neurosci.,* 16, 214–218, 1993.
35. Anderson, R.W. et al., Two-dimensional saccade-related population activity in superior colliculus in monkey, *J. Neurophysiol.,* 80, 798–817, 1998.
36. Kang, I. and Lee, C., Properties of saccade-related neurons in the cat superior colliculus: patterns of movement fields and discharge timing, *Exp. Brain Res.,* 131, 149–164, 2000.
37. Soetedjo, R., Kaneko, C.R., and Fuchs, A.F., Evidence against a moving hill in the superior colliculus during saccadic eye movements in the monkey, *J. Neurophysiol.,* 87, 2778–2789, 2000.
38. Freedman, E.G. and Sparks, D.L., Response properties of saccade-related burst cells in the superior colliculus of the monkey during large, head un-restrained, gaze shifts, *Soc. Neurosci. Abstr.,* 20, 142, 1994.
39. Bozis, A. and Moschovakis, A.K., Neural network simulations of the primate oculomotor system III. An one-dimensional, one-directional model of the superior colliculus, *Biol. Cybern.,* 79, 215–230, 1998.
40. Goossens, H.H. and van Opstal, A.J., Blink-perturbed saccades in monkey. II. Superior colliculus activity, *J. Neurophysiol.,* 83, 3420–3452, 2000.
41. Keller, E.L. and Edelman, J.A., Use of interrupted saccade paradigm to study spatial and temporal dynamics of saccadic burst cells in superior colliculus in monkey, *J. Neurophysiol.,* 72, 2754–2770, 1994.

42. Stanford, T.R., Freedman, E.G., and Sparks, D.L., The site and parameters of microstimulation determine the properties of eye movements evoked from the primate superior colliculus: Evidence for independent collicular signals of saccade displacement and velocity, *J. Neurophysiol.,* 76, 3360–3381, 1996.
43. du Lac, S. and Knudsen, E.I., Neural maps of head movement vector and speed in the optic tectum of the barn owl, *J. Neurophysiol.,* 63, 131–146, 1990.
44. King, S.M., Dean, P., and Redgrave, P., Bypassing the saccadic pulse generator: possible control of head movement trajectory by rat superior colliculus, *Eur. J. Neurosci.,* 3, 790–801, 1991.
45. Paré, M., Crommelinck, M., and Guitton, D., Gaze shifts evoked by stimulation of the superior colliculus in the head-free cat conform to the motor map but also depend on stimulus strength and fixation activity, *Exp. Brain Res.,* 101, 123–139, 1994.
46. Tehovnic, E.J., Head and body movements evoked electrically from the caudal superior colliculus of rats: pulse frequency effects, *Behav. Brain Res.,* 34, 71–78, 1989.
47. Hikosaka, O. and Wurtz, R.H., Saccadic eye movements following injections of lidocaine into the superior colliculus, *Exp. Brain Res.,* 61, 531–539, 1986.
48. Hikosaka, O. and Wurtz, R.H., Modification of saccadic eye movements by GABA-related substances. I. Effect of muscimol and bicuculline in monkey superior colliculus, *J. Neurophysiol.,* 53, 266–291, 1985.
49. van Opstal, A.J., van Gisbergen, J.A., and Smit, A.C., Comparison of saccades evoked by visual stimulation and collicular electrical stimulation in the alert monkey, *Exp. Brain Res.,* 79, 299–312, 1990.
50. Stryker, M.P. and Schiller, P.H., Eye and head movements evoked by electrical stimulation of monkey superior colliculus, *Exp. Brain Res.,* 23, 211–227, 1975.
51. Robinson, D.L. and Jarvis, C.D., Superior colliculus neurons studied during head and eye movements of the behaving monkey, *J. Neurophysiol.,* 37, 533–540, 1974.
52. Freedman, E.G., Stanford, T.R., and Sparks, D.L., Combined eye-head gaze shifts produced by electrical stimulation of the superior colliculus in rhesus monkeys, *J. Neurophysiol.,* 76, 927–952, 1996.
53. Freedman, E.G. and Sparks, D.L., Activity of cells in the deeper layers of the superior colliculus of rhesus monkey: Evidence for a gaze displacement command, *J. Neurophysiol.,* 78, 1669–1690, 1997.
54. Isa, T. and Sasaki, S., Brainstem control of head movements during orienting; organization of the premotor circuits, *Progress in Neurobiology,* 66, 205–241, 2002.
55. Corneil, B.D., Olivier, E., and Munoz, D.P., Neck muscle responses to stimulation of monkey superior colliculus. I. Topography and manipulation of stimulation parameters, *J. Neurophysiol.,* 88, 1980–1999, 2002.
56. Freedman, E.G. and Sparks, D.L., Eye-head coordination during head unrestrained gaze shifts in rhesus monkeys. manipulation of stimulation parameters, *J. Neurophysiol.,* 77, 2328–2348, 1997.
57. Lestienne, E., Vidal, P.P., and Berthoz, A., Gaze changing behaviour in head restrained monkey, *Exp. Brain Res.* 53, 349–356, 1984.
58. Roucoux, A., Vidal, P.P., Veraart, C., and Crommelinck, M., The relation of neck muscle activity to horizontal eye position in the alert cat. I. Head fixed, in *Physiological and Pathological Aspects of Eye Movements*, Roucoux, A. and Crommelinck, M. Eds., Dr.W. Junk Publishers, The Hague, 1982, pp. 371–378.
59. Vidal, P.P., Roucoux, A., and Berthoz, A., Horizontal eye-position related activity in neck muscles of the alert cat, *Exp. Brain Res.,* 46, 448–453, 1982.
60. Laurutis, V. and Robinson, D., The vestibulo-ocular reflex during human saccadic eye movements, *J. Physiol., (Lond)* 373, 2293–2301, 1986.

61. Pelisson, D. and Prablanc, C., Vestibulo-ocular reflex (VOR) induced by passive head rotation and goal-directed saccadic eye movements do not simply add in man, *Brain Res.,* 380, 397–400, 1986.
62. Tomlinson, R.D. and Bahra, P.S., Combined eye-head gaze shifts in the primate I. Metrics, *J. Neurophysiol.,* 56, 1542 -1557, 1986.
63. Tomlinson, R.D., Combined eye-head gaze shifts in the primate III. Contributions to the accuracy of gaze saccades, *J. Neurophysiol.,* 64, 1873 -1891, 1990.
64. Guitton, D. and Volle, M., Gaze control in humans: eye-head coordination during orienting movements to targets within and beyond the oculomotor range, *J. Neurophysiol.,* 58, 427–459, 1987.
65. Tabak, S., Smeets, J.B.J., and Collewijn, H., Modulation of the human vestibuloocular reflex during saccades: probing by high frequency oscillation and torque pulses to the head, *J. Neurophysiol.,* 76, 3249–3263, 1996.
66. Freedman, E.G., Interactions between eye and head control signals can account for movement kinematics, *Biol. Cybern.,* 84, 453–462, 2001.
67. Galiana, H.L. and Guitton, D., Central organization and modeling of eye-head coordination during orienting gaze shifts, in *Sensing and Controlling Motion*, Cohen, B., Tomko, D.L., and Guedry, F.E. Eds., NYAS, New York, 1992, pp. 452–471.
68. Guitton, D., Munoz, D.P., and Galiana, H.L., Gaze control in the cat: studies and modeling of the coupling between orienting eye and head movements in different behavioral tasks, *J. Neurophysiol.,* 64, 509–531, 1990.

Index

A

B

C

D

E

F

G

H

R

S

T

U

V

W

X

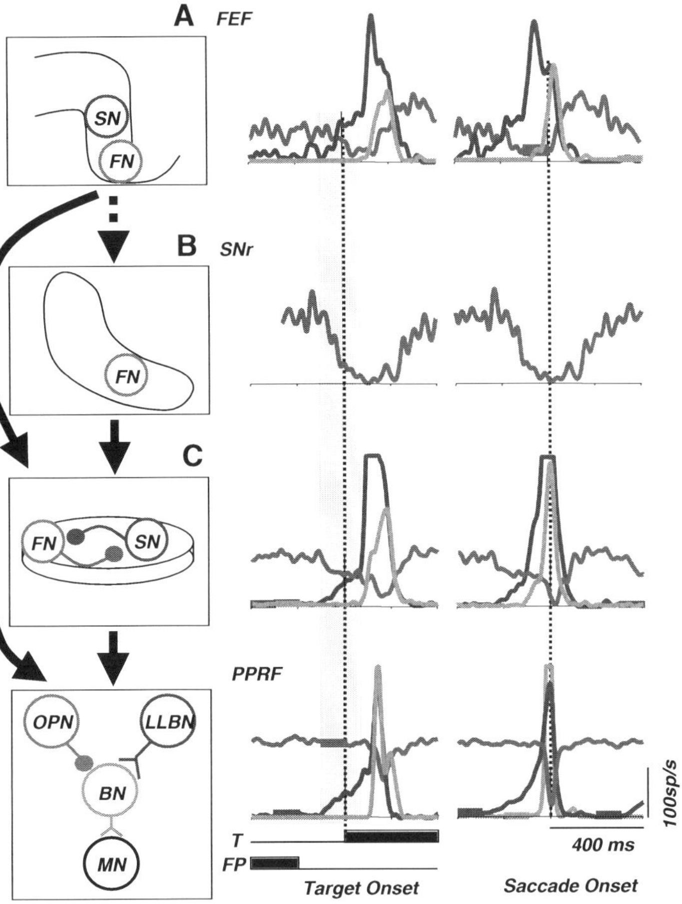

FIGURE 3.3 Discharge of neurons in the frontal eye fields (A), substantia nigra pars reticulata (B), intermediate layers of the superior colliculus (C), and paramedian pontine reticular formation (D) during the gap-saccade paradigm. The spike density waveforms are aligned on target appearance (left column) and saccade onset (right column). Cells at all three levels are modulated by the gap period. Saccade neurons are silent during visual fixation of the fixation point and then increase their discharge for saccades. Of these saccade neurons, some become active before target appearance (blue traces), while others remain silent until after target appearance (green traces). Neurons with tonic activity during fixation have a drop in discharge rate during the gap and saccade initiation (red traces). Vertical gray bar denotes the end of gap epoch, highlighting neurons that change their activity during the gap period. Abbreviations: BN: burst neuron; FN: fixation neuron; LLBN: long-lead burst neuron; MN: motoneuron; OPN: omnipause neuron; SN: saccade neuron.

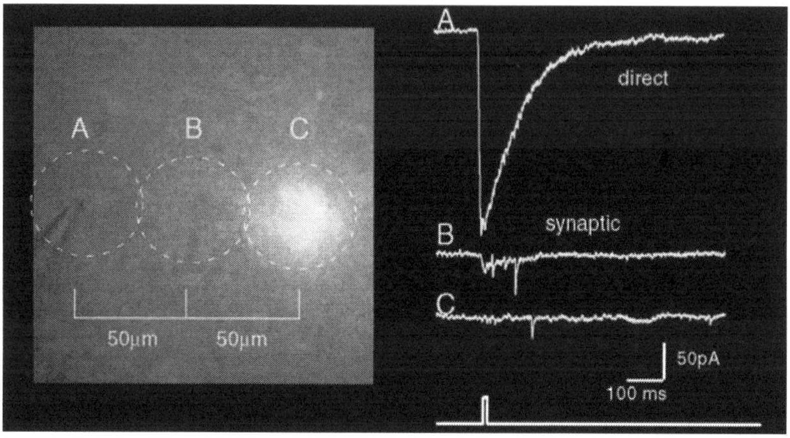

FIGURE 6.3 (A) Photostimulation with uncaged glutamate directly activates glutamate receptors on the soma and dendrites of the recorded cell and evokes a large, slow response. In this photograph of the living slice, the micropipette can be seen in contact with the cell body. The dashed circle indicates the size of the laser spot, which was photographed using a long exposure time after the experiment in C. (B) In contrast to stimulation at site A, stimulation at site B, which is only 50 μm from A, evoked brief synaptic events, but little or no direct response. (C) Stimulation at this site failed to evoke a response.

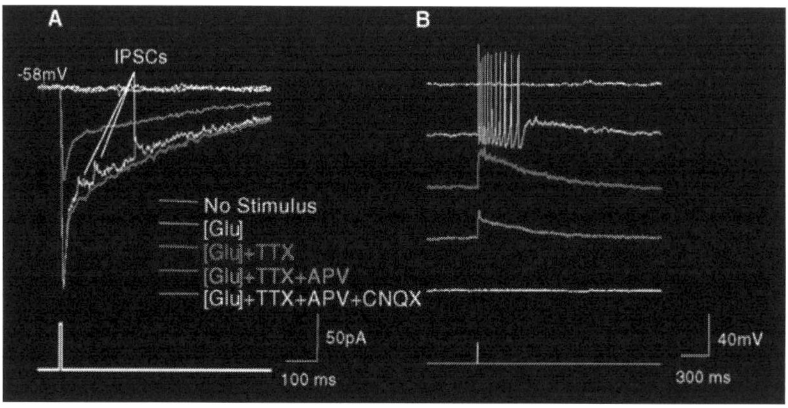

FIGURE 6.4 (A) Photostimulation over the cell body elicits a direct response with outward synaptic currents superimposed (IPSCs, yellow). These IPSCs are evoked by inputs from local inhibitory interneurons that are presynaptic to the patch-clamped cell and are excited by the photostimulus to fire action potentials. The synaptic mediation of the IPSCs is indicated by the observation that they are blocked by bath application of TTX, which eliminates action potentials generated by presynaptic cells (red). The response that remains after application of TTX is evoked by direct activation of receptors on the recorded cell, and therefore can be blocked by application of NMDA and AMPA receptor blockers (gray). (B) Current clamp responses evoked by uncaging glutamate directly over the soma illustrate the efficacy of photostimulation in producing action potentials in slice neurons. Photostimulation at the soma is sufficient to elicit a high-frequency train of action potentials in this neuron (yellow). TTX blocks the action potentials, and the glutamatergic blockers APV and CNQX eliminate the remaining direct response.

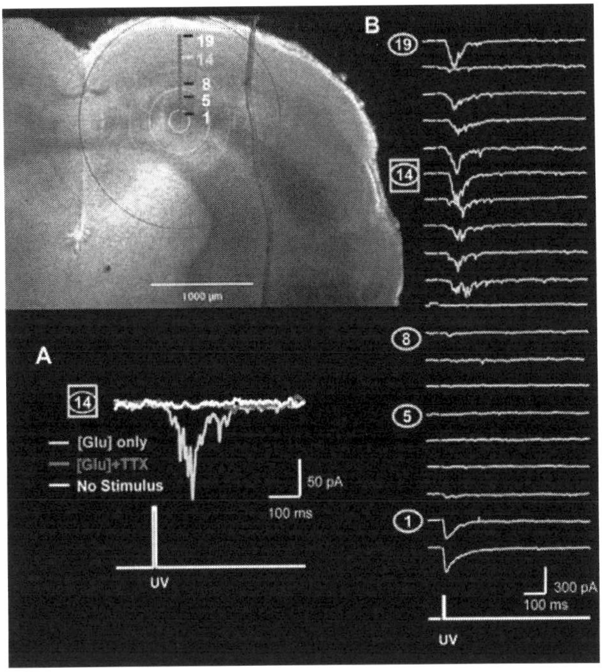

FIGURE 6.5 Synaptic circuitry between superficial and intermediate layers is demonstrated by short and long distance (6 to 1000 µm) photostimulation. The patch-clamped cell is located in SGI, at the yellow dot just below stimulation site #1. (A) The responses evoked in the SGI cell can be blocked by TTX, indicating that they are mediated through the activity of presynaptic cells that are activated by the photostimulus in SGS. (B) Robust EPSCs can be evoked from stimulation sites as far away from the patch clamped cell as the upper superficial gray layer (site #19).

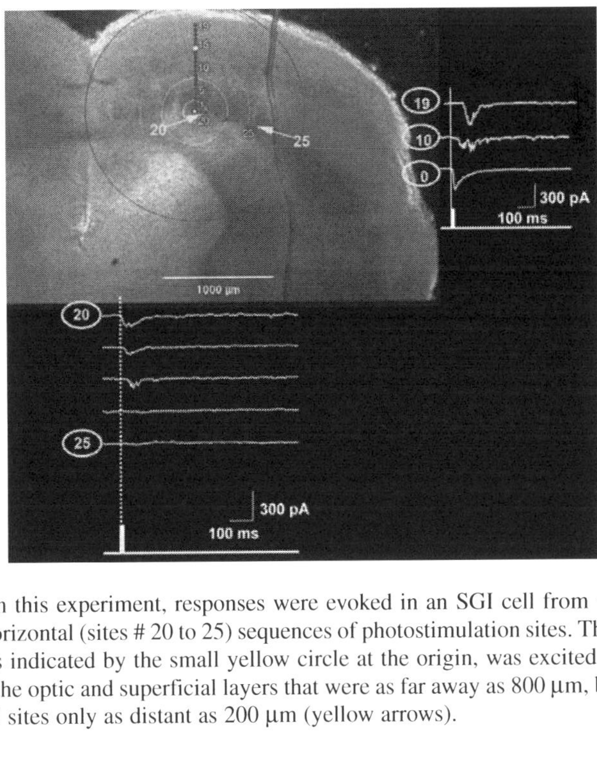

FIGURE 6.6 In this experiment, responses were evoked in an SGI cell from vertical (sites # 0 to 19) and horizontal (sites # 20 to 25) sequences of photostimulation sites. The cell, whose soma location is indicated by the small yellow circle at the origin, was excited by overlying vertical sites in the optic and superficial layers that were as far away as 800 µm, but was excited by horizontal sites only as distant as 200 µm (yellow arrows).

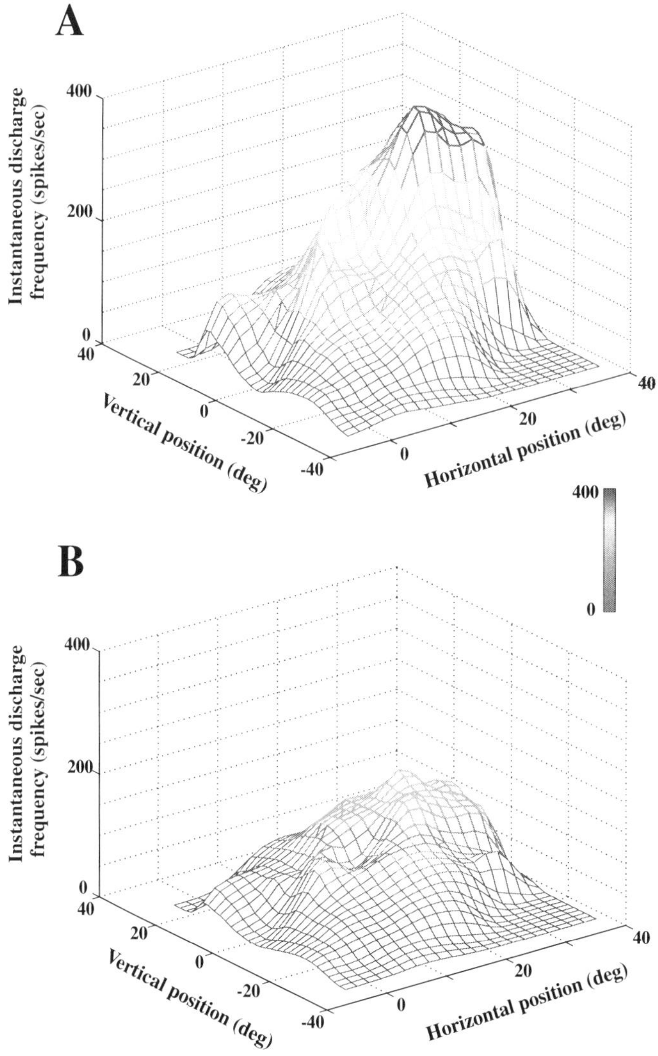

FIGURE 11.1 Movement field (MF) fit for a sample SRBN in the left SC. The estimation of the two-dimensional shape of the field was made at two perisaccadic time epochs using a model-free radial basis function technique. The MF fit was computed from discharge data recorded while the animal made saccades to more than 100 locations scattered in the right visual hemifield. (A) Discharge at 8 ms before saccade onset when activity was maximal. (B) Discharge at saccade end. An attempt to fit the two-dimensional discharge with a log-Gaussian model in amplitude and a Gaussian in direction did not converge.

9780849300974